THIRD EDITION

THE ART OF SYSTEMS ARCHITECTING

THE ART OF SYSTEMS ARCHITECTING

MARK W. MAIER
EBERHARDT RECHTIN

CRC Press
Taylor & Francis Group
Boca Raton   London   New York

CRC Press is an imprint of the
Taylor & Francis Group, an **informa** business

CRC Press
Taylor & Francis Group
6000 Broken Sound Parkway NW, Suite 300
Boca Raton, FL 33487-2742

International Standard Book Number-13: 978-1-4200-7913-5 (Hardcover)

Library of Congress Cataloging-in-Publication Data

Maier, Mark.
    The art systems of architecting / Mark W. Maier. -- 3rd ed.
       p. cm.
    Includes bibliographical references and index.
    ISBN 978-1-4200-7913-5 (alk. paper)
    1. Systems engineering. I. Title.

TA168.M263 2009
620.001'171--dc22
                                            2008044161

Visit the Taylor & Francis Web site at
http://www.taylorandfrancis.com

and the CRC Press Web site at
http://www.crcpress.com

*To Eberhardt Rechtin,
who opened new vistas to
so many of us, and inspired
us to go out and find more.*

———————

Mark Maier

Contents

Part II: New Domains, New Insights

Preface

The Continuing Development of Systems Architecting

> Architecting, the planning and building of struc-
> tures, is as old as human societies — and as modern
> as the exploration of the solar system.

So began this book's original 1991 predecessor.* The earlier work was based on the premise that architectural methods, similar to those formulated centuries before in civil works, were being used, albeit unknowingly, to create and build complex aerospace, electronic, software, command, control, and manufacturing systems. If so, then still other civil works architectural tools and ideas — such as qualitative reasoning and the relationships between client, architect, and builder — should be found even more valuable in today's more recent engineering fields. Five and ten years later, at the time of the first and second editions of this book, judging from several hundred retrospective studies at the University of Southern California of dozens of post–World War II systems, the original premise was validated. Since then the use of architectural concepts has become common in systems engineering discussions. A central premise of the application of the civil architecture metaphor, that creating and building systems too complex to be treated by engineering analysis alone can be addressed through structured methods at the level of heuristics, has been further validated.

Of great importance for the future, the new fields have been creating architectural concepts and tools of their own and at an accelerating rate. This book includes a number of the more broadly applicable ones, among them heuristic tools, progressive design, intersecting waterfalls, feedback architectures, spiral-to-circle software acquisition, technological

---

* Rechtin, E., *Systems Architecting, Creating and Building Complex Systems.* Englewood Cliffs, NJ: Prentice Hall, 1991, hereafter called Rechtin 1991.

innovation, architecture and business strategy, and the rules of the political process as they affect system design.

Arguably, these developments could, even should, have occurred sooner in this modern world of systems. Why now?

Architecting in the Systems World

A strong motivation for expanding the architecting process into new fields has been the retrospective observation that success or failure of today's widely publicized (and unpublicized) systems often seems preordained — that is, traceable to their beginnings. Some system development projects start doomed, and no downstream engineering efforts are likely to rescue them. Other projects seem fated for success almost in spite of poor downstream decisions. The initial concept is so "right" that its success is inevitable, even if not necessarily with the first group that tries to execute it. This is not a new realization. It was just as apparent to the ancient Egyptians, Greeks, and Romans who originated classical architecting in response to it. The difference between their times and now is in the extraordinary complexity and technological capability of what could then and now be built.

Today's architecting must handle systems of types unknown until very recently, for example, systems that are very high quality, real time, closed loop, reconfigurable, interactive, software intensive, and, for all practical purposes, autonomous. New domains like personal computers, intersatellite networks, health services, and joint service command and control are calling for new architectures — and for architects specializing in those domains. Their needs and lessons learned are in turn leading to new architecting concepts and tools and to the acknowledgment of a new formalism — and evolving profession — called systems architecting, a combination of the principles and concepts of both systems and of architecting. However, for all the new complexity, many of the roots of success and failure are nearly constant over time. By examining a series of case studies, interwoven with a discussion of the particular domains to which they belong, we can see how relatively timeless principles (for example, technical and operational coupled revolution, strategic consistency) largely govern success and failure.

The reasons behind the general acknowledgment of architecting in the new systems world are traceable to that remarkable period immediately after the end of the Cold War in the mid-1980s. Abruptly, by historical standards, a 50-year period of continuity ended. During the same period, there was a dramatic upsurge in the use of smart, real-time systems, both civilian and military, that required much more than straightforward refinements of established system forms. Long-range management strategies and design rules, based on years of continuity, came under challenge.

That challenge was not short-lived; instead, it has resorted itself repeatedly in the years between editions of this book. It is now apparent that the new era of global transportation, global communications, global competition, and global security turmoil is not only different in type and direction; it is unique technologically and politically. It is a time of restructuring and invention, of architecting new products and processes, and of new ways of thinking about how systems are created and built.

Long-standing assumptions and methods are under challenge. For example, for many engineers, architectures were a given; automobiles, airplanes, and even spacecraft had the same architectural forms for decades. What need was there for architecting? Global competition soon provided an answer. Architecturally different systems were capturing markets. Consumer product lines and defense systems are well-reported examples. Other questions remained: How can software architectures be created that evolve as fast as their supporting technologies? How deeply should a systems architect go into the details of all the system's subsystems? What are the relationships between the architectures of systems and the human organizations that design, build, support, and use them?

Distinguishing between Architecting, Engineering, and Project Management

Because it is the most asked by engineers in the new fields, the first issue to address is the distinction between architecting and engineering in general — that is, regardless of engineering discipline. Although civil engineers and civil architects, even after centuries of debate, have not answered that question in the abstract, they have in practice. Generally speaking, engineering deals almost entirely with measurables using analytic tools derived from mathematics and the hard sciences; that is, engineering is a deductive process. Architecting deals largely with unmeasurables using nonquantitative tools and guidelines based on practical lessons learned; that is, architecting is an inductive process. Architecting embraces the world of the user/sponsor/client, with all the ambiguity and imprecision that may entail. Architecting seeks to communicate across the gap from the user/sponsor/client to the engineer/developer, and architecting is complete (at least its initial phase) when a system is well-enough defined to engage developers. At a more detailed level, engineering is concerned more with quantifiable costs, architecting more with qualitative worth. Engineering aims for technical optimization, architecting for client satisfaction. Engineering is more of a science, and architecting is more of an art. Although the border between them is often fuzzy, the distinction at the end is clear.

Table P.1 Characteristics of the Roles on the Architecting and Engineering Continuum

Characteristic	Architecting	Architecting and Engineering	Engineering
Situation/goals	Ill-structured	Constrained	Understood
	Satisfaction	Compliance	Optimization
Methods	Heuristics	⟵————————⟶	Equations
	Synthesis	⟵————————⟶	Analysis
	Art and science	Art **and** science	**Science** and art
Interfaces	Focus on "mis-fits"	Critical	Completeness
System integrity maintained through	"Single mind"	Clear objectives	Disciplined methodology and process
Management issues	Working for client	Working with Client	Working for builder
	Conceptualization and certification	Whole waterfall	Meeting project requirements
	Confidentiality	Conflict of interest	Profit versus cost

In brief, the practical distinction between engineering and architecting is in the problems faced and the tools used to tackle them. This same distinction appears to apply whether the branch involved is civil, mechanical, chemical, electrical, electronic, aerospace, software, or systems.* Both architecting and engineering can be found in every one of the established disciplines and in multidisciplinary contexts. Architecting and engineering are roles, distinguished by their characteristics. They represent two edges of a continuum of systems practice. Individual engineers often fill roles across the continuum at various points in their careers or on different systems. The characteristics of the roles, and a suggestion for an intermediate role, are shown in Table P.1.

As the table indicates, architecting is characterized by dealing with ill-structured situations, situations where neither goals nor means are known with much certainty. In systems engineering terms, the requirements for the system have not been stated more than vaguely, and the architect cannot appeal to the client for a resolution, as the client has engaged the architect precisely to assist and advise in such a resolution. The architect engages in a joint exploration of requirements and design, in contrast to the classic engineering approach of seeking an optimal design solution to a clearly defined set of objectives.

---

* The systems branch, possibly new to some readers, is described in Rechtin 1991 and in Chapter 1 of this book.

Because the situation is ill structured, the goal cannot be optimization. The architect seeks satisfactory and feasible problem-solution pairs. Good architecture and good engineering are both the products of art and science, and a mixture of analysis and heuristics. However, the weight will fall on heuristics and "art" during architecting.

An "ill-structured" problem is a problem where the statement of the problem depends on the statement of the solution. In other words, knowing what you can do changes your mind about what you want to do. A solution that appears correct based on an initial understanding of the problem may be revealed as wholly inadequate with more experience. Architecting embraces ill-structured problems. A basic tenet of architecting is to assume that one will face ill-structured problems and to configure one's processes so as to allow for it.

One way to clearly see the distinction between architecting and engineering is in the approach to interfaces and system integrity. When a complex system is built (say one involving 10,000 person-years of effort), only absolute consistency and completeness of interface descriptions and disciplined methodology and process will suffice. When a system is physically assembled, it matters little whether an interface is high tech or low tech; if it is not exactly correct the system does not work. In contrast, during architecting, it is necessary only to identify the interfaces that cannot work — the mis-fits. Mis-fits must be eliminated during architecting, and then interfaces should be resolved in order of criticality and risk as development proceeds into engineering.

One important point is that the table represents management in the classical paradigm of how architecting is done, not necessarily how it actually is done. Classically, architecting is performed by a third party working for the client. In practice, the situation is more complex as the architecting might be done by the builder before a client is found, might be mixed into a competitive procurement, or might be done by the client. These variations are taken up in chapters to come.

As for project management, architecting clearly exists within the larger project cycle. If we examine the development of systems very holistically, looking from the earliest to the latest phases, we see architecting existing within that large picture. But, at a practical level, what is usually taught as project management has a narrower focus, as does what is usually taught as systems engineering. The narrower focus assumes that definite requirements (in the unambiguous, orthogonal, measurable, and testable senses) exist and can be documented, that budgets and schedules exist and must be managed, and that specific end points are defined through contracts or other agreements. For a given organization (a contract developer, a government project office), that narrower focus may be all that matters, and may encompass billions of dollars. Often, by the time that narrower focus has been arrived at, the architecting is over. Often, by the time that narrower focus has been arrived at, the project is already doomed to failure or well on its way to success.

Table P.1 implies an important distinction in architecting as currently practiced. The table, and this book, emphasize architecting as decision making. Architecting has been accomplished when the fundamental structural decisions about a system have been made, regardless of what sort of architecture description document has been produced. In contrast, many "architecture" projects currently being conducted are description-centric. Their basis is producing an architecture framework compliant description document about a system or system-of-systems that typically already exists. These are sometimes called "as-is" or "baseline" architecture documents. This book has relatively little to say about such projects. The authors' emphasis, and the emphasis of this book, is on the structural decisions that underlie the "as-is" system. The methods of this book could be useful applied to making an assessment of those decisions, and reevaluating those decisions.

Architecting as Art and Science

Systems architecting is the subject of this book, and the art of it in particular, because, being the most interdisciplinary, its tools can be most easily applied in the other branches. Good architecting is not just an art, and virtually all architects of high-technology systems, in the authors' experience, have strong science backgrounds. But, the science needed for systems architecting already is the subject of many publications, but few address the art systematically and in depth. The overriding objective of this book is to bring the reader a toolbox of techniques for handling ill-structured, architectural problems that are different from the engineering methods already taught well and widely published.

It is important in understanding the subject of this book to clarify certain expressions. The word "architecture" in the context of civil works can mean a structure, a process, or a profession; in this text, it refers only to the structure, although we will often consider "structures" that are quite abstract. The word "architecting" refers only to the process. Architecting is an invented word to describe how architectures are created, much as engineering describes how "engines" and other artifacts are created. In another, subtler, distinction from conventional usage, an "architect" is meant here to be an individual engaged in the process of architecting, regardless of domain, job title, or employer. By definition and practice, from time to time an architect may perform engineering and an engineer may perform architecting — whatever it takes to get the job done.

Clearly, both processes involve elements of the other. Architecting requires top-level quantitative analysis to determine feasibility and quantitative measures to certify readiness for use. Engineering can and occasionally does require the creation of architecturally different alternatives to resolve otherwise intractable design problems. Good engineers are armed with an array of heuristics to guide tasks ranging from structuring a mathematical analysis to debugging a piece of electronic hardware. For complex systems, both engineering and architecting are essential.[*] In practice, it is usually necessary to draw a sharp line between them only when that sharp line is imposed by business or legal requirements.

Criteria for Mature and Effective Systems Architecting

An increasingly important need of project managers and clients is for criteria to judge the maturity and effectiveness of systems architecting in their projects — criteria analogous to those developed for software development by Carnegie Mellon's Software Engineering Institute. Based upon experience to date, criteria for systems architecting appear to be, in rough order of attainment:

- A recognition by clients and others of the need to architect complex systems.
- An accepted discipline to perform that function; in particular, the existence of architectural methods, standards, and organizations.
- A recognized separation of value judgments and technical decisions between client, architect, and builder.
- A recognition that architecture is an art as well as a science; in particular, the development and use of nonanalytic as well as analytic techniques.
- The effective utilization of an educated professional cadre — that is, of masters-level, if not doctorate-level, individuals and teams engaged in the process of systems-level architecting.

By those criteria, systems architecting is in its adolescence, a time of challenge, opportunity, and controversy. History and the needs of global competition would seem to indicate adulthood is close at hand.

---

[*] For further elaboration on the related questions of the role of the architect, see Rechtin 1991, pp. 11–14; on the architect's tools, Parts I and III of this book; on architecting as a profession, Part IV of this book and *Systems Engineering*, the Journal of the International Council on Systems Engineering.

The Architecture of This Book

The first priority of this book has been to restate and extend into the future the retrospective architecting paradigm of Rechtin 1991.* An essential part of both retrospective and extended paradigms is the recognition that systems architecting is part art and part science. Part I of this book further develops the art and extends the central role of heuristics. Part II introduces five important domains that contribute to the understanding of that art. We buttress the retrospective lessons of the original book by providing some detailed stories on some of the case studies that motivated the original work, and use those case studies to introduce each chapter in Part II. Part III helps bridge the space between the science and the art of architecting. In particular, it develops the core architecting process of modeling and representation. Part IV concentrates on architecting as a profession: its relationship to business strategy and activities, the political process and its part in system design, and the professionalization of the field through education, research, and peer-reviewed journals.

The architecture of Part II deserves an explanation. Without one, the reader may inadvertently skip some of the domains — builder-architected systems, manufacturing systems, social systems, software systems, and collaborative systems — because they are outside the reader's immediate field of interest. These chapters, instead, recognize the diverse origins of heuristics, illustrating and exploiting them. Heuristics often first surface in a specialized domain where they address an especially prominent problem. Then, by abstraction or analogy, they are carried over to others and become generic. Such is certainly the case in the selected domains. In these chapters, the usual format of stating a heuristic and then illustrating it in several domains is reversed. Instead it is stated, but in generic terms, in the domain where it is most apparent. Readers are encouraged to scan all the chapters of Part II. The chapters may even suggest domains, other than the reader's, where the reader's experience can be valuable in these times of vocational change. References are provided for further exploration. For professionals already in one of the domains, the description of each is from an architectural perspective, looking for those essentials that yield generic heuristics and providing in return other generic ones that might help better

---

* This second book is an extension of Rechtin 1991, not a replacement for it. However, this book reviews enough of the fundamentals that it can stand on its own. If some subjects, such as examples of specific heuristics, seem inadequately treated, the reader can probe further in the earlier work. There are also a number of areas covered there that are not covered here, including the challenges of ultraquality, purposeful opposition, economics, and public policy; biological architectures and intelligent behavior; and assessing architecting and architectures. A third book, Rechtin, E., *Systems Architecting of Organizations, Why Eagles Can't Swim*, Boca Raton, FL: CRC Press, 1999, introduces a part of systems architecting related to, but different from, the first two.

understand those essentials. In any case, the chapters most emphatically are not intended to advise specialists about their specialties.

Architecting is inherently a multidimensional subject, difficult to describe in the linear, word-follows-word, format of a book. Consequently, it is occasionally necessary to repeat the same concept in several places, internally and between books. A good example is the concept of systems. Architecting can also be organized around several different themes or threads. Rechtin 1991 was organized around the well-known waterfall model of system procurement. As such, its applicability to software development was limited. This book, more general, is by fundamentals, tools, tasks, domains, models, and vocation. Readers are encouraged to choose their own personal theme as they go along. It will help tie systems architecting to their own needs.

Exercises are interspersed in the text, designed for self-test of understanding and critiquing the material just presented. If the reader disagrees, then the disagreement should be countered with examples and lessons learned — the basic way that mathematically unprovable statements are accepted or denied. Most of the exercises are thought problems, with no correct answers. Read them, and if the response is intuitively obvious, charge straight ahead. Otherwise, pause and reflect a bit. A useful insight may have been missed. Other exercises are intended to provide opportunities for long-term study and further exploration of the subject. That is, they are roughly the equivalent of a master's thesis.

Notes and references are organized by chapter. Heuristics by tradition are boldfaced when they appear alone, with an appended list of them completing the text.

Changes Since the Second Edition

Since the publication of the second edition, it has become evident that some materials available to the authors are not generally available (case studies) and some subjects have been extensively developed in the years since publication. The authors have benefited from extensive feedback from working systems architects through teaching courses, seminars, and professional application. Where appropriate, that feedback has been incorporated into the book in the form of clearer explanations, useful case studies, better examples, and corrections to misunderstandings.

In several areas, we have added new material. A new chapter covers the relationships between architecting and the larger business (whether commercial or government) in which it is embedded. This subject has taken on great importance as it becomes apparent how deeply business strategy and architecture interrelate. We argue in this chapter that architecture can be seen as the physical (or technical) embodiment of strategy. Conversely, architecture without strategy is, essentially by definition, incoherent. Many of the common problems encountered in attempting to improve architecting practices can be linked directly to problems in organizational strategy. Moreover, this linkage provides fertile ground for looking at intellectual links with other engineering-related subjects, such as decision theory.

The chapter on architecture description frameworks has been revised in the light of developments since the second edition. As the importance of architectures has become more broadly accepted, standards have been promulgated and in some cases mandated. Most of these standards are related to architecture description, the equivalent of blueprint standards. The standards are roughly similar in intellectual approach, but they use distinctly different terminology and make quite different statements about what features are important. There is now enough experience in the community to identify common problems, and to recommend techniques drawn from the metaphor that motivates this book to address them.

We have also folded case study material into the book. The cases studied here formed part of the basic story used by the authors in a number of educational settings, but many of their details were either hard to find in print or became completely out of print. The generally available case study materials are also mostly historical and do not try to architecturally interpret the decisions that went into the systems. As a result, we have compiled some of the most interesting material that fits readily into book format here, and interleaved their presentation with the discussion of the related system categories.

Readership and Usage of This Book

This book is written for present and future systems architects, for experienced engineers interested in expanding their expertise beyond a single field, and for thoughtful individuals concerned with creating, building, or using complex systems. It is intended either for simple reading, for reference in professional practice, or in classroom settings. From experience with its predecessor, the book can be used as a reference work for graduate studies, for senior capstone courses in engineering and architecture, for executive training programs, and for the further education of consultants and systems acquisition and integration specialists, and as background for legislative staffs.

The book is a basic text for courses in systems architecture and engineering at several universities and in several extended professional courses. Best practice in using this book in such courses appears to be to combine it with selected case studies and an extended case exercise. Because architecting is about having skills, not about having rote knowledge, it can only be demonstrated in the doing. The author's courses have been built around course-long case exercises, normally chosen in the student's individual field. In public courses, such as at universities, the case studies presented here are appropriate for use. The source materials are reasonably available, and students can expand on what is presented here and create their own interpretations. In professional education settings, it is preferable to replace the case studies in class with case studies drawn directly from the student's home organizations.

Everything in this book represents the opinions of the authors and does not represent the official position of The Aerospace Corporation or its customers. All errors are the responsibility of the authors.

Acknowledgments

Eberhardt Rechtin, who originated and motivated so much of the thinking here, passed away in 2006. Although no longer with us, his spirit, and words, pervade this book. The first edition of this book was formulated while Rechtin taught at the University Southern California (USC). He treasured his interactions with his students there and believed that the work was enormously improved through the process of teaching them. At least a dozen of them deserve special recognition for their unique insights and penetrating commentary: Michael Asato, Kenneth Cureton, Susan Dawes, Norman P. Geis, Douglas R. King, Kathrin Kjos, Jonathan Losk, Ray Madachy, Archie W. Mills, Jerry Olivieri, Tom Pieronek, and Marilee Wheaton. The quick understanding and extension of the architecting process by all the students was been a joy to behold and a privilege to acknowledge.

Several members of the USC faculty were instrumental in finding a place for this work, and the associated program. In particular, there was Associate Dean Richard Miller, now President of Olin College; Associate Dean Elliot Axelband, who originally requested this book and directed the USC Masters Program in Systems Engineering and Architecture; and two members of the School of Engineering staff, Margery Berti and Mary Froehlig, who architected the Master of Science in Systems Architecture and Engineering out of an experimental course and a flexible array of multidisciplinary courses at USC. Particular thanks go to Charles Weber, who greatly encouraged Eb Rechtin in creating the program, and then encouraged his then graduate student, Mark Maier, to take the first class offered in systems architecting as part of his Ph.D. in Electrical Engineering Systems. Brenda Forman, then of USC, now retired from the Lockheed Martin Corporation and the author of Chapter 12, accepted the challenge of creating a unique course on the "facts of life" in the national political process and how they affect — indeed often determine — architecting and engineering design.

Our colleagues at The Aerospace Corporation have been instrumental in the later development of the ideas that have gone into this book.

Mark Maier has taught many versions of this material under the auspices of the Aerospace Systems Architecting Program and its derivatives. That program was dependent on the support of Mal Depont, William Hiatt, Dave Evans, and Bruce Gardner of the Aerospace Institute. The program in turn had many other collaborators, including Kevin Kreitman, Andrea Amram, Glenn Buchan, and James Martin. Also of great importance to the quality of the presentation has been the extensive editing and organization of the materials in the Aerospace Systems Architecting Program by Bonnie Johnston and Margaret Maher.

Manuscripts may be written by authors, but publishing them is a profession and contribution unto itself requiring judgment, encouragement, tact, and a considerable willingness to take risk. For all of these we thank Norm Stanton, a senior editor of Tayor & Francis/CRC Press and editor of the first edition of this book, who has understood and supported the field beginning with the publication of Frederick Brooks' classic architecting book, *The Mythical Man-Month*, more than two decades ago; and Cindy Carelli for her support of subsequent editions of this book.

Of course, a particular acknowledgment is due to the Rechtin and Maier families for the inspiration and support they have provided over the years, and their continuing support in revising this book.

Mark Maier

part I

Introduction

A Brief Review of Classical Architecting Methods

Architecting: The Art and Science of Designing and Building Systems[1]

The four most important methodologies in the process of architecting are characterized as normative, rational, participative, and heuristic[2] (Table I.1). As might be expected, like architecting itself, they contain both science and art. The science is largely contained in the first two, normative and rational, and the art in the last two, participative and heuristic.

The normative technique is solution based; it prescribes architecture as it "should be" — that is, as given in handbooks, civil codes, and pronouncements by acknowledged masters. Follow it and the result will be successful by definition.

Limitations of the normative method — such as responding to major changes in needs, preferences, or circumstances — led to the rational method, scientific and mathematical principles to be followed in arriving at a solution to a stated problem. It is method based or rule based. Both the normative and rational methods are analytic, deductive, experiment

Table I.1 Four Architecting Methodologies

Normative (solution based)
Examples: building codes and communications standards
Rational (method based)
Examples: systems analysis and engineering
Participative (stakeholder based)
Examples: concurrent engineering and brainstorming
Heuristic (lessons learned)
Examples: Simplify. Simplify. Simplify. and SCOPE!

based, easily certified, well understood, and widely taught in academia and industry. Moreover, the best normative rules are discovered through engineering science (think of modern building codes) — truly a formidable set of positives.

However, although science-based methods are absolutely necessary parts of architecting, they are not the focus of this book. They are already well treated in a number of architectural and engineering texts. Most people who are serious practitioners of systems architecting, or who aspire to be serious practitioners, come from an engineering and science background. They already realize the necessity of applying scientific and quantitative thinking to the design of complex systems. Equally necessary, and the focus of this part of the book, is the art, or practice, needed to complement the science for highly complex systems.

In contrast with science-based methodologies, the art or practice of architecting — like the practices of medicine, law, and business — is nonanalytic, inductive, difficult to certify, less understood, and, at least until recently, is seldom taught formally in either academia or industry. It is a process of insights, vision, intuitions, judgment calls, and even "taste."[3] It is key to creating truly new types of systems for new and often unprecedented applications. Here are some of the reasons.

For unprecedented systems, past data are of limited use. For others, analysis can be overwhelmed by too many unknowns, too many stakeholders, too many possibilities, and too little time for data gathering and analysis to be practical. To cap it off, many of the most important factors are not measurable. Perceptions of worth, safety, affordability, political acceptance, environmental impact, public health, and even national security provide no realistic basis for numerical analyses — even if they were not highly variable and uncertain. Yet, if the system is to be successful, these perceptions must be accommodated from the first, top-level, conceptual model down through its derivatives.

The art of architecting, therefore, complements its science where science is weakest: in dealing with immeasurables, in reducing past experience and wisdom to practice, in conceptualization, in inspirationally putting disparate things together, in providing "sanity checks," and in warning of likely but unprovable trouble ahead. Terms like reasonable assumptions, guidelines, indicators, elegant design, and beautiful performance are not out of place in this art, nor are lemon, disaster, snafu, or loser. These terms are hardly quantifiable, but are as real in impact as any science.

The participative methodology recognizes the complexities created by multiple stakeholders. Its objective is consensus. As a notable example, designers and manufacturers need to agree on a multiplicity of details if an end product is to be manufactured easily, quickly, and profitably.

In simple but common cases, only the client, architect, and contractor have to be in agreement. But as systems become more complex, new and different participants have to agree as well.

Concurrent engineering, a recurrently popular acquisition method, was developed to help achieve consensus among many participants. Its greatest values, and its greatest contentions, are for systems in which widespread cooperation is essential for acceptance and success, for example, systems that directly impact on the survival of individuals or institutions. Its well-known weaknesses are undisciplined design by committee, diversionary brainstorming, the closed minds of "groupthink," and members without power to make decisions but with unbridled right to second guess. Arguably, the greatest mistake that can be made in concurrent engineering is to attempt to quantify it. It is not a science. It is a very human art.

The heuristics methodology is based on "common sense" — that is, on what is sensible in a given context. Contextual sense comes from collective experience stated in as simple and concise a manner as possible. These statements are called heuristics, the subject of Chapter 2, and are of special importance to architecting because they provide guides through the rocks and shoals of intractable, "wicked" system problems. *Simplify!* is the first and probably most important of them. They exist in the hundreds if not thousands in architecting and engineering, yet they are some of the most practical and pragmatic tools in the architect's kit of tools.

Different Methods for Different Phases of Architecting

The nature of classical architecting changes as the project moves from phase to phase. In the earliest stages of a project, it is a structuring of an unstructured mix of dreams, hopes, needs, and technical possibilities when what is most needed has been called an inspired synthesizing of feasible technologies. It is a time for the art of architecting. Later on, architecting becomes an integration of, and mediation among, competing subsystems and interests — a time for rational and normative methodology. And finally, there comes certification to all that the system is suitable for use, when it may take all the art and science to that point to declare the system as built is complete and ready for use.

Not surprisingly, architecting is often individualistic, and the end results reflect it. As Frederick P. Brooks put it in 1983[4] and Robert Spinrad stated in 1987,[5] the greatest architectures are the product of a single mind — or of a very small, carefully structured team. To which should be added in all fairness: a responsible and patient client, a dedicated builder, and talented designers and engineers.

Notes

1. *Webster's II, New Riverside University Dictionary.* Boston, MA: Riverside 1984. As adapted for systems by substitution of "building systems" for "erecting buildings."
2. For a full discussion of these methods, see Lang, Jon, *Creating Architectural Theory, The Role of the Behavioral Sciences in Environmental Design.* New York: Van Nostrand Reinhold, 1987; Rowe, Peter G., *Design Thinking.* Cambridge, MA: MIT Press, 1987. They are adapted for systems architecting in Rechtin 1991, pp. 14–22.
3. Spinrad, Robert J., in a lecture at the University of Southern California, 1988.
4. Brooks, Frederick P., *The Mythical Man-Month, Essays on Software Engineering.* Reading, MA: Addison Wesley, 1983.
5. Spinrad, Robert J., at a Systems Architecting lecture at the University of Southern California, Fall 1987.

chapter 1

Extending the Architecting Paradigm

Introduction: The Classical Architecting Paradigm

The recorded history of classical architecting, the process of creating architectures, began in Egypt more than 4,000 years ago with the pyramids, the complexity of which had been overwhelming designers and builders alike. This complexity had at its roots the phenomenon that as systems became increasingly more ambitious, the number of interrelationships among the elements increased far faster than the number of elements. These relationships were not solely technical. Pyramids were no longer simple burial sites; they had to be demonstrations of political and religious power, secure repositories of god-like rulers and their wealth, and impressive engineering accomplishments. Each demand, of itself, would require major resources. When taken together, they generated new levels of technical, financial, political, and social complications. Complex interrelationships among the combined elements were well beyond what the engineers' and builders' tools could handle.

From that lack of tools for civil works came classical or civil architecture. Millennia later, technological advances in shipbuilding created the new and complementary fields of marine engineering and naval architecture. In this century, rapid advances in aerodynamics, chemistry, materials, electrical energy, communications, surveillance, information processing, and software have resulted in systems whose complexity is again overwhelming past methods and paradigms. One of those is the classical architecting paradigm. But, if we are to understand and respond to the complexity overwhelming the classical paradigm, we must first understand that classical paradigm.

Responding to Complexity

Complex: Composed of interconnected or interwoven parts.[1]
System: A set of different elements so connected or related as to perform a unique function not performable by the elements alone.[2]

It is generally agreed that increasing complexity* is at the heart of the most difficult problems facing today's systems architecting and engineering. When architects and builders are asked to explain cost overruns and schedule delays, by far the most common, and quite valid, explanation is that the system is much more complex than originally thought. The greater is the complexity, the greater the difficulty. It is important, therefore, to understand what is meant by system complexity if architectural progress is to be made in dealing with it.

The definitions of *complexity* and *systems* given at the beginning of this section are remarkably alike. Both speak to interrelationships (interconnections, interfaces, and so forth) among parts or elements. As might be expected, the more elements and interconnections, the more complex the architecture and the more difficult the system-level problems.

Less apparent is that qualitatively different problem-solving techniques are required at high levels of complexity than at low ones. Purely analytical techniques, powerful for the lower levels, can be overwhelmed at the higher ones. At higher levels, architecting methods, experience-based heuristics, abstraction, and integrated modeling must be called into play.[3] The basic idea behind all of these techniques is to simplify problem solving by concentrating on its essentials. Consolidate and simplify the objectives. Focus on the things with the highest impact, things that determine other things. Put to one side minor issues likely to be resolved by the resolution of major ones. Discard the nonessentials. Model (abstract) the system at as high a level as possible, then progressively reduce the level of abstraction. In short: *Simplify!*

It is important in reading about responses to complexity to understand that they apply throughout system development, not just to the conceptual phase. The concept that a complex system can be progressively partitioned into smaller and simpler units, and hence into simpler problems, omits an inherent characteristic of complexity — interrelationships among the units. As a point of fact, poor aggregation and partitioning during development can *increase* complexity, a phenomenon all too apparent in the organization of work breakdown structures.

This primacy of complexity in system design helps explain why a single "optimum" seldom if ever exists for such systems. There are just too many variables. There are too many stakeholders and too many conflicting interests. No practical way may exist for obtaining information critical in making a "best" choice among quite different alternatives.

---

* A system need not be large or costly to be complex. The manufacture of a single mechanical part can require over 100 interrelated steps. A $10 microchip can contain millions of interconnected active elements.

The High Rate of Advances in the Computer and Information Sciences

Unprecedented rates of advance in the computer and information sciences have further exacerbated an already complex picture. The advent of smart, software-intensive systems is producing a true paradigm shift in system design. Software, long treated as the glue that tied hardware elements together, is becoming the center of system design and operation. We see it in consumer electronic devices of all types. The precipitous drop in hardware costs has generated a major design shift — from "keep the computer busy" to "keep the user busy." Designers happily expend hardware resources to save redesigning either hardware or software. We see it in automobiles, where software increasingly determines the performance, quality, cost, and feel of cars and trucks. We see it in aircraft, where controls are coming to drive aerodynamic and structural design, and military system designers discuss a shift to designing the airframe around the sensors instead of designing the sensors around the airframe.

We see the paradigm shift in the design of spacecraft and personal computers, where complete character changes can be made in minutes. In effect, such software-intensive systems "change their minds" on demand. It is no longer a matter of debate whether machines have "intelligence"; the only real questions are of what kinds of intelligence and how best to use each one. And, because its software largely determines what and how the user perceives the system as a whole, its design will soon control and precede hardware design much as hardware design controls software today. This shift from "hardware first" to "software first" will force major changes on when and how system elements are designed, and who, with what expertise, will design the system as a whole. The impact on the value of systems to the user has been and will continue to be enormous.

One measure of this phenomenon is the proportion of development effort devoted to hardware and software for various classes of product. Anecdotal reports from a variety of firms in telecommunications and consumer electronics commonly show a reversal of the proportion from 70% hardware and 30% software to 30% hardware and 70% software. This shift has created major challenges and destroyed some previously successful companies. When the cost of software development dominates total development, systems should be organized to simplify software development. But good software architectures and good hardware architectures are often quite different. Good architectures for complex software usually emphasize layered structures that cross many physically distinct hardware entities. Good software architectures also emphasize information hiding and close parallels between implementation constructs and domain concepts at the upper layers. These are in contrast to the emphasis on hierarchical decomposition, physical locality of communication, and interface transparency in good hardware architectures. Organizations find trouble when their workload moves from hardware to software dominated, but their management and development skills no longer "fit" the systems they should support.

Particularly susceptible to these changes are systems that depend upon electronics and information systems and that do not enjoy the formal partnership with architecting that structural engineering has long enjoyed. This book is an effort to remedy that lack by showing how the historical principles of classical architecting can be extended to modern systems architecting.

The Foundations of Modern Systems Architecting

Although the day-to-day practice may differ significantly,[4] the foundations of modern systems architecting are much the same across many technical disciplines. Generally speaking, they are a systems approach, a purpose orientation, a modeling methodology, ultraquality, certification, and insight.[5] Each will be described in turn.

A Systems Approach

A systems approach is one that focuses on the system as a whole, specifically linking value judgments (what is desired) and design decisions (what is feasible). A true systems approach means that the design process includes the "problem" as well as the solution. The architect seeks a joint problem–solution pair and understands that the problem statement is not fixed when the architectural process starts. At the most fundamental level, *systems are collections of different things that together produce results unachievable by the elements alone.* For example, only when all elements are connected and working together do automobiles produce transportation, human organs produce life, and spacecraft produce information. These system-produced results, or "system functions," derive almost solely from the interrelationships among the elements, a fact that largely determines the technical role and principal responsibilities of the systems architect.

Systems are interesting because they achieve results, and achieving those results requires different things to interact. From much experience with it over the last decade, it is difficult to underestimate the importance of this specific definition of systems to what follows, literally on a word-by-word basis. Taking a systems approach means paying close attention to results, the reasons we build a system. Architecture *must* be grounded in the client's/user's/customer's purpose. Architecture is not just about the structure of components. One of the essential distinguishing features of architectural design versus other sorts of engineering design is the degree to which architectural design embraces results from the perspective

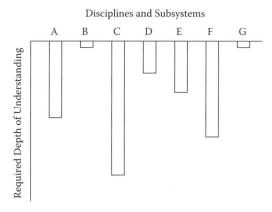

Figure 1.1 The architect's depth of understanding of subsystem and disciplinary details.

of the client/user/customer. The architect does not assume some particular problem formulation, as "requirements" is fixed. The architect engages in joint exploration, ideally directly with the client/user/customer, of what system attributes will yield results worth paying for.

It is the responsibility of the architect to know and concentrate on the critical few details and interfaces that really matter and not to become overloaded with the rest. It is a responsibility that is important not only for the architect personally but for effective relationships with the client and builder. To the extent that the architect must be concerned with component design and construction, it is with those specific details that critically affect the system as a whole.

For example, a loaded question often posed by builders, project managers, and architecting students is, "How deeply should the architect delve into each discipline and each subsystem?" A graphic answer to that question is shown in Figure 1.1, exactly as sketched by Bob Spinrad in a 1987 lecture at the University of Southern California. The vertical axis is a relative measure of how deep into a discipline or subsystem an architect must delve to understand its consequences to the system as a whole. The horizontal axis lists the disciplines, such as electronics or stress mechanics, and the subsystems, such as computers or propulsion systems. Depending upon the specific system under consideration, a great deal, or a very little depth, of understanding may be necessary.

But that leads to another question, "How can the architect possibly know before there is a detailed system design, much less before system test, what details of what subsystem are critical?" A quick answer is: only through experience, through encouraging open dialog with subsystem specialists, and by being a quick, selective, tactful, and effective student of the system and its needs. Consequently, and perhaps more than any other specialization, architecting is a continuing, day-to-day learning process. No two systems are exactly alike. Some will be unprecedented, never built before.

> *Exercise*: Put yourself in the position of an architect asked to help a client build a system of a new type whose general nature you understand (a house, a spacecraft, a nuclear power plant, or a system in your own field) but which must considerably outperform an earlier version by a competitor. What do you expect to be the critical elements and details and in what disciplines or subsystems? What elements do you think you can safely leave to others? What do you need to learn the most about? Reminder: You will still be expected to be responsible for all aspects of the system design.

Critical details aside, the architect's greatest concerns and leverage are still, and should be, with the systems' connections and interfaces: First, because they distinguish a system from its components; second, because their addition produces unique system-level functions, a primary interest of the systems architect; and third, because subsystem specialists are likely to concentrate most on the core and least on the periphery of their subsystems, viewing the latter as (generally welcomed) external constraints on their internal design. Their concern for the system as a whole is understandably less than that of the systems architect; if not managed well, the system functions can be in jeopardy.

A Purpose Orientation

Systems architecting is a process driven by a client's purpose or purposes. A president wants to meet an international challenge by safely sending astronauts to the moon and back. Military services need nearly undetectable strike aircraft. Cities call for pollutant-free transportation.

Clearly, if a system is to succeed, it must satisfy a useful purpose at an affordable cost for an acceptable period of time. Note the explicit value judgments in these criteria: a *useful* purpose, an *affordable* cost, and an *acceptable* period of time. Every one is the client's prerogative and responsibility, emphasizing the criticality of client participation in all phases of system acquisition. But of the three criteria, satisfying a *useful* purpose is predominant. Without it being satisfied, all others are irrelevant. Architecting therefore begins with, and is responsible for maintaining, the integrity of the system's utility or purpose.

For example, the Apollo manned mission to the moon and back had a clear purpose, an agreed cost, and a no-later-than date. It delivered on all three. Those requirements, kept up front in every design decision, determined the mission profile of using an orbiter around the moon and not an earth-orbiting space station, and on developing electronics for a lunar orbit rendezvous instead of developing an outsize propulsion system for a direct approach to the lunar surface.

As another example, NASA Headquarters, on request, gave the NASA/JPL Deep Space Network's huge ground antennas a clear set of priorities: first performance, then cost, then schedule, even though the primary missions they supported were locked into the absolute timing of planetary arrivals. As a result, the first planetary communication systems were designed with an alternate mode of operation in case the antennas were not yet ready. As it turned out, and as a direct result of the NASA risk-taking decision, the antennas were carefully designed, not rushed, and satisfied all criteria not only for the first launch but for all launches for the next 40 years or so.

The Douglas Aircraft DC-3, though originally thought by the airline (later TWA) to require three engines, was rethought by the client and the designers in terms of its underlying purpose — to make a profit on providing affordable long-distance air travel over the Rocky and Sierra Nevada mountains for paying commercial passengers. The result was the two-engine DC-3, the plane that introduced global air travel to the world.

When a system fails to achieve a useful purpose, it is doomed. When it achieves some purpose but at an unfavorable cost, its survival is in doubt, but it may survive. The purpose for which the Space Shuttle was conceived and sold, low-cost transport to low earth orbit, has never been achieved. However, its status as the sole U.S. source of manned space launch has allowed its survival. Many will argue that the Space Shuttle was a tremendous technical achievement, and there is little doubt it was. The success of architecting is not measured by technical success, but by success in mission. In a similar fashion, it has proven impossible to meet the original purpose of the space station at an acceptable cost, but its role in the U.S. manned space program and international space diplomacy has assured minimum survival. In contrast, the unacceptable cost/benefit ratios of the supersonic transport, the space-based ballistic missile defense system, and the superconducting supercollider terminated all these projects before their completion.

Curiously, the end use of a system is not always what was originally proposed as its purpose. The F-16 fighter aircraft was designed for visual air-to-air combat, but in practice it has been most used for ground support. The ARPANET-INTERNET communication network originated as a government-furnished computer-to-computer linkage in support of university research; it is now most used, and paid for, by individuals for e-mail and information accessing. Both are judged as successful. Why? Because, as circumstances changed, providers and users redefined the meaning of useful, affordable, and acceptable. A useful heuristic comes to mind: *Design the structure with "good bones."* It comes from the architecting of buildings, bridges, and ships, where it refers to structures that are resilient to a wide range of stresses and changes in purpose. It could just as well come from physiology and the remarkably adaptable spinal column and appendages of all vertebrates — fishes, amphibians, reptiles, birds, and mammals.

> *Exercise*: Identify a system whose purpose is clear and unmistakable. Identify, contact, and if possible, visit its architect. Compare notes and document what you learned.

Technology-driven systems, in notable contrast to purpose-driven systems, tell a still different story. They are the subject of Chapter 3.

A Modeling Methodology

Modeling is the creation of abstractions or representations of the system to predict and analyze performance, costs, schedules, and risks and to provide guidelines for systems research, development, design, manufacture, and management. Modeling is the centerpiece of systems architecting — a mechanism of communication to clients and builders, of design management with engineers and designers, of maintaining system integrity with project management, and of learning for the architect, personally.

> *Examples*: The balsa wood and paper scale models of a residence, the full-scale mockup of a lunar lander, the rapid prototype of a software application, the computer model of a communication network, or the mental model of a user.

Modeling is of such importance to architecting that it is the sole subject of Part III. Modeling is the fabric of architecting because architecting is at a considerable distance of abstraction from actual construction. The architect does not manipulate the actual elements of construction. The architect builds models that are passed into more detailed design processes. Those processes lead, eventually, to construction drawings or the equivalent and actual system fabrication or coding.

Viewing architecting and design as a continuum of modeling refinement leads naturally to the "stopping question." Where does architecting stop and engineering or design begin? Or, when should we stop any design activity and move onto the next stage? From a modeling perspective, there is no stopping. Rather modeling is seen to progress and evolve, continually solving problems from the beginning of a system's acquisition to its final retirement. There are of course conceptual models, but there are also engineering models and subsystem models; models for simulation, prototypes, and system test; demonstration models, operational models and mental models by the user of how the system behaves. From another perspective, careful examination of the "stopping question" leads us to a better understanding of the purpose of any particular architecting or design phase. Logically, they stop when their purpose is fulfilled.

Models are in fact created by many participants, not just by architects. These models must somehow be made consistent with overall system imperatives. It is particularly important that they be consistent with the architect's system model, a model that evolves, becoming more and more concrete and specific as the system is built. It provides a standard against which consistency can be maintained and is a powerful tool in maintaining the larger objective of system integrity. And finally, when the system is operational and a deficiency or failure appears, a model — or full-scale simulator if one exists — is brought into play to help determine the causes and cures of the problem. The more complete the model, the more accurately possible failure mechanisms can be duplicated until the only cause is identified.

In brief, modeling is a multipurpose, progressive activity, evolving and becoming less abstract and more concrete as the system is built and used.

Ultraquality Implementation

Ultraquality is defined as a level of quality so demanding that it is impractical to measure defects, much less certify the system prior to use.[6] It is a limiting case of quality driven to an extreme, a state beyond acceptable quality limits (AQLs) and statistical quality control. It requires a zero defect approach not only to manufacturing but also to design, engineering, assembly, test, operation, maintenance, adaptation, and retirement — in effect, the complete life cycle.

Some examples include a new-technology spacecraft with a design lifetime of at least 10 years, a nuclear power plant that will not fail catastrophically within the foreseeable future, and a communication network of millions of nodes, each requiring almost 100% availability. In each case, the desired level of quality cannot, even in principle, be directly measured; or, only the absence of the quality desired can be directly measured. Ultraquality is a recognition that the more components there are in a system, the more reliable each component must be to a point where, at the element level, defects become impractical to measure within the time and resources available. Or, in a variation on the same theme, the operational environment cannot be created during test at a level or for a duration that allows measurement at the system level. Yet, the reliability goal of the system as a whole must still be met. In effect, it reflects the reasonable demand that a system — regardless of size or complexity — should not fail to perform more than about 1% or less of the time. An intercontinental ballistic missile (ICBM) should not. A space shuttle, at least 100 times more complex, should not. An automobile should not. A passenger airliner, at least 100 times more complex, should not; as a matter of fact, we expect the airliner to fail far, far less than the family car.

> *Exercise*: Trace the histories of commercial aircraft and passenger buses over the last 50 years in terms of the number of trips that a passenger would expect to make without an accident. What does that mean to vehicle reliability as trips lengthen and become more frequent, as vehicles get larger, faster, and more complex? How were today's results achieved? What trends do you expect in the future? Did more software help or hinder vehicle safety?

The subject would be moot if it were not for the implications of this "limit state" of zero defects to design. Zero defects, in fact, originated as long ago as World War II, largely driven by patriotism. As a motivator, the zero defects principle was a prime reason for the success of the Apollo mission to the moon.

To show the implications of ultraquality processes, if a manufacturing line operated with zero defects, there would be no need, indeed it would be worthless, to build elaborate instrumentation and information-processing support systems. This would reduce costs and time by 30%. If an automobile had virtually no design or production defects, then sales outlets would have much less need for large service shops with their high capital and labor costs. Indeed, the service departments of the finest automobile manufacturers are seldom fully booked, resembling something like the famous Maytag commercial. Very little repair or service, except for routine maintenance, is required for 50,000 to 100,000 miles. Not coincidentally, these shops invariably are spotlessly clean, evidence of both the professional pride and discipline required for sustaining an ultraquality operation. Conversely, a dirty shop floor is one of the first and best indicators to a visitor or inspector of low productivity, careless workmanship, reduced plant yield, and poor product performance. The rocket, ammunition, solid-state component, and automotive domains all bear witness to that fact.

As another example, microprocessor design and development has maintained the same per-chip defect rate even as the number and complexity of operations increased by factors of thousands. The corresponding failure rate per individual operation is now so low as to be almost unmeasurable. Indeed, for personal computer applications, a microprocessor hardware failure more than once a year is already unacceptable.

Demonstrating this limit state in high quality is not a simple extension of existing quality measures, though the latter may be necessary in order to get within range of it. In the latter there is a heuristic: [Measurable] *acceptance tests must be both complete and passable.* How then can inherently unmeasurable ultraquality be demanded or certified? The answer is a mixture of analytical and heuristic approaches, forming a set of surrogate procedures, such as zero defects programs. Measurements play an important role but are always indirect because of the immeasurability of the core quality factors of interest.

In looking at procedural approaches, a powerful addition to pre-1990 ultraquality techniques was the concept, introduced in the last few years, that each participant in a system acquisition sequence is both a buyer and a supplier. The original application, apparently a Japanese idea, was that each worker on a production line was a buyer from the preceding worker in the production line as well as a supplier to the next. Each role required a demand for high quality — that is, a refusal to buy a defective item and a concern not to deliver a defective one likely to be refused.[7] In effect, the supplier–buyer concept generates a self-enforcing quality program with built-in inspection. There would seem to be no reason why the same concept should not apply throughout system acquisition — from architect to engineer to designer to producer to seller to end user. As with all obvious ideas, the wonder is why it was not self-evident earlier.

When discussing ultraquality, it may seem odd to be discussing heuristics. After all, is not something as technologically demanding as quality beyond measure, the performance of things like heavy space boosters, not the domain of rigorous, mathematical engineering? In part, of course, it is. But experience has shown that rigorous engineering is not enough to achieve ultraquality systems. Ultraquality is achieved by a mixture of analytical and heuristic methods. The analytical side is represented by detailed failure analysis and even the employment of proof techniques in system design. In some cases, these very rigorous techniques have been essential in allowing certain types of ultraquality systems to be architected.

Flight computers are a good example of the mixture of analytical and heuristic considerations in ultraquality systems. Flight control computers for statically unstable aircraft are typically required to have a mean time between failures (where a failure is one that produces incorrect flight control commands) on the order of 10 billion hours. This is clearly an ultraquality requirement because the entire production run of a given type of flight computer will not collectively run for 10 billion hours during its operational lifetime. The requirement certainly cannot be proved by measurement and analysis. Nevertheless, aircraft administration authorities require that such a reliability requirement be certified.

Achieving the required reliability would seem to necessitate a redundant computer design as individual parts cannot reach that reliability level. The problem with redundant designs is that introducing redundancy also introduces new parts and functions, specifically the mechanisms that manage the redundancy, and must lock out the signals from redundant sections that have failed. For example, in a triple redundant system, the redundant components must be voted to take the majority position (locking out a presumptive single failure). The redundancy management components are subject to failure, and it is possible that a redundant system is actually more likely to fail than one without redundancy. Further, "fault tolerance" depends upon the fault to be tolerated. Tolerating mechanical failure is of limited value if the fault is human error.

Creating redundant computers has been greatly helped by better analysis techniques. There are proof techniques that allow pruning of the unworkable failure trees by assuming "Byzantine" failure* models. These techniques allow strong statements to be made about the redundancy properties of designs. The heuristic part is trying to verify the absence of "common-mode-failures," or failures in which several redundant and supposedly independent components fail at the same time for the same reason.

The Ariane 5 space launch vehicle was destroyed on its initial flight in a classic common mode failure. The software on the primary flight control computer caused the computer to crash shortly after launch. The dual redundant system then switched to the backup flight control computer, which had failed as well moments before for exactly the same reason that the primary computer failed. Ironically, the software failure was due to code leftover from the Ariane 4 that was not actually necessary for the phase of flight in which it was operating. Arguably, in the case of the Ariane 5, more rigorous proof-based techniques of the mixed software and systems design might have found and eliminated the primary failure. But, the failure is a classical example of a "common mode failure," where redundant systems are simultaneously carried away by the same reason. Greater rigor in tracing how an implemented system meets the assumptions it was built to can never eliminate the failures that are inherent in the original assumptions.

Thus, the analytical side is not enough for ultraquality. The best analysis of failure probabilities and redundancy can only verify that the system

---

* In a Byzantine failure, the failed component does the worst possible thing to the system. It is as if the component were possessed by a malign intelligence. The power of the technique is that it lends itself to certification, at least within the confines of well-defined models.

as built agrees with the model analyzed, and that the model possesses desired properties. It cannot verify that the model corresponds to reality. Well-designed ultraquality systems fail, but they typically fail for reasons not anticipatable in the reliability model.

Certification

Certification is a formal statement by the architect to the client or user that the system, as built, meets the criteria both for client acceptance and for builder receipt of payment; that is, it is ready for use (to fulfill its purposes). Certification is the grade on the "final exams" of system test and evaluation. To be accepted, it must be well supported, objective, and fair to client and builder alike.

> *Exercise*: Pick a system for which the purposes are reasonably clear. What tests would you, as a client, demand be passed for you to accept and pay for the system? What tests would you, as a builder, contract to pass in order to be paid? Whose word would each of you accept that the tests had or had not been passed? When should such questions be posed? (Hopefully, quite early, before the basic concept has been decided upon!)

Clearly, if certification is to be unchallenged, then there must be no perception of conflict of interest of the architect. This imperative has led to three widely accepted, professionally understood, constraints[8] on the role of the architect:

1. *A disciplined avoidance of value judgments* — that is, of intruding in questions of worth to the client; questions of what is satisfactory, what is acceptable, affordable, maintainable, reliable, and so on. Those judgments are the imperatives, rights, and responsibilities of the client. As a matter of principle, the client should judge on desirability and the architect should decide (only) on feasibility. To a client's question of "What would you do in my position?" the experienced architect responds only with further questions until the client can answer the original one. To do otherwise makes the architect an advocate and, in some sense, the "owner" of the end system, preempting the rights and responsibilities of the client. It may make the architect famous, but the client will feel used. Residences, satellites, and personal computers have all suffered from such

preemption (Frank Lloyd Wright houses, low earth-orbiting satellite constellations, and the Lisa computer, respectively).*

2. *A clear avoidance of perceived conflict of interest* through participation in research and development, including ownership or participation in organizations that can be, or are, building the system. The most evident conflict here is the architect recommending a system element that the architect will supply and profit from. This constraint is particularly important in builder-architected systems (Chapter 3).†

3. *An arms-length relationship with project management* — that is, with the management of human and financial resources other than of the architect's own staff. The primary reason for this arrangement is the overload and distraction of the architect created by the time-consuming responsibilities of project management. A second conflict, similar to that of participating in research and development, is created whenever architects give project work to themselves. If clients, for reasons of their own, nonetheless ask the architect to provide project management, it should be considered as a separate contract for a different task requiring different resources.

Insight and Heuristics

A picture is worth a thousand words.

Chinese Proverb, 1000 b.c.

One insight is worth a thousand analyses.

Charles Sooter, April 1993

Insight, or the ability to structure a complex situation in a way that greatly increases understanding of it, is strongly guided by lessons learned from one's own or others' experiences and observations. Given enough lessons, their meaning can be codified into succinct expressions called "heuristics," a Greek term for guide. Heuristics are an essential complement to analytics, particularly in situations where analysis alone cannot provide either insights or guidelines.[9] In many ways, they resemble what are called principles in other arts; for example, the importance of balance and proportion in a painting, a musical composition, or the ensemble of a string quartet. Whether as heuristics or principles, they encapsulate the

---

* That said, when we break away from the classical architecting paradigm, we will see how responsibilities may change, and the freedom and risks inherent in doing so.

† Precisely this constraint led the Congress to mandate the formation in 1960 of a nonprofit engineering company, The Aerospace Corporation, out of the for-profit TRW Corporation, a builder in the aerospace business.

insights that have to be attained and practiced before a masterwork can be achieved.

Both architecting and the fine arts clearly require insight and inspiration as well as extraordinary skill to reach the highest levels of achievement. Seen from this perspective, the best systems architects are indeed artists in what they do. Some are even artists in their own right. Renaissance architects like Michaelangelo and Leonardo da Vinci were also consummate artists. They not only designed cathedrals, they executed the magnificent paintings in them. The finest engineers and architects, past and present, are often musicians; Simon Ramo and Ivan Getting, famous in the missile and space field, and, respectively, a violinist and pianist, are modern-day examples.

The wisdom that distinguishes the great architect from the rest is the insight and the inspiration, that combined with well-chosen methods and guidelines and fortunate circumstances, creates masterworks. Unfortunately, wisdom does not come easily. As one conundrum puts it:

- *Success comes from wisdom.*
- *Wisdom comes from experience.*
- *Experience comes from mistakes.*

Therefore, because success comes only after many mistakes, something few clients would willingly support, one might think it is either unlikely or must follow a series of disasters.

This reasoning might well apply to an individual. But applied to the profession as a whole, it clearly does not. The required mistakes and experience and wisdom gained from them can be those of one's predecessors, not necessarily one's own. Organizations that care about successful architecting consider designing their program portfolios to generate experience. When staged experience is understood as important, staged experience can be designed into an organization.

And from that understanding comes the role of education. It is the place of education to research, document, organize, codify, and teach those lessons so that the mistakes need not be repeated as a prerequisite for future successes. Chapter 2 is a start in that direction for the art of systems architecting.

The Architecture Paradigm Summarized

This book uses the terms *architect, architecture,* and *architecting* with full consciousness of the "baggage" that comes with their use. Civil architecture is a well-established profession with its own professional societies, training programs, licensure, and legal status. Systems architecting borrows from it its basic attributes:

1. The architect is principally an agent of the client, not the builder. Whatever organization the architect is employed by, the architect must act in the best interests of the client for whom the system is being developed.
2. The architect works jointly with the client and builder on problem and solution definition. System requirements are an output of architecting, not really an input. Of course, the client will provide the architect some requirements, but the architect is expected to jointly help the client determine the ultimate requirements to be used in acquiring the system. An architect who needs complete and consistent requirements to begin work, though perhaps a brilliant builder, is not an architect.
3. The architect's product, or "deliverable," is an architecture representation, a set of abstracted designs of the system. The designs are not (usually) ready to use to build something. They have to be refined, just as the civil architect's floor plans, elevations, and other drawings must be refined into construction drawings.
4. The architect's product is not just physical representations. As an example, the civil architect's client certainly expects a "ballpark" cost estimate as part of any architecture feasibility question. So, too, in systems architecting, where an adequate system architecture description must cover whatever aspects of physical structure, behavior, cost, performance, human organization, or other elements are needed to clarify the clients' priorities.
5. An initial architecture is a vision. An architecture description is a set of specific models. The architecture of a building is more than the blueprints, floor plans, elevations, and cost estimates; it includes elements of ulterior motives, belief, and unstated assumptions. This distinction is especially important in creating standards. Standards for architecture, like community architectural standards, are different from blueprint standards promoted by agencies or trade associations.

Architecting takes place within the context of an acquisition process. The traditional way of viewing hardware acquisitions is known as the waterfall model. The waterfall model captures many important elements of architecting practice, but it is also important in understanding other acquisition models, particularly the spiral for software, incremental development for evolutionary designing, and collaborative assembly for networks.

The Waterfall Model of Systems Acquisition

As with products and their architectures, no process exists by itself. All processes are part of still larger ones. And all processes have subprocesses.

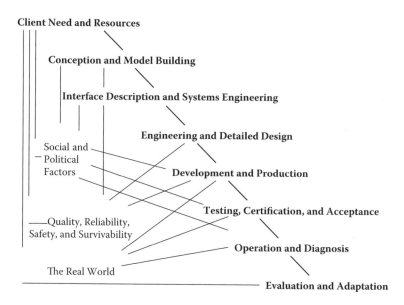

Figure 1.2 The expanded waterfall.

As with the product of architecture, so also is the process of architecting a part of a still larger activity, the acquisition of useful things.

Hardware acquisition is a sequential process that includes design, engineering, manufacturing, testing, and operation. This larger process can be depicted as an expanded waterfall, Figure 1.2.[10] The architect's functional relationship with this larger process is sketched in Figure 1.3. Managerially, the architect could be a member of the client's or the builder's organization, or of an independent architecting partnership in which perceptions of conflict of interest are to be avoided at all costs. In any case and wherever the architect is physically or managerially located, the relationships to the client and the acquisition process are essentially as shown. The strongest (thickest line) decision ties are with client need and resources, conception and model building, and with testing, certification, and acceptance. Less prominent are the monitoring ties with engineering and manufacturing. There are also important, if indirect, ties with social and political factors, the "illities" and the "real world."

This waterfall model of systems acquisition has served hardware systems acquisition well for centuries. However, as new technologies create new, larger-scale, more complex systems of all types, others have been needed and developed. The most recent ones are due to the needs of software-intensive systems, as will be seen in Chapters 4 and 6 and in Part III. Although these models change the roles and methods of the

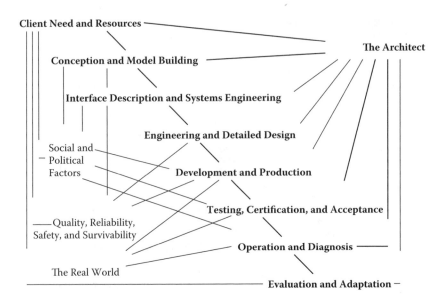

Figure 1.3 The architect and the expanded waterfall. (Adapted from Rechtin, E., *Systems Architecting, Creating and Building Complex Systems.* Englewood Cliffs, NJ: Prentice Hall, 1991. With permission from Prentice Hall.)

architecting process, the basic functional relationships shown in Figure 1.3 remain much the same.

In any case, the relationships in Figure 1.3 are more complex than simple lines might suggest. As well as indicating channels for two-way communication and upward reporting, they infer the tensions to be expected between the connected elements, tensions caused by different imperatives, needs, and perceptions.

Some of competing technical factors are shown in Figure 1.4.[11] This figure was drawn such that directly opposing factors pull in exactly opposite directions on the chart. For example, continuous evolution pulls against product stability; a typical balance is that of an architecturally stable, evolving product line. Low-level decisions pull against strict process control, which can often be relieved by systems architectural partitioning, aggregation, and monitoring. Most of these trade-offs can be expressed in analytic terms, which certainly helps, but some cannot, as will become apparent in the social systems world of Chapter 5.

> *Exercise:* Give examples from a specific system of what information, decisions, recommendations, tasks, and tensions might be expected across the lines of Figure 1.4.

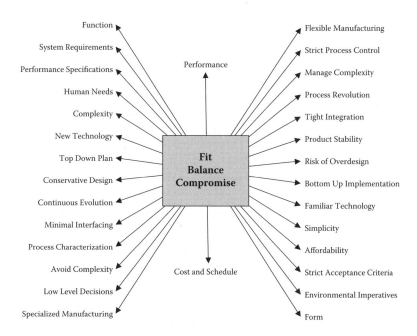

Figure 1.4 Tensions in systems architecting. (From Rechtin, E., *Systems Architecting, Creating and Building Complex Systems.* Englewood Cliffs, NJ: Prentice Hall, 1991. With permission from Prentice Hall.)

Spirals, Increments, and Collaborative Assembly

Software developers have long understood that most software-intensive projects are not well suited to a sequential process but to a highly iterative one such as the spiral. There is a strong incentive to iteratively modify software in response to user experience. As the market, or operational environment, reveals new desires, those desires are fed back into the product. One of the first formalizations of iterative development is due to Boehm and his famous spiral model. The spiral model envisions iterative development as a repeating sequence of steps. Instead of traversing a sequence of analysis, modeling, development, integration, and test just once, software may return over and over to each. The results of each are used as inputs to the next. This is depicted in Figure 1.5.

The original spiral model is intended to deliver one, hopefully stable, version of the product, the final of which is delivered at the end of the last spiral cycle. Multiple cycles are used for risk control. The nominal approach is to set a target number of cycles at the beginning of development, and partition the whole time available over the target number of cycles. The objective of each cycle is to resolve the most risky thing remaining. So, for example, if user acceptance was adjudged as the most

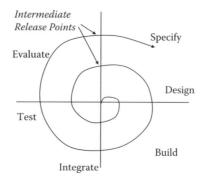

Figure 1.5 The "classic" spiral development model employs multiple cycles through the waterfall model's steps to reach a final release point.

risky at the beginning of the project, the first spiral would concentrate on those parts of the system that produce the greatest elements of user experience. Even the first cycle tests would focus on increasing user acceptance. Similarly, if the most risky element was adjudged to be some internal technical performance issue, the product of the initial cycle would focus on technical feasibility.

Many software products, or the continuing software portion of many product lines, are delivered over and over again. A user may buy the hardware once and expect to be offered a steady series of software upgrades that improve system functionality and performance. This alters a spiral development process (which has a definite end) to an incremental process, which has no definite end. The model is now more like a spiral spiraling out to circles, which represent the stable products to be delivered. After one circle is reached, an increment is delivered, and the process continues. Actually, the notion of incremental delivery appears in the original spiral model where the idea is that the product of spirals before the last can be an interim product release if, for example, the final product is delayed.

Finally, there are a number of systems in use today that are essentially continuously assembled, and where the assembly process is not directly controlled. The canonical example is the Internet, where the pieces evolve with only loose coupling to the other pieces. Control over development and deployment is fundamentally collaborative. Organizations, from major corporations to individual users, choose which product versions to use and when. No governing body exists (at least, not yet) that can control the evolution of the elements. The closest thing at present to a governing body, the Internet Society and its engineering arm, the Internet Engineering Task Force (IETF), can affect other behavior only through persuasion. If member organizations do not choose to support IETF standards, the IETF has no authority of compel compliance, or to block noncomplying implementations.

We call systems like this "collaborative systems." The development process is collaborative assembly. Whether or not such an uncontrolled process can continue for systems like the Internet as they become central to daily life is unknown, but the logic and heuristics of such systems now is the subject of Chapter 7. In Chapter 12 we address again different models of programs as examples of "program architecture" or patterns for designing a development program. Many strategic goals a client has require addressing in the structure of the program rather than in the structure of the system.

Scopes of Architecting

What is the scope of systems architecting? By scope, we mean what things are inside the architect's realm of concern and responsibility and which are not? In the classic architecting paradigm (what we have discussed so far in this chapter), the client has a problem and wants to build a system in response. The system is the response to the client's needs, as constrained by the client's resources and any other outside constraints. The concern of architecting is finding a satisfactory and feasible system in response to the client's problem. A primary difference with other conceptualizations of similar situations is that architecting does not assume that the client's problem is well structured, or that the client fully understands it. It is quite likely a full understanding of the problem will have to emerge from the client–architect interaction.

As we look beyond the classic scope of problem and system, we see several other issues of scope. First, a system is built within a "program," here defined as the collection of arrangements of funding, contracts, builders, and other elements necessary to actually build and deploy a system, whether a single-family house or the largest defense system. The program has a structure; we can say the program has an architecture. The architecture of the program will consist of strategic decisions, like is the system delivered once or many times? Is the system incrementally developed from breadboard to brassboard, or is it incrementally developed through fully deployable but reduced functionality deliveries? How is the work parceled out to different participants?

Who is responsible for the programmatic architectural decisions? In some cases, it may be important to integrate programmatic structure into the technical structure of the system. For example, if the system is to be partitioned over particular subsystem manufacturers, the system must possess a structure in subsystem compatible with what the suppliers can produce, and those subsystems must be specifiable in ways that allow for eventual integration. Whether or not the programmatic architectural

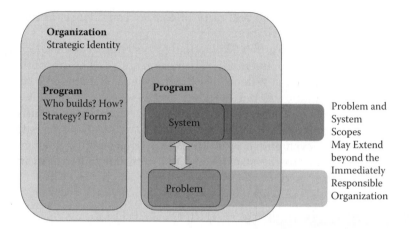

Figure 1.6 The relationship between contexts for architecting. Our core concern is with the relationship between problem and system. But, the structure of the development program and the identity and portfolio of the responsible organization are additional concerns.

decisions are in the scope of the system architect's responsibilities, there is a scope of programmatic architecture that somebody must carry out.

Moving upward and outward one more layer, the client presumably has an organization, even if the organization is only him- or herself. That organization may have multiple programs and be concerned with multiple systems. The client's organization also has a structure, which many would call an architecture. The principal concerns at the organizational scope are the organization's strategic identity, or how does the organization give itself a mission? The organization exists to do something, what is that something? Is it to make money in particular markets, to advance a particular technology, or to carry out part of a military mission?

From the perspective of the system architect, it is unlikely that the strategic identity of the client's organization is in-play in anything like the sense that the basic problem description is. However, the strategic identity of the client is important to the system architect. If that strategic identity is unclear, or poorly articulated, or simply unrealistic, then it will be very difficult for the client to make effective value judgments.

These scopes are illustrated in Figure 1.6. Figure 1.6 also illustrates one more issue of scope. Back at the scope of immediate concern to the system architect, both the solution and problem may apply well outside the immediate program and the client's organization. Other clients may have the same or similar problems. A system developed for one client may apply elsewhere. Part of architecting is the one-to-one system-to-client orientation, and individual customization; but this does not mean that others many not also be served by similar systems. Depending on the architect, it is quite likely that the issues of scope in problems and systems applying outside the immediate client's realm will be important. We return to these issues in detail in Chapter 12.

Conclusion

A system is a collection of different things that together produce results unachievable by themselves alone. The value added by systems is in the interrelationships of their elements.

Architecting is creating and building structures—that is, "structuring." *Systems* architecting is creating and building *systems*. It strives for fit, balance, and compromise among the tensions of client needs and resources, technology, and multiple stakeholder interests.

Architecting is both an art and a science — both synthesis and analysis, induction and deduction, and conceptualization and certification — using guidelines from its art and methods from its science. As a process, it is distinguished from systems engineering in its greater use of heuristic reasoning, lesser use of analytics, closer ties to the client, and particular concern with certification of readiness for use.

The foundations of systems architecting are a systems approach, a purpose orientation, a modeling methodology, ultraquality, certification, and insight. To avoid perceptions of conflict of interest, architects must avoid value judgments, avoid perceived conflicts of interest, and keep an arms-length relationship with project management.

A great architect must be as skilled as an engineer and as creative as an artist or the work will be incomplete. Gaining the necessary skills and insights depends heavily on lessons learned by others, a task of education to research and teach.

The role of systems architecting in the systems acquisition process depends upon the phase of that process. It is strongest during conceptualization and certification, but never absent. Omitting it at any point, as with any part of the acquisition process, leads to predictable errors of omission at that point to those connected with it.

Notes and References

1. *Webster's II, New Riverside University Dictionary.* Boston, MA: Riverside, 1984, p. 291.
2. Rechtin, E., *Systems Architecting, Creating and Building Complex Systems.* Englewood Cliffs, NJ: Prentice Hall, 1991, p. 7.
3. Another factor in overruns and delays is uncertainty, the "unknowns" and "unknown unknowns." Uncertainty is highest during conceptualization, less in design, still less in redesign, and least in an upgrade. As with complexity, the higher the level, the more important become experience-based architecting methods. See Carpenter, Robert Glenn, System Architects' Job

Characteristics and Approach to the Conceptualization of Complex Systems. Doctoral dissertation in Industrial and Systems Engineering, University of Southern California, Los Angeles, 1995.

4. Cuff, Dana, *Architecture: The Story of Practice.* Cambridge, MA: MIT Press, 1991.

5. Rechtin, Eberhardt, Foundations of Systems Architecting, Systems Engineering, *Journal of the National Council on Systems Engineering*, Vol. 1, Number 1, pp. 35–42, July/September 1994.

6. Discussed extensively in Juran, J. M., *Juran on Planning for Quality.* New York: The Free Press, 1988; Phadke, Madhav S., *Quality Engineering Using Robust Design.* Englewood Cliffs, NJ: Prentice Hall, 1989; Rechtin, E., *Systems Architecting, Creating and Building Complex Systems.* Englewood Cliffs, NJ: Prentice Hall, 1991, pp. 160–187.

7. See also Chapter 4.

8. From Cantry, Donald, What Architects Do and How to Pay Them, *Architectural Forum*, Vol. 119, pp. 92–95, September 1963; and from discussions with Roland Russell, AIA.

9. See King, Douglas R., The Role of Informality in System Development or, a Critique of Pure Formality, University of Southern California (USC), 1992 (unpublished but available through USC).

10. Rechtin, E., *Systems Architecting, Creating and Building Complex Systems.* Englewood Cliffs, NJ: Prentice Hall, 1991, p. 4.

11. Rechtin, E., *Systems Architecting, Creating and Building Complex Systems.* Englewood Cliffs, NJ: Prentice Hall, 1991, p. 156.

chapter 2

Heuristics as Tools

Introduction: A Metaphor

Mathematicians are still smiling over a gentle self-introduction by one of their famed members. "There are *three* kinds of mathematicians," he said, "those that know how to count and those that don't." The audience waited in vain for the third kind until, with laughter and appreciation, they caught on. Either the member could not count to three — ridiculous — or he was someone who believed that there was more to mathematics than numbers, important as they were. The number theorists appreciated his acknowledgement of them. The "those that don'ts" quickly recognized him as one of their own, the likes of a Gödel who, using thought processes alone, showed that no set of theorems can ever be complete.

Modifying the self-introduction only slightly to the context of this chapter: There are three kinds of people in our business, those who know how to count and those who do not — including the authors.

Those who know how to count (most engineers) approach their design problems using analysis and optimization, powerful and precise tools derived from the scientific method and calculus. Those who do not (most architects) approach their qualitative problems using guidelines, abstractions, and pragmatics generated by lessons learned from experience — that is, heuristics. As might be expected, the tools each use are different because the kinds of problems they solve are different. We routinely and accurately describe an individual as "thinking like an engineer" — or architect, or scientist, or artist. Indeed, by their tools and works you will know them.

Of course, we exaggerate to make a point. The reality is that architects often compute (must compute), and engineers use many heuristics. Both are complex amalgams of art and science. To be one who uses heuristics does not mean avoiding being systematic and quantitative. Consider how people who are very good at debugging hardware or software go about their work. Being systematic and quantitative in the search is an essential practice, but the search is guided by heuristic. But, the complexity of integrating the art and science can wait. For now we want to understand those things that are squarely part of the "art" of systems architecting.

This chapter, metaphorically, is about architects' heuristic tools. As with the tools of carpenters, painters, and sculptors, there are literally hundreds of them — but only a few are needed at any one time and for a specific job at hand. To continue the metaphor, although a few tool users make their own, the best source is usually a tool supply store — whether it be for hardware, artists' supplies, software — or heuristics. Appendix A, Heuristics for Systems-Level Architecting, is a heuristics store, organized by task, just like any good hardware store. Customers first browse, and then select a kit of tools based on the job, personal skill, and knowledge of the origin and intended use of each tool.

Heuristic has a Greek origin, *heuriskein*, a word meaning "to find a way" or "to guide" in the sense of piloting a boat through treacherous shoals. Architecting is a form of piloting. Its rocks and shoals are the risks and changes of technology, construction, and operational environment that characterize complex systems. Its safe harbors are client acceptance and safe, dependable, long life. Heuristics are guides along the way — channel markings, direction signs, alerts, warnings, and anchorages — tools in the larger sense. But they must be used with judgment. No two harbors are alike. The guides may not guarantee safe passage, but to ignore them may be fatal. The stakes in architecting are just as high — reputations, resources, vital services, and, yes, lives. Consonant with their origin, the heuristics in this book are intended to be trusted, time-tested guidelines for serious problem solving.

Heuristics as so defined are narrower in scope, subject to more critical test and selection, and intended for more serious use than other guidelines, for example, conventional wisdom, aphorisms, maxims, rules of thumb, and the like. For example, a pair of mutually contradictory statements like (1) *look before you leap* and (2) *he who hesitates is lost*, are hardly useful guides when encountering a cliff while running for your life. In this book, neither of these pairs would be a valid heuristic because they offer contradictory advice for the same problem.

The purpose of this chapter is therefore to help the reader — whether architect, engineer, or manager — find or develop heuristics that can be trusted, organize them according to need, and use them in practice. The first step is to understand that heuristics are abstractions of experience.

Heuristics as Abstractions of Experience

One of the most remarkable characteristics of the human race is its ability not only to learn, but to pass on to future generations sophisticated abstractions of lessons learned from experience. Each generation knows more, learns more, plans more, tries more, and succeeds more than the previous one because it does not need to repeat the time-consuming process of re-living the prior experiences. Think of how extraordinarily efficient are such quantifiable abstractions as $F = ma$, $E = mc^2$ and $x = F(y,z,t)$; of algorithms, charts, and graphs; and of the basic principles of economics. This kind of efficiency is essential if large, lengthy, complex systems and long-lived product lines are to succeed. Few architects ever work on more than two or three complex systems in a lifetime. They have neither the time nor opportunity to gain the experience needed to create first-rate architectures from scratch. By much the same process, qualitative heuristics, condensed and codified practical experience, came into being to complement the equations and algorithms of science and engineering in the solving of complex problems. Passed from architect to architect, from system to system, they worked. They helped satisfy a real need.

In contrast to the symbols of physics and mathematics, the format of heuristics is words expressed in the natural languages. Unavoidably, they reflect the cultures of engineering, business, exploration, and human relations in which they arose. The birth of a heuristic begins with anecdotes and stories, hundreds of them, in many fields which become parables, fables, and myths,[1] easily remembered for the lessons they teach. Their impact, even at this early stage, can be remarkable not only on politics, religion, and business but also on the design of technical systems and services. The lessons that have endured are those that have been found to apply beyond the original context, extended there by analogy, comparison, conjecture, and testing.* At their strongest they are seen as self-evident truths requiring no proof.

There is an interesting human test for a good heuristic. An experienced listener, on first hearing one, will know within seconds that it fits that individual's model of the world. Without having said a word to the speaker, the listener almost invariably affirms its validity by an unconscious nod of the head, and then proceeds to recount a personal experience that strengthens it. Such is the power of the human mind.

Selecting a Personal Kit of Heuristic Tools

> The art in architecting lies not in the wisdom of the heuristics, but in the wisdom of knowing which heuristics apply, a priori, to the current project.[2]

All professions and their practitioners have their own kits of tools, physical and heuristic, selected from their own and others' experiences to

---

* This process is one of inductive reasoning, "a process of truth estimation in the face of incomplete knowledge which blends information known from experience with plausible conjecture" (Klir, George J., *Architecture of Systems Problem Solving*. New York: Plenum Press, 1985, p. 275). More simply, it is an extension or generalization from specific examples. It contrasts with deductive reasoning, which derives solutions for specific cases from general principles.

match their needs and talents. But, in the case of architecting prior to the late 1980s, selections were limited and, at best, difficult to acquire. Many heuristics existed, but they were mainly in the heads of practitioners. No efforts had been made to articulate, organize, and document a useful set. The heuristics in this book were codified largely through the University of Southern California graduate course in Systems Architecting. The students and guest instructors in the course, and later program, were predominantly experienced engineers who contributed their own lessons learned throughout the West Coast aerospace, electronics, and software industries. Both as class exercises, and through the authors' writings, they have been expressed in heuristic form and organized for use by architects, educators, researchers, and students.

An initial collection[3] of about 100 heuristics was soon surpassed by contributions from over 200 students, reaching nearly 1,000 heuristics within 6 years.[4] Many, of course, were variations on single, central ideas — just as there are many variations of hammers, saws, and screwdrivers — repeated in different contexts. The four most widely applicable of these heuristics were as follows, in decreasing order of popularity:

1. *Do not assume that the original statement of the problem is necessarily the best, or even the right one.*

   Example: The original statement of the problem for the F-16 fighter aircraft asked for a high-supersonic capability, which is difficult and expensive to produce. Discussions with the architect, Harry Hillaker, brought out that the reason for this statement was to provide a quick exit from combat, something far better provided by a high thrust-to-weight, low supersonic design. In short, the original high speed statement was replaced by a high acceleration one, with the added advantage of exceptional maneuverability.

2. *In partitioning, choose the elements so that they are as independent as possible; that is, choose elements with low external complexity and high internal complexity.*

   Example: One of the difficult problems in the design of microchips is the efficient use of their surface area. Much of that area is consumed by connections between components — that is, by communications rather than by processing. Carver Mead of Caltech has now demonstrated that a design based on minimum communications between process-intensive nodes results in much more efficient use of space, with the interesting further result that the chip "looks elegant" — a sure sign of a fine architecture and another confirmation of the heuristic: ·

3. *The eye is a fine architect. Believe it.*
   Simplify. Simplify. Simplify.

   Example: One of the best techniques for increasing reliability while decreasing cost and time is to reduce the piece part count of a device. Automotive engineers, particularly recently, have produced remarkable results by substituting single castings for multiple assemblies and by reducing the number of fasteners and their associated assembly difficulties by better placement.

4. *Build in and maintain options as long as possible in the design and implementation of complex systems. You will need them.*

   Example: In the aircraft business they are called "scars." In the software business they are called "hooks." Both are planned breaks or entry points into a system that can extend the functions the system can provide. In aircraft, they are used for lengthening the fuselage to carry more passengers or freight. In software, they are used for inserting further routines, or to allow integration of data with other programs.

Though these four heuristics do not make for a complete tool kit, they do provide good examples for building one. All are aimed at reducing complexity, a prime objective of systems architecting. All have been trusted in one form or another in more than one domain. All have stood the test of time for decades if not centuries.

The first step in creating a larger kit of heuristics is to determine the criteria for selection. The following were established to eliminate unsubstantiated assertions, personal opinions, corporate dogma, anecdotal speculation, mutually contradictory statements, and the like. As it turned out, they also helped generalize domain-specific heuristics into more broadly applicable statements. The strongest heuristics passed all the screens easily. The criteria were as follows:

- The heuristic must make sense in its original domain or context. To be accepted, a strong correlation, if not a direct cause and effect, must be apparent between the heuristic and the successes or failures of specific systems, products, or processes. Academically speaking, both the rationale for the heuristic and the report that provided it were subject to peer and expert review. As might be expected, a valid heuristic seldom came from a poor report.
- The general sense, if not the specific words, of the heuristic should apply beyond the original context. That is, the heuristic should be useful in solving or explaining more than the original problem from which it arose. An example is the preceding *do not assume* heuristic. Another is *Before the flight it is opinion; after the flight it is obvious.* In the latter, the word "flight" can sensibly be replaced by test, experiment, fight, election, proof, or trial. In any case, the heuristic should not be wrong or contradictory in other domains where it could lead to serious misunderstanding and error. This heuristic applies in general to ultraquality systems. When they fail, and they usually fail after all the tests are done and they are in actual use, the cause of

the failure is typically a deterministic consequence of some incorrect assumptions; and we wonder how we missed such an obvious failure of our assumptions.

- The heuristic should be easily rationalized in a few minutes or on less than a page. As one of the heuristics states, *If you can't explain it in five minutes, either you don't understand it or it doesn't work* (Darcy McGinn 1992 from David Jones). With that in mind, the more obvious the heuristic is on its face, and the fewer the limitations on its use, the better. *Example: A model is not reality.*

- The opposite statement of the heuristic should be foolish, clearly not "common sense." For example: The opposite of *Murphy's Law — If it can fail, it will* — would be "If it can fail, it won't," which is patent nonsense.

- The heuristic's lesson, though not necessarily its most recent formulation, should have stood the test of time and earned a broad consensus. Originally this criterion was that the heuristic *itself* had stood the test of time, a criterion that would have rejected recently formulated heuristics based on retrospective understanding of older or lengthy projects. *Example: The beginning is the most important part of the work* (Plato 4th Century B.C.), reformulated more recently as *All the serious mistakes are made in the first day.**

It is probably true that heuristics can be even more useful if they can be used in a set, like wrenches and screwdrivers, hammers and anvils, or files and vises. The taxonomy grouping in a subsequent section achieves that possibility in part.

It is also probably true that a proposed action or decision is stronger if it is consistent with several heuristics rather than only one. A set of heuristics applicable to acceptance procedures substantiates that proposition.

And it would certainly seem desirable that a heuristic, taken in a sufficiently restricted context, could be specialized into a design rule, a quantifiable, rational evaluation, or a decision algorithm. If so, heuristics of this type would be useful bridges between architecting, engineering, and design. There are many cases where we have such progressive extensions, from a fairly abstract heuristic that is broadly applicable to a set of more narrowly applicable, but directly quantifiable, design rules.

Using Heuristics

Virtually everybody, after brief introspection, sees that heuristics play an important role in their design and development activities. However, even if we accept that everyone uses heuristics, it is not obvious that those heuristics can be communicated and used by others. This book takes the

---

* Spinrad, Robert, Lecture at the University of Southern California, 1988.

approach that heuristics can be effectively communicated to others. One lesson from student use of Rechtin 1991,[5] and previous editions of this book, is that heuristics do transfer from one person to another, but not always in simple ways. It is useful to document heuristics and teach from them, but learning styles differ.

People typically use heuristics in three ways. First, they can be used as evocative guides. They work as guides if they evoke new thoughts in the reader. Some readers have reported that they use the catalog of heuristics in the appendices at random when faced with a difficult design problem. If one of the heuristics seems suggestive, they follow up by considering how that heuristic could describe the present situation, what solutions it might suggest, or what new questions it suggests.

The second usage is as codifications of experience. In this usage, the heuristic is like an outline heading, a guide to the detailed discussion that follows. In this case, the stories behind the heuristics can be more important than the basic statement. The heuristic is a pedagogical tool, a way of teaching lessons not well captured in other engineering teaching methods.

The third usage is the most structured. It is when heuristics are integrated into development processes. This means that the heuristics are attached to an overall design process. The design process specifies a series of steps and models to be constructed. The heuristics are attached to the steps as specific guidelines for accomplishing those steps.

A good example is in software. A number of software development methods have a sequence of models, from relatively abstract to code in a programming language. Object-oriented methods, for example, usually begin with a set of textual requirements, build a model of classes and objects, and then refine the class/object model into code in the target programming environment. There are often intermediate steps in which the problem-domain-derived objects are augmented with objects and characteristics from the target environment. A problem in all such methods is knowing how to construct the models at each step. The transformation from a set of textual requirements to classes and objects is not unique, but it involves extensive judgment by the practitioner. Some methods provide assistance to the practitioner by giving explicit, prescriptive heuristics for each step.

A Process Framework for Architecting Heuristics

In Part III of this book, we will present a basic process framework for system architecting. The process framework will define activities repeatedly required in effective architecting, and discuss how those activities can be arranged relative to each other. We will also place those activities in a larger architecture project framework. This process framework is illustrated in Figure 2.1. As noted above, one method for using heuristics is to attach them to steps in a design process. By doing so, the heuristics

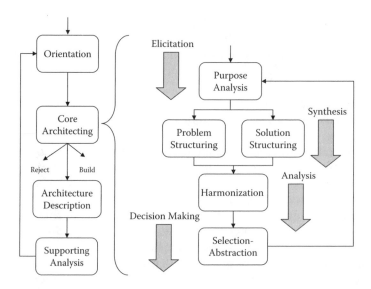

Figure 2.1 Activities in an architecting process model.

become local guides to each aspect of the process. A complete process with step-by-step designated models and transformation heuristics is not appropriate for general systems architecting. There is simply too much variation from domain to domain, too many unique domain aspects, and too many important domain-specific tools. Even so, it is useful to recognize the basic structure of the process framework and how the heuristics relate to that framework.

It is important to distinguish between the activity cycle for an entire development program and the activity cycle for an architecture project. The goal of a development program is to build and deliver a system. The goal of an architecture project is something else. In the simplest case, the goal of the project is to initiate a development program. Even in the simple case we recognize that development programs go in fits and starts. There might be several discrete architecture projects, simultaneously or sequentially developing architectural concepts for every actual development program. Some architecture projects do not have a specific system development as their goal, as in the architecture projects that concern collaborative systems (which we take up in Chapter 7).

The beginning of an architecting project is "orientation" or determining where you are and where you want to go. This refers both to the architecture project as well as the underlying, assumed but not yet existing, system development project. Orientation is less technical and more business. Its intent is to ensure that the architecture effort can proceed for at least one iterative cycle in an organized fashion. Heuristics associated with orientation relate to topics like identifying the driving characteristics of a project, finding leading stakeholders, and clarifying relationships between the architect, sponsors, and downstream users. Orientation is about scoping and planning, and so the heuristics of Appendix A and in Chapter 9 under the associated topics apply most strongly. Orientation leads to core architecting, which is characterized by purpose analysis, problem structuring, solution structuring, harmonization, and selection-abstraction.

Purpose analysis is a broad-based study of why the capability or system of interest has value. It works from an understanding of the client strategy and expands to all stakeholders with significant power over the eventual construction, deployment, and operation of the system. Purpose analysis is an elicitation activity, and so all heuristics that relate to elicitation apply most strongly here.

Problem structuring is where we organize elements of the problem space with a primary focus on a "value model." The value model is an explicit model of the most important stakeholder's preferences, and it is intended to capture them without regard to consistency. That is, we want to be able to assess alternatives in the value system of each major stakeholder, realizing that the resulting preference orderings will not be the same. Any reconciliation necessary among them will be conducted later. Its concern is on the problem side of the problem–system tension. It is a synthesis activity in the sense that we are synthesizing problem descriptions, preferably several, with somewhat different scopes. In terms of Appendix A and Chapter 9, the associated heuristics are drawn mostly from modeling and prioritizing.

In solution structuring, we synthesize models of solutions, again multiple solutions that should differ in scope and scale. The heuristics that apply are drawn from those that cover modeling, aggregating, and partitioning.

Harmonization is a dominantly analytical activity in which we integrate problem and solution descriptions and assess value. Harmonization is a preparation for selection-abstraction. Selection is easy to understand; it is picking answers. An important distinction between the approach of systems architecting and most decision analysis texts is that we do not assume when we enter selection that there is a unitary, exclusive decision to make. At some point in the process, if the overall goal is to build a system, we must clearly make a decision about a preferred configuration. But we might travel down this process road many times before reaching such a unitary decision. Along the way, we may wish to hold onto multiple solution configurations, classes of solution configurations, and multiple problem descriptions. We make no decision before its time. As it was put in Rechtin 1991, *Hold onto the agony of decision as long as possible.* We also introduce the notion of abstraction for those cases where architecting has been completed even though no single configuration has been selected.

As an example of abstraction over selection, consider the case of a family-of-systems, say the collection of printers made by a single company.

There are shared properties or components across the whole family (for example, interfaces, software rendering engines, supply chains). These shared elements are the concern of the family-of-systems architect, and are abstractions of the entire family. It is inaccurate to talk about selecting the whole family (though we might select the market-niche structure of the whole family), but it is accurate to consider selection of properties of the whole family abstract into a family-of-systems architecture. We refer to that form of selection as "abstraction."

Architectural projects ultimately produce architecture descriptions, a document. We illustrate this as a following step to Core Architecting in Figure 2.1. In reality, architecture descriptions are developed at least partially in parallel with the architectural decision making. But, it is helpful to illustrate the separation of the two activities to emphasize that architecting is about decision making, and architectures are about decisions. Architecture descriptions can only document those decisions. The quality (or lack thereof) of those decisions must stand on its own. An excellently drawn description will not make up for poor architecture decisions.

Finally, in practice, architects discover in the process where they need additional knowledge. Straightforward progress through an architecture study may be interrupted by the discovery that we do not know critical numbers related to the cost or performance of a key system element, or we do not understand the technicalities of a particular stakeholder problem, or we lack clear input on preferences from a stakeholder. In most cases, it is more effective to put such issues aside by making suitable assumptions, returning to the issues after completing an end-to-end pass through architectural analysis, resolving those detailed issues in studies, and returning to another iterative cycle through the architecting process.

Heuristics on Heuristics

A phenomenon observed as heuristics discovered by the USC graduate students is that the discoverers themselves began thinking heuristically. They found themselves creating heuristics directly from observation and discussion, and then trying them out on professional architects and engineers, some of whose experiences had suggested them. (Most interviewees were surprised and pleased at the results.) The resultant provisional heuristics were then submitted for academic review as parts of class assignments.

Kenneth L. Cureton, carrying the process one step further, generated a set of heuristics on how to generate and apply heuristics,[6] from which the following were chosen.

Generating Useful Heuristics

- Humor [and careful choice of words] in a heuristic provides an emotional bite that enhances the mnemonic effect [Karklins].

- Use words that transmit the "thrill of insight" into the mind of the beholder.
- For maximum effect, try embedding both descriptive and prescriptive messages in a heuristic.
- Many heuristics can be applied to heuristics [e.g., *Simplify! Scope!*].
- Do not make a heuristic so elegant that it only has meaning to its creator, and thus loses general usefulness.
- Rather than adding a conditional statement to a heuristic, consider creating a separate but associated heuristic that focuses on the insight of dealing with that conditional situation.

Applying Heuristics

- If it works, then it is useful.
- Knowing when and how to use a heuristic is as important as knowing what and why.
- Heuristics work best when applied early to reduce the solution space.
- Strive for balance — too much of a good thing or complete elimination of a bad thing may make things worse, not better!
- Practice, practice, practice!
- Heuristics are not reality, either!

A Taxonomy of Heuristics

The second step after finding or creating individual heuristics is to organize them for easy access so that the appropriate ones are at hand for the immediate task. The collection mentioned earlier in this chapter was accordingly refined and organized by architecting task.* In some ways, the resultant list — presented in Appendix A — was self-organizing. Heuristics tended to cluster around what became recognized as basic architecting tasks. For example, although certifying is shown last and is one of the last formal phases in a waterfall, it actually occurs at many milestones as "sanity checks" are made along the way and subsystems are assembled. The tasks, elaborated in Chapter 9, are as follows:

- Scoping and planning
- Modeling
- Prioritizing
- Aggregating

* The original 100 of Rechtin 1991 were organized by the phases of a waterfall. The list in Appendix A of this book recognizes that many heuristics apply to several phases, that the spiral model of system development would in any case call for a different categorization, and that many of the tasks described here occur over and over again during systems development.

- Partitioning
- Integrating
- Certifying
- Assessing
- Evolving and rearchitecting

The list is further refined by distinguishing between two forms of heuristic. One form is descriptive; that is, it describes a situation but does not indicate directly what to do about it. Another is prescriptive; that is, it prescribes what might be done about the situation. An effort has been made in the appendix to group prescriptions under appropriate descriptions with some, but not complete, success. Even so, there are more than enough generally applicable heuristics for the reader to get started.

And then there are sets of heuristics that are domain-specific to aircraft, spacecraft, software, manufacturing, social systems, and so on. Some of these can be deduced or specialized from more general ones given here. Or, they can be induced or generalized from multiple examples in specialized subdomains. Still more fields are explored in Part III, adding further heuristics to the general list.

You are encouraged to discover still more, general and specialized, in much the same way the more general ones here were — by spotting them in technical journals, books,[7] project reports, management treatises, and conversations.

The Appendix A taxonomy is not the only possible organizing scheme, any more than all tool stores are organized in the same way. In Appendix A one heuristic follows another, one-dimensionally, as in any list. But some are connected to others in different categories, or could just as easily be placed there. Some are "close" to others and some are further away. Ray Madachy, then a graduate student, using hypertext linking, converted the list into a two-dimensional, interconnected "map" in which the main nodes were architecting themes: conception and design; the systems approach; quality and safety; integration, test, and certification; and disciplines.[8] To these were linked each of the 100 heuristics in the first systems architecting text,[9] which in turn were linked to each other. The ratio of heuristic-to-heuristic links to total links was about 0.2; that is, about 20% of the heuristics overlapped into other nodes.

The Madachy taxonomy, however, shared a limitation common to all hypertext methods — the lack of upward scalability into hundreds of objects — and consequently was not used for Appendix A. Nonetheless, it could be useful for organizing a modest-sized personal tool kit or for treating problems already posed in object-oriented form, for example, computer-aided design of spacecraft.[10]

New Directions

Heuristics are a popular topic in systems and software engineering, though they do not often go by that name. A notable example is the pattern language. The idea of patterns and pattern languages comes from Christopher Alexander and has been adapted to other disciplines by other writers. Most of the applications are to software engineering.

A pattern is a specific form of prescriptive heuristic. A number of forms have been used in the literature, but all are similar. The basic form is a pattern name, a statement of a problem, and a recommended form of solution (to that problem). So, for example, a pattern in civil architecture has the title "Masters and Apprentices," the problem statement describes the need for junior workers to learn while working from senior master workers, and the recommended solution consists of suitable arrangements of work spaces.

When a number of patterns in the same domain are collected together, they can form a pattern language. The idea of a pattern language is that it can be used as a tool for synthesizing complete solutions. The architect and client use the collected problem statements to choose a set that is well-matched to the client's concerns. The resulting collection of recommended solutions is a collection of fragments of a complete solution. It is the job of the architect to harmoniously combine the fragments into a whole.

In general, domain-specific, prescriptive heuristics are the easiest for apprentices to explain and use. So, patterns on coding in programming are relatively easy to teach and learn to use. This is borne out by the observed utility of coding pattern books in university programming courses. Similarly, an easy entry to the use of heuristics is when they are attached as step-by-step guides in a structured development process. At the opposite end, descriptive heuristics on general systems architecting are the hardest to explain and use. They typically require the most experience and knowledge to apply successfully. The catalog of heuristics in Appendix A has heuristics across the spectrum.

Conclusion

Heuristics, as abstractions of experience, are trusted, nonanalytic guidelines for treating complex, inherently unbounded, ill-structured problems. They are used as aids to decision making, value judgments, and assessments. They are found throughout systems architecting, from earliest conceptualization through diagnosis and operation. They provide bridges between client and builder, concept and implementation, synthesis and analysis, and system and subsystem. They provide the successive transitions from qualitative, provisional needs to descriptive and prescriptive guidelines, and thence to rational approaches and methods.

This chapter has introduced the concept of heuristics as tools — how to find, create, organize, and use them for treating the qualitative problems of systems architecting. Appendix A provides a ready source of them organized by architecting task — in effect, a tool store of systems architecting heuristic tools.

Notes and References

1. For more of the philosophical basis of heuristics, see Asato, Michael, The Power of the Heuristic, University of Southern California, 1988, and Rowe, Alan J., The Meta Logic of Cognitively Based Heuristics, in Watkins, P. R. and L. B. Eliot, eds. *Expert Systems in Business and Finance: Issues and Applications.* New York: John Wiley & Sons, 1988.
2. Williams, Paul L., 1992 Systems Architecting Report, University of Southern California (unpublished).
3. Rechtin, E., *Systems Architecting, Creating and Building Complex Systems.* Englewood Cliffs, NJ: Prentice Hall, 1991, pp. 312–319.
4. Rechtin, E. ed., Collection of Student Heuristics in Systems Architecting, 1988–93, University of Southern California (unpublished), 1994.
5. Rechtin, E., *Systems Architecting, Creating and Building Complex Systems.* Englewood Cliffs, NJ: Prentice Hall, 1991. Note that throughout the rest of this chapter, this reference is referred to as Rechtin 1991.
6. Cureton, Kenneth L., Metaheuristics, USC graduate report, December 9, 1991.
7. Many of the ones in Appendix A come from a similar appendix in Rechtin 1991.
8. Madachy, Ray, Thread Map of Architecting Heuristics, University of Southern California, April 21, 1991 and Formulating Systems Architecting Heuristics for Hypertext, University of Southern California, April 29, 1991.
9. Rechtin 1991, Appendix A, pp. 311–319.
10. As an example, see Asato, Michael, Final Report. Spacecraft Design and Cost Model, University of Southern California, 1989.

part II

New Domains, New Insights

Part II explores from an architectural point of view five domains beyond those of aerospace and electronics, the sources of most examples and writings to date. The chapters can be read for several purposes. For a reader familiar with a domain, there are broadly applicable heuristics for more effective architecting of its products. For ones unfamiliar with it, there are insights to be gained from understanding problems differing in the degree but not in kind from one's own. To coin a metaphor, if the domains can be seen as planets, then this part of the book corresponds to comparative planetology, the exploration of other worlds to benefit one's own. The chapters can be read for still another purpose, as a template for exploring other, equally instructive, domains. An exercise for that purpose can be found at the end of Chapter 7, "Collaborative Systems."

Each of the chapters is preceded by a brief case study. Each of the case studies is chosen to be relevant to the chapter to which it is attached. Many students and readers have asked about case studies of real systems to assist in understanding the application of the materials. Unfortunately, really good engineering and architecting case studies are notoriously hard to obtain. The stories and details are rarely published. Books published on major systems are more likely to focus on the people involved than on the technical decision making. Many of the most interesting stories are buried behind walls of proprietary information. By the time the full story can be published, it is often old. We, the authors, think the older stories carry timeless lessons, so we have included several here. Each includes some references back to the original literature, where it is readily available, so the interested reader can follow up with further investigation of his or her own. In a few cases, we abstracted several cases into one where the original stories have not yet been published, and the combination makes the lessons clearer.

From an educational point of view, this part is a recognition that one of the best ways of learning is by example, even if the example is in a different field or domain. One of the best ways of understanding another discipline is to be given examples of problems it solves. And one of the best ways of learning architecting is to recognize that there are architects in every domain and at every level from which others can learn and with whom all can work. At the most fundamental level, all speak the same language and carry out the same process, systems architecting. Only the examples are different.

Chapter 3 explores systems for which form is predetermined by a builder's perceptions of need. Such systems differ from those that are driven by client purposes by finding their end purpose only if they succeed in the marketplace. The uncertainty of end purpose has risks and consequences that it is the responsibility of architects to help reduce or exploit. Central to doing so are the protection of critical system parameters and the formation of innovative architecting teams. These systems can be either evolutionary or revolutionary. Not surprisingly, there are important differences in the architectural approach. The case study is an old one, but an excellent one, on the development of the DC-3 airplane.

Chapter 4 highlights the fact that manufacturing has its own waterfall, quasi-independent of the more widely discussed product waterfall, and that these two waterfalls must intersect properly at the time of production. A spiral-to-circle model is suggested to help understand the integration of hardware and software. Ultraquality and feedback are shown to be the keys to both lean manufacturing and flexible manufacturing, with the latter needing a new information flow architecture in addition. The case study is on the development of mass production, particularly its development at Ford and later Toyota.

Chapter 5 on sociotechnical systems introduces a number of new insights to those of the more technical domains. Economic questions and value judgments play a much stronger role here, even to the point of outright veto of otherwise worthwhile systems. A new tension comes to center stage, one central to social systems but too often downplayed in others until too late — the tension between facts and perceptions. It is so powerful in defining success that it can virtually mandate system design and performance, solely because of how that architecture is perceived. The case study is on architecting intelligent transportation systems.

Chapter 6 serves to introduce the domain of software as it increasingly becomes the center of almost all modern systems designs. Consequently, whether stand-alone or as part of a larger system, software systems must accommodate to continually changing technologies and product usage. In very few other domains is annual, much less monthly, wholesale replacement of a deployed system economically feasible or even considered. In point of fact, it is considered normal in software systems, precisely because of software's unique ability to continuously and rapidly evolve in response to changes in technology and user demands. Software has another special property; it can be as hard or as soft as needed. It can be hard-wired if certification must be precise and unchanging. Or it can be as soft as a virtual environment molded at the will of a user. For these and other reasons, software practice is heavily dependent on heuristic guidelines and organized, layered modeling. It is a domain in which architecting development is very active, particularly in progressive modeling and rapid prototyping. The case study is on the transition from hierarchical to layered systems, a major point of contention in software systems. It is abstracted from several real cases familiar to the authors.

Chapter 7 introduces an old but newly significant class of systems, collaborative systems. Collaborative systems exist only because the participants actively and continuously work to keep it in existence. A collaborative system is a dynamic assemblage of independently owned and operated components, each one of which exists and fulfills its owner's purposes whether or not it is part of the assemblage. These systems have been around for centuries in programs of public works. But today we find wholly new forms in communications (the Internet and World Wide Web), transportation (intelligent transportation systems), militaries (multinational reconnaissance-strike and defensive systems), and software (open source software). The architecting paradigm begins to shift in collaborative systems because the architect no longer has a single client who can make and execute decisions. The architect must now deal with more complex relationships and must find architectures in less familiar structures, such as architecture through communication or command protocol specification. The case study is on the Global Positioning System (GPS), which did not start as a collaborative system, but which is rapidly evolving into one.

The nature of modern software and information-centric systems, and their central role in new complex systems makes a natural lead into Part III, "Models and Modeling."

Case Study 1: DC-3

Even though the DC-3 airplane was designed and built in the 1930s, it is not uncommon for someone today to have flown on one. Seventy years after its origination, the DC-3 is still flying effectively and profitably, albeit mostly only in remote areas. The DC-3 is commonly cited as the most successful airplane ever built. What accounts for the extraordinary success of the DC-3 airplane? The history of the DC-3's development extensively illustrates many of the key lessons of systems architecting, especially the following:

1. The role of the very small architecting team in bringing vision and coherence to the system concept.
2. The cooperative nature of the effective architect–client relationship, even when the architect belongs to the builder organization.
3. The role of coupled technological and operational change in creating revolutionarily successful systems.
4. The role of evolutionary development in enabling revolutionary development.

Because of the extraordinary success of the DC-3, there is a broad literature on its history and on the history of other airplanes at that time. One of the most valuable sources for the architectural history of the DC-3, and an exceptional source of architecting heuristics, is the paper "The Well Tempered Aircraft" by Arthur Raymond.[1] A more extensive history is presented in the online book provided by the DC-3 history society.[2]

The History

In a room of the Smithsonian Air and Space Museum devoted to flight between World Wars I and II, three key airplanes can be seen together. They are the Ford Trimotor, the Boeing 247, and the DC-3. Of these, only the DC-3 can be seen outside of an air museum or historical air show. In 1930 the Ford Trimotor was state-of-the-art in passenger and cargo aircraft. It carried eight passengers and enabled passenger and cargo service across the United States. But, by modern standards, the airplane was barely usable. The large reciprocating motor on the nose coupled noise and vibration (and sometimes exhaust) directly into the passenger and cargo areas. The framed fuselage put large spars directly through the passenger and cargo area, with obvious inconvenience for both types of service. Reliability and safety were far from modern standards. Regardless, it was such an improvement over its predecessor, and delivered such value, that 199 were built (see Figure CS1.1).

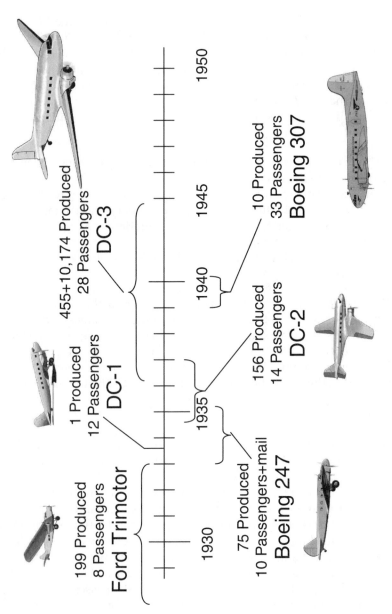

Figure CS1.1 Timeline of the DC-3 and related aircraft.

In 1930, aeronautical technology was changing quickly. Engines were improving very rapidly in power and power-to-weight ratio, new structural concepts were being tested, and understanding of aerodynamics was improving very rapidly (from prototype airplanes, theoretical study, and the first generation of capable wind tunnels). Two young companies riding the early boom in aeronautics, Boeing and Douglas Aircraft, were developing new airplane concepts exploiting these new technologies. But, the two companies faced very different business situations and clients.

At the time, Boeing and United Airlines were very closely related. As a result, as Boeing looked into how to exploit the emerging aeronautical technologies, they did so with extensive knowledge of United operations and sources of revenue. The key insight that came from that knowledge was that essentially all of the profit from operations came from carrying government subsidized airmail. What we now think of as the regular business of airlines, carrying passengers and general freight, was financially ancillary to the airmail. As a result, when Boeing conducted design studies for how to best exploit the new technology in engines, aerodynamics, and structures, they focused on an aircraft that was optimized for the routing structure imposed by the U.S. Postal Service. The result was the Boeing 247. There is no doubt that the Boeing 247 was a revolutionary airplane technologically. And from the perspective of passengers, it was far more comfortable than the Ford Trimotor, and much faster. United quickly ordered sixty, a large leap in production capacity for the Boeing of the time. But, it was not revolutionary from a business–operational perspective. The 247 was intended to do business the way it was being done, just much better.

At the same time, Douglas Aircraft, working with the airline TWA (and later American), also began design studies for airplanes incorporating the newly available technology. TWA was originally interested in the 247 but was unable to obtain any deliveries because of the long backlog to United. Unlike Boeing, Douglas and their airline partners were thinking well beyond the immediate profit source of airmail. As a result, they began designing airplanes larger than necessary for the airmail role. The DC-1 was produced contemporaneously with the Boeing 247, and was roughly the same size. Douglas and their customers realized the advantages of the new overall design given the new technologies but believed the airplane was too small. They proceeded quickly to the DC-2.

The DC-1 was essentially a proof-of-concept airplane. Douglas and the airlines intended it to be a production representative airplane, but it served mainly to prove the concept and demonstrate the way forward. It also demonstrated what TWA had imposed as a key requirement, that the airplane be able to survive a single-engine-out condition anywhere in flight, most notably over the highest-altitude mountain points of TWA's routes. This was successfully demonstrated during DC-1 tests. The extensive DC-1 tests also revealed a wide variety of issues with the very new design, and the need for significant redesign (significant enough for the redesign to be a new airplane) before full production.

The DC-2 was larger, and was commercially successful, as witnessed by its production run of 156 aircraft (see Figure CS1.1 for the times and figures). The production run of the DC-2 was already larger than the Boeing 247, and nearly the size of the Ford Trimotor's, the previously most successful airplane. American airlines, after some experience with the DC-2, approached Douglas about a further upsizing, with intent to use the airplane in cross-country sleeper service. Douglas and their team began design work immediately on the DC-3. It was much larger, with a passenger capacity double that of the DC-2. Its production run was much larger than any previous airplane, reflecting its revolutionary success in the commercial airline business. Even though Douglas was confident of the excellence of the DC-3, the magnitude of the success was a surprise. The company chose an initial production to produce tooling with a design life of 50 units (which Raymond regarded as "rather daring"). That tooling lasted through hundreds of aircraft. With 455 of the initial commercial model produced, it was the foundation of the modern, then rapidly growing airline business.

Of course, the story does not end here. Boeing saw the success of the DC-3 and moved to counter with an even larger and higher-performance aircraft, albeit after some delay. Boeing was well placed to continue to move up in aircraft size and performance because of the simultaneous work for the U.S. Army Air Corps on the large four engine bombers (among them the B-17 and later the B-29 of World War II fame). Boeing countered with the Boeing 307, with a capacity of thirty-three passengers, larger than the DC-3.

Here history intervenes in the story. The 307 was produced from 1939 to 1940. At this point, U.S. industry was already converting to war production. After the attack on Pearl Harbor in 1941, essentially all airplane production was converted to war production, but in large measure, the conversion had already begun. The U.S. Army Air Corps needed transports, bombers, fighters, and all types of aircraft. Boeing, with its advantages in large bombers, moved its production primarily to bombers. The DC-3 was an obvious choice as a transport. It was a proven, mature design with proven utility and reliability. Enormous contracts for producing military variants of the DC-3 came rapidly, and more than 10,000 were produced in various military configurations. This huge production base became the foundation for the aircraft to fly productively for decades after production ended.

After World War II, the competition in commercial airplanes resumed, but from a new point. The DC-3 existed in such large numbers there was hardly room for a direct competitor. The technology for building and operating much larger aircraft had been extensively developed. The transports produced after World War II were larger still, mostly four-engine aircraft. And soon after, the transition to jet engines would revolutionize the architecture of commercial aircraft, and the airline industry, once again.

Architecture Interpretation

As interesting as the capsule history of the DC-3 may be, this history is not the primary focus here in this book. The reader may find many extensive histories of the DC-3 and its competitors. But, we are interested here in understanding and interpreting its architecture, not just on its own, but in relationship to its competitors and in the context of its builders, sponsors, and users.

Three Story Variations

Three different but related contexts can be considered in the DC-3 story. The first way of seeing the story is as one of architectural revolution fueled by technology. In this way, we see the DC-3 as a technology-enabled architectural jump over the Ford Trimotor. The moral of this story is that technological advance combined with architectural vision creates a revolutionary system. This story is, of course, true; but it is also incomplete. The DC-3 was a revolutionary advance over the Ford Trimotor, and it was a combination of technological advance and architectural vision. But, it did not happen in one step, it did not happen in only one place, and it did not happen all at once. If the DC-3 was a technology-driven jump, than so was the Boeing 247. To understand the success of the DC-3 over the 247, we need to look beyond the first story of a technology-driven jump.

In the second story we see the Boeing 247 and the DC-3 as a story in the hazards of optimality. The moral of the second story is that being optimal with respect to the problem as currently or originally understood is not always the best choice. The DC-3 achieved enormous success because it did not optimally serve existing markets; instead, it leapfrogged and enabled new markets. The revolution was not just in technology of airplanes, it was in the coupling of technological change with operational change. The DC-3 became a huge success when its owners changed their business model in response to its capabilities. In this story, we can see the Boeing 247 as a cautionary tale to not look too narrowly, especially in times of rapid change.

The third story expands the second by seeing what Boeing did after the appearance of the DC-3. When the DC-3 opened new markets, Boeing did not stand still. They had already invested in the 247, and it was being used on airmail routes, but they did not continue to build it in the face of the greater success of the DC-3. Instead, they followed where the DC-3 had revealed the market to be (larger, faster, higher-capacity aircraft) by building the 307. The 307 might have been a highly successful aircraft, except that World War II intervened and upset the competition with the forced conversion to war production.

The third story must color our perception of success and failure, and Boeing versus Douglas' decision making. Boeing started the revolution with the 247. Boeing was eclipsed by the DC-3, but that has to be viewed in the larger context of builder's strategic positions and capabilities. Boeing started with a stronger business position and a direct relationship with the leading customer, United Air Lines. Boeing also held a "real option"* on moving to even larger aircraft in a way that Douglas did not. Thus, Boeing could logically make a more conservative decision for the competitive and technological positioning of the 247 than made sense for Douglas. This leads naturally to the next question.

Was the Boeing 247 Successfully Architected?

It seems obvious that the DC-3 was very successfully architected. It is generally regarded as the most successful aircraft of all time, and beautifully combined technical and operational innovation. The combination was deliberate, if not entirely foreseen.[3] The natural follow-on question is to ask how successfully was the Boeing 247 architected? Obviously, it was a much less successful aircraft. But, it was the aircraft its sponsors requested. It did effectively exploit the new technology, and it did what was asked. The general question is, if a sponsor gets the system he or she asks for, and as a result loses in a competitive environment, did the architects perform either job effectively?

There is no universal answer to this question. The answer depends very much on how the development environment structures the relationship between the architect and sponsor. In the classical architecting paradigm, the architect must be careful not to substitute his or her own value judgments for those of the client. So, if the system reflects the client's value judgments, and the system is ultimately unsuccessful because those value judgments do not reflect reality, the architecting job has still been done well. But, it is also traditionally well within the architect's responsibility to warn the client of the certain or likely consequences of proposed courses of action. If it is evident to the architect that the design process is leading to something that can be easily opposed, this must be made plain to the client. The client may choose to proceed anyway, but the consequences should be clear.

---

* A real option in this context is the ability to build alternative systems at relatively low cost because of other investments. In this case, Boeing held a real option on larger aircraft through its involvement in building large bombers for the U.S. Army Air Corps.

In some cases, the architect may have an ethical or even legal responsibility beyond that of the responsibility to the client. Public buildings must be built in accordance with public safety. A system architect working for a government has some responsibility beyond just the immediate acquisition program to the national interest. In our DC-3 story, the architect was part of the builder organization and so had a great stake in ultimate success or failure. A builder-architect cannot shrug off poor client decision making as the builder-architect is also the client and rises or falls on the result. The builder-architect should have a level of ownership of the problem a third-party architect need not.

What Is the "Architecture" of the DC-3?

Asking "What is the architecture of the DC-3" illustrates the contrast between architecture as *physical design* and architecture as *concept development* points of view. Both the Boeing 247 and DC-3 shared the same essential structural, technical features. Both were two-engine, hollow fuselage, modern configuration transport aircraft. From the outside, both look quite similar. Both used very similar technology. In the sense of overall physical design, they are quite similar.

However, in performance attributes and in operational placement, they are quite different. The DC-3 is considerably larger and, more importantly, is enough larger for the performance margin to have great operational significance. The DC-3 performs missions the Boeing 247 cannot, and enables business models that the Boeing 247 cannot. In a larger context, the design of the DC-3 embodies a different business strategy than the Boeing 247. If we think of architecture as the technical embodiment of strategy, we see the distinct difference between the architectures of the two systems.

Art Raymond's Principles

One of the attractions of the DC-3 story is the excellent Art Raymond paper previously referenced. Raymond's paper provides a set of eight timeless principles for architecting that hold as well today as they did when first articulated[1]:

1. *Proper environment*: This includes the physical facilities in which designers work, but Raymond's focus was on the confidence and enthusiasm of the sponsors and adequate financing. In Raymond's words:

> The thing above all else that makes a project go
> is the enthusiasm of its backers; not false enthusiasm put on for effect — sooner or later this is seen

through — but rather the enthusiasm that comes from the conviction that the project is sound, worthwhile, and due to succeed.

2. *Good initial choice*: In Raymond's terms, a good initial choice is one that neatly combines value and feasibility. He particularly emphasizes the role of elegant compromise between conflicting factors and clearly identifying the need or mission for the aircraft. The biggest failures come not from systems that are technological failures, but from those that fail to meet any need well enough to generate demand.

3. *Excellence of detail design*: Although this book is focused on architecture as the initial concept, detailed design is likewise important. An excellent initial concept can be ruined by poor detailed design (although a poor initial concept is very unlikely to be saved by excellence in detailed design).

4. *Thorough development*: Raymond's perspective on thorough development emphasizes design refinement after the first test flight. In Raymond's era, the refinement of flying qualities of airplanes was quite important, and occurred mostly after the first flight. Calculations and wind tunnel tests were sufficient for basic performance, but refining handling qualities to a point of excellence required extensive flight testing.

5. *Follow-through*: Follow-through refers to the system life cycle after delivery to the operator. In the case of a commercial aircraft, some of the important elements include operator and maintainer training, maintenance and service facilities, development of spare parts, design updates in response to service data, and technical manuals. The value of the system to its customers/operators is directly related to the quality of follow-through. From the perspective of systems architecting, the follow-through elements may be inside the boundaries of the initial concept development. The quality of the initial concept may be determined by its amenability to effective follow-through.

6. *Thorough exploitation*: All successful aircraft are extensively modified during their operational lifetimes. The DC-3 was produced in an enormous number of variations, and even today there are firms that adapt modern avionics to the remaining DC-3 airframes. Successful systems are designed to accommodate a range of modifications. This is familiar in modern commercial aircraft where many interior configurations are available, usually several different choices of engine, freighter and passenger versions, and extended-range or capacity versions.

7. *Correct succession*: No matter how successful a system is, there comes a time when it is more effective to break away and re-architect. Conversely, breaking away when the time is not ripe incurs high cost to little effect. The essential judgment here is projection of technical and operational trends. There is an opportunity for succession when either (or better yet both) will move substantially over the time required to develop the successor system.

8. *Adaptiveness*: The DC-1, 2, 3 sequence is the best illustration of adaptiveness. Adaptiveness really means responsiveness to the future environment as it unfolds, rather than as it was projected. Projections are the foundation of planning, and real strategy is the ability to adapt to the environment as it unfolds. In this story, we see several examples of adaptiveness in architecture. Douglas did not settle for the DC-1, even though it met the contractual specifications provided by TWA. Instead, they adapted to the operational environment as it developed, first with the improved DC-2 and then with the much upsized DC-3. Likewise, Boeing illustrated effective adaptiveness in the sense of retaining (and then exercising) real options for larger aircraft. When their first attempt at a revolutionary aircraft was insufficient, they used large aircraft technology from their military aircraft to upsize their flagship commercial aircraft to the Boeing 307.

Notes and References

1. Raymond, A., The Well Tempered Aircraft, 39th Wilbur Wright Memorial Lecture, *Journal of the Royal Aeronautical Society*, September 1951.
2. A history of the DC-3, referenced from www.dc3history.org/chapters/chapter_1.
3. Raymond, op cit.

chapter 3

Builder-Architected Systems

> No system can survive that doesn't serve a useful purpose.
>
> **Harry Hillaker***

Introduction: The Form-First Paradigm

The classical architecting paradigm is not the only way to create and build large complex systems, nor is it the only regime in which architects and architecting is important. A different architectural approach, the "form first," begins with a builder-conceived architecture in mind, rather than with a set of client-accepted purposes. Its architects are generally members of the technical staff of the company. Their client is the company; although the intention is to reach a customer base in the market.

Incremental Development for an Existing Customer

Most builder-initiated architectures are variations of existing ones; as examples, consider jet aircraft, personal computers, smart automobiles, and follow-on versions of existing software applications. The original architectures having proved by use to be sound, variations and extensions should be of low risk. Extensive reuse of existing modules should be expected because design assumptions, system functions, and interfaces are largely unchanged.

The architect's responsibilities remain much the same as under the classical paradigm, but with an important addition: the identification of proprietary architectural features deemed critical to maintaining competitive advantage in the marketplace. Lacking this identification, the question "who owns what?" can become so contentious for both builder and customer that product introduction can be delayed for years.

Far more important than these relatively low risks is the paradigm shift from function-to-form (purpose driven) to one of form-to-function (form driven). Unlike the classical paradigm, in form-first architecting,

---

* Chief architect, General Dynamics F-16 Fighter. As stated in a University of Southern California (USC) Systems Architecting lecture, November 1989.

one's customers judge the value of the product after rather than before the product has been developed and produced. In the classical paradigm, the customer is responsible for the value judgments, and so should expect to be satisfied with the resultant system. In a form-first, builder-architected system, the architect hopes the customer will find it satisfactory, but there are no guarantees. The judgment of success begins only after the system is built and delivered.

The resultant risk has spawned several risk-reduction strategies. The simplest is an early prototype demonstration to present customers, with its associated risks of premature rejection. The more rapidly prototypes can be developed and delivered, the more rapidly feedback can be gained from customers. Another recent strategy is the open source method for designing software, a process in which customers become developers, or at least active participants with developers. Anyone interested can participate, comment, submit ideas, develop software, and use the system, all at no cost to the participant. The project being tied together by the Internet (and some unique social conventions), everyone — and particularly the builder and potential clients — knows and can judge its utility. The risk of rejection is sharply reduced at the possible cost of control of design. The open source community is a principal example of collaborative system assembly. We discuss that topic specifically in Chapter 7.

New Markets for Existing Products

The next level of architecting intensity is reached when the builder's motivation is to reach uncertain or "latent" markets in which the unknown customer must acquire the product before judging its value. Almost certainly, the product will have to be at least partially rearchitected in cost, performance, availability, quantities produced, and so forth. To succeed in the new venture, architecting must be particularly alert, making suggestions or proposing options without seriously violating the constraints of an existing product line. Hewlett-Packard in the 1980s developed this architecting technique in a novel way. Within a given product line, say that of a "smart" analytic instrument, a small set of feasible "reference" architectures are created, each of which is intended to appeal to a different kind of customer. Small changes in that architecture then enable tailoring to customer-expressed priorities. Latent markets discovered in the process can then be quickly exploited by expansion of the product line.

The original product line architecture can be maintained with few modifications or risks until a completed system is offered to the market. Ideally, the architectural features of the product line are largely invariant, but the architectural features of individual products change rapidly. The product line sets out constraints and resources, and the individual products use them to produce valued features. The architecture of the product line is dominantly, though not exclusively, the intersection of the architectures of the circumscribed products. The architecture of the product line is dominated by the common features, the things that bring value to taking a product-line approach. In one sense only can the architecture of the product line be thought of as the union of the architectures of the products, which is the sense in which the product line defines the collection of niches into which each product will fit. The product line makes global decisions about where individual products can be developed, and where they cannot.

New Products, New Markets

Of greatest risk are those form-first, technology-driven systems that create major qualitative changes in system-level behavior, changes in kind rather than of degree. Systems of this type almost invariably require across-the-board new starts in design, development, and use. They most often arise when radically new technologies become available, such as jet engines, new materials, microprocessors, lasers, software architectures, and intelligent machines. Although new technologies are infamous for creating unpleasant technological and even sociological surprises, by far the greatest single risk in these systems is one of timing. Even if the form is feasible, introducing a new product either too early or too late can be punishing. Douglas Aircraft Company was too late into jet aircraft, losing out for years to The Boeing Company. Innumerable small companies have been too early, unable to sustain themselves while waiting for the technologies to evolve into engineered products. High-tech defense systems, most often due to a premature commitment to a critical new technology, have suffered serious cost overruns and delays.

Technological Substitutions within Existing Systems

The second greatest risk is in not recognizing that before they are completed, technology-driven architectures will require much more than just replacing, one at a time, components of an older technology for those of a newer one. Painful experience shows that without widespread changes in the system and its management, technology-driven initiatives seldom meet expectations and too often cost more for less value. As examples, direct replacements of factory workers with machines,[1] of vacuum tubes with transistors, of large inventories with just-in-time deliveries, and of experienced analysts with computerized management information systems, all collapsed when attempted by themselves in a system that was otherwise unchanged. They succeeded only when incorporated in concert with other matched and planned changes. It is not much of an exaggeration to say that the latter successes were well architected, the former failures were not.

In automobiles, the most recent and continuing change is the insertion of ultraquality electronics and software between the driver and the mechanical subsystems of the car. This remarkably rapid evolution removes the driver almost completely from contact with, or direct physical control of, those subsystems. It considerably changes such overall system characteristics as fuel consumption, aerodynamic styling, driving performance, safety, and servicing and repair — as well as the design of such possibly unexpected elements as engines, transmissions, tires, dashboards, seats, passenger restraints, and freeway exits. As a point of fact, the automotive industry expected that by the turn of the century more than 93% of all automotive equipment would be computer controlled,[2] a trend evidently welcomed and used by the general public or it would not have been done. A telling indicator of the public's perception of automotive performance and safety was the virtually undisputed increase in national speed limits. Safe, long-distance, highway travel at 70 mph (117 km/hr) was rare, even dangerous, two decades ago. Even if the highways were designed for it, conventional cars and trucks were not. It is now common, safe, and legal. Perhaps the most remarkable fact about this rapid evolution is that most customers were never aware of it. This result came from a commitment to quality so high that a much more complex system could be offered that, contrary to the usual experience, worked far better than its simpler predecessor.

In aircraft, an equivalent, equally rapid, technology-driven evolution is "fly by wire," a change that, among other things, is forcing a social revolution in the role of the pilot and in methods of air traffic control. More is involved than the form-fit-function replacement of mechanical devices with a combination of electrical, hydraulic, and pneumatic units. Aerodynamically stable aircraft, which maintain steady flight with nearly all controls inoperative, are steadily being replaced with ones that are less stable, more maneuverable, and computer controlled in all but emergency conditions. The gain is more efficient, potentially safer flight. But the transition has been as difficult as that between visual and instrument-controlled flight.

In inventory control, a remarkable innovation has been the very profitable combination in one system of point-of-sale terminals, of a shift of inventory to central warehouses and of just-in-time deliveries to the buyer. Note the word *combination*. None of the components has been particularly successful by itself. The risk here is greater susceptibility to interruption of supply or transportation during crises.

In communications, satellites, packet switching, high-speed fiber-optic lines, e-mail, the World Wide Web, and electronic commerce have combined for easier access to a global community, but with increasing concerns about privacy and security. The innovations now driving the communications revolution were not, individually, sufficient to create this revolution. It has been the interaction of the innovations, and the changes in business processes and personal habits connected to them, that have made the revolution.

In all of these examples, far more is affected than product internals. Affected also are such externals as manufacturing management, equity financing, government regulations, and the minimization of environmental impact, to name but a few. These externals alone could explain the growing interest by innovative builders in the tools and techniques of systems architecting. How else could well-balanced, well-integrated, financially successful, and socially acceptable total systems be created?

Consequences of Uncertainty of End Purpose

Uncertainty of end purpose, no matter what the reason, can have serious consequences. The most serious is the likelihood of serious error in decisions affecting system design, development, and production. Builder-architected systems are often solutions looking for a problem and hence are particularly vulnerable to the infamous "error of the third kind": working on the wrong problem.

Uncertainty in system purposes also weakens them as criteria for design management. Unless a well-understood basis for configuration control exists and can be enforced, system architectures can be forced off course by accommodations to crises of the moment. Some of the most expensive cases of record have been in attempts to computerize management information systems. Lacking clear statements of business purposes and market priorities, irreversible ad hoc decisions were made which so affected their performance, cost, and schedule that the systems were scrapped. Arguably, the best prevention against "system drift" is to decide on provisional or baseline purposes and stick to them. But what if those baseline purposes prove to be wrong in the marketplace?

Architecture and Competition

In the classical architecting paradigm, there is little or no role for competition. The client knows what he or she wants, or learns through interaction with the architect. When a system is delivered that is consonant with the client's values, the client should be satisfied. In many other cases, builder-architected systems prominent among them, success is judged more on competitive performance than on adherence to client values.

To reconcile how architecting and architecture relates to competition, we must set the context of the organization's overall competitive strategy.

Architecting cannot be talked about in the abstract; it has to be grounded in the strategies of the organization conducting it. In builder-architected systems, this means the competitive posture of the builder. Broadly speaking, we can identify three major competitive strategies with architectural consequences: disrupt and dominate, agile response, and attrition.

Disrupt and Dominate

This strategy is based on creating systems that disrupt existing operational patterns or markets, and building barriers to prevent others from taking advantage of those disruptions. In "Case Study 1", the DC-3 was a disruptive system in that it caused systematic change to how airlines did business. However, Douglas was unable to raise a strong barrier to prevent Boeing from entering the market space (although Douglas had a valuable lead of several years). The Apple iPod and iTunes music store combination is an example, where patents, copyrights, secrecy of proprietary technologies, and exclusive contractual arrangements have successfully formed barriers to competitive entry.

The architectural challenges in supporting this strategy are twofold. First, the quality of the architecting must be exceptional, as the architect must create beyond the boundaries of current systems. Great imagination is required, while simultaneously maintaining sufficient options (see the next section) to adapt to the inevitable failures of imagination. Second, the approach must allow protection from competitors who will employ an agile response strategy.

Agile Response

This strategy emphasizes the organization's capability to react more quickly and effectively than the competition. We emphasize both speed and effectiveness, because an ineffective response quickly delivered is still ineffective. A key distinction between the disrupt and dominate strategy and agile response is that agile response seeks to exploit the underlying flux in markets or military situations without disrupting their overall structure. An agile responder in a commercial environment produces new products within established markets faster and more effectively than the competition but does not try to create entirely new markets. The agile response strategy is especially effective in immature markets where changes in consumer preference and technology create many new opportunities.

From an architectural perspective, the challenges for agile response are again twofold. First, to carry this strategy out effectively, the organization must be able to very rapidly conceive, develop, and deliver new systems. This means that architecting must be fast and must support a very compressed development cycle. Second, at one higher level of abstraction, the architecture of the organization and its product lines must support agility. The organization and product lines must be structured to facilitate agility. Typically the product-line architecture evolves much more slowly than the products, and the product-line architecture sets out critical invariants allowing rapid development and deployment.

Attrition

The classic example on the military side of the attrition strategy is to win by having more firepower, manpower, logistic power, and willingness to suffer than your opponent. A business equivalent strategy is to prevail through access to large amounts of low-cost capital, low-wage labor, and large distribution channels. When coupled to a strong organizational capability for learning and improvement, this is a powerful strategy, especially in mature markets where consumer preference changes slowly.

Architecting in the attrition strategy is relatively slow and deliberate. The key architecture is the one embodied in the organization. Successful conduct of the attrition strategy is dependent on access to the requisite resources, cheaply and at a large scale. The strategy is likely to fail either when encountering a still larger and more fit competitor, or when the underlying environment (markets, operations, and technology) has an inherent rate of change high enough so that an agile response strategy becomes more effective, or when the change is sufficient to be open to disruption.

Reducing the Risks of Uncertainty of End Purpose

A powerful architecting guide to protect against the risk of uncertain purposes is to *build in and maintain options*. With options available, early decisions can be modified or changed later. Other possibilities include the following: Build in options to stop at known points to guarantee at least partial satisfaction of user purposes without serious losses in time and money, for example, in databases for accounting and personnel administration. Create architectural options that permit later additions, a favorite strategy for automobiles and trucks. Provisions for doing so are hooks in software to add applications and peripherals, scars in aircraft to add range and seats, shunts in electrical systems to isolate troubled sections, contingency plans in tours to accommodate cancellations, and forgiving exits from highways to minimize accidents.

In software, a general strategy is: *Use open architectures. You will need them once the market starts to respond.* As will be seen, a further refinement of this domain-specific heuristic will be needed, but this simpler version makes the point for now.

And then there is the always welcome heuristic: every once in a while, *Pause and reflect*. Reexamine the cost-effectiveness of system features such as high precision pointing for weather satellites or cross-talk levels for tactical communication satellites.* Review why interfaces were placed where they were. Check for unstated assumptions such as the Cold War continuing indefinitely† or the 1960s generation turning conservative as it grew older.

Risk Management by Intermediate Goals

Another strategy to reduce risk in the development of system-critical technologies is by scheduling a series of intermediate goals to be reached by precursor or partial configurations. For example, build simulators or prototypes to tie together and synchronize otherwise disparate research efforts.[3] Build partial systems, demonstrators, or models to help assess the sensitivity of customer acceptance to the builder's or architect's value judgments,[4] a widely used market research technique. And, as will be seen in Chapter 7, if these goals result in stable intermediate forms, they can be powerful tools for integrating hardware and software.

Clearly, precursor systems have to be almost as well architected as the final product. If not, their failure in front of a prospective customer can play havoc with future acceptance and ruin any market research program. As one heuristic derived from military programs warns, *The probability of an untimely failure increases with the weight of brass in the vicinity.* If precursors and demonstrators are to work well "in public," they better be well designed and well built.

Even if a demonstration of a precursor succeeds, it can generate excessive confidence, particularly if an untested requirement is critical. In one case, a U.S. Air Force (USAF) satellite control system successfully and very publicly demonstrated the ability to manage one satellite at a time; the critical task, however, was to control multiple, different satellites, a test it subsequently flunked. Massive changes in the system as a whole were required. In another similar case, a small launch vehicle, arguably successful as a high-altitude demonstrator of single-stage-to-orbit, could not be scaled up to full size or full capability for embarrassingly basic mechanical and materials reasons.

These kinds of experiences led to the admonition: *Do the hard parts first,* an extraordinarily difficult heuristic to satisfy if the hard part is a unique function of the system as a whole. Such has been the case for a near-impenetrable missile defense system, a stealthy aircraft, a general aviation air traffic control system, a computer operating system, and a national tax

---

* In real life, both features proved to be unnecessary but could not be eliminated by the time that truth was discovered.
† A half-joking question in defense planning circles in the early 1980s used to be, "What if peace broke out?" Five years later, it had.

reporting system. The only credible precursor, to demonstrate the hard parts, had to be almost as complete as the final product.

In risk management terms, if the hard parts are, perhaps necessarily, left to last, then the risk level remains high and uncertain to the very end. The justification for the system therefore must be very high and the support for it very strong or its completion will be unlikely. For private businesses, this means high-risk venture capital. For governments, it means support by the political process, a factor in system acquisition for which few architects, engineers, and technical managers are prepared. Chapter 13 is a primer on the subject.

The "What Next?" Quandary

One of the most serious long-term risks faced by a builder of a successful, technology-driven system is the lack of, or failure to win a competition for, a successor or follow-on to the original success.

The first situation is well exemplified by a start-up company's lack of a successor to its first product. Lacking the resources in its early, profitless, years to support more than one research and development effort, it could only watch helplessly as competitors caught up and passed it by. Ironically, the more successful the initial product, the more competition it will attract from established and well-funded producers anxious to profit from a sure thing. Soon the company's first product will be a "commodity," something that many companies can produce at a rapidly decreasing cost and risk. Unable to repeat the first success, soon enough the start-up enterprise fails or is bought up at fire-sale prices when the innovator can no longer meet payroll. Common. Sad. Avoidable? Possibly.

The second situation is the all-too-frequent inability of a well-established company that had been successfully supplying a market-valued system to win contracts for its follow-on. In this instance, the very strength of the successful system, a fine architecture matched with an efficient organization to build it, can be its weakness in a time of changing technologies and shifting market needs. The assumptions and constraints of the present architecture can become so ingrained in the thinking of participants that options simply do not surface.

In both situations, the problem is largely architectural, as is its alleviation.

For the innovative company, it is a matter of control of critical architectural features. For the successful first producer, it is a matter of knowing, well ahead of time, when purposes have changed enough that major rearchitecting may be required. Each situation will be considered in turn.

Controlling the Critical Features of the Architecture

The critical part of the answer to the start-up company's "what next" quandary is control of the architecture of its product through proprietary ownership of its basic features.[5] This is the second half of a disrupt and dominate strategy. Examples of such features are computer operating systems, interface characteristics, communication protocols, microchip configurations, proprietary materials, patents, exclusive agreements with critical suppliers or distributors, and unique and expensive manufacturing capabilities. Good products, although certainly necessary, are not sufficient. They must also arrive on the market as a steadily improving product line, one that establishes, de facto, an architectural standard.

Surprisingly, one way to achieve that objective is to use the competition instead of fighting it. Because success invites competition, it may well be better for a start-up to make its competition dependent, through licensing, upon a company-proprietary architecture rather than to have it incentivized to seek architectural alternatives. Finding architectural alternatives takes time. But licensing encourages the competition to find new applications, add peripherals, and develop markets, further strengthening the architectural base, adding to the source company's profits and its own development base.[6] Heuristically: *Successful architectures are proprietary, but open.**

This strategy was well exemplified by Microsoft in opening and licensing its personal computer (PC) operating system while Apple refused to do so for its Macintosh. The resultant widespread cloning of the PC expanded not only the market as a whole, but Microsoft's share of it. The Apple share dropped. The dangers of operating in this kind of open environment, however, are also illustrated in the case of PC hardware. The PC standard proved much more open than IBM intended. Where it was assumed they could maintain a price advantage through the economies of scale, the advantage disappeared. The commoditization of the PC also drove down profit margins until even a large share proved substantially unprofitable, at least for a company structured as IBM. IBM struggled for years (unsuccessfully) to move the PC market in a direction that would allow it to retain some degree of proprietary control and return profits. In contrast, Microsoft and Intel have struck a tremendously profitable balance between proprietary protection and openness. The Intel instruction set architecture has been copied, but no other company has been able to achieve a market share close to Intel's. Microsoft has grown both through proprietary and open competition, the former in operating systems and the latter in application programs.

Apple was not entirely closed. Apple was "open enough" to create a substantial market in software and peripheral devices. Opening up

* "Open" here means adaptable, friendly to add-ons, and selectively expandable in capability.

too far can destroy any possibility of maintaining a competitive advantage. Staying too closed prevents the creation of a synergistic market. Architecture and strategy need to be consistent.

A different kind of architectural control is exemplified by the Bell telephone system with its technology generated by the Bell Laboratories, its equipment produced largely by Western Electric, and its architectural standards maintained by usage and regulation. Others include Xerox in copiers, Kodak in cameras, and Hewlett-Packard in instruments. All these product-line companies began small, controlled the basic features, and prospered. But, as each of these also demonstrated, success is not forever.

Thus, for the innovator, the essentials for continued success are not only a good product, but also the generation, recognition, and control of its basic architectural features. Without these essentials, there may never be a successor product. With them, many product architectures, as architecturally controlled product lines, have lasted for years following the initial success. Which adds even more meaning to: *There's nothing like being the first success.*[7]

Abandonment of an Obsolete Architecture

A different risk reduction strategy is needed for the company that has established and successfully controlled a product-line architecture[8] and its market, but is losing out to a successor architecture that is proving to be better in performance, cost, or schedule. There are many ways that this can happen. Perhaps the purposes that original architecture has satisfied can better be done in other ways. Typewriters have largely been replaced by personal computers. Perhaps the conceptual assumptions of the original architecture no longer hold. Energy may no longer be cheap. Perhaps competitors found a way of bypassing the original architectural controls with a different architecture. Personal computers destroyed the market for Wang word processors and eventually for proprietary workstations. And, as a final example, cost risk considerations precluded building larger and larger spacecraft for the exploration of the solar system.

To avoid being superceded architecturally requires a strategy, worked out well ahead of time, to set to one side or cannibalize that first architecture, *including the organization matched with it,* and to take preemptive action to create a new one. The key move is the well-timed establishment of an innovative architecting team, unhindered by past success and capable of creating a successful replacement. Just such a strategy was undertaken by Xerox in a remake of the corporation as it saw its copier architecture start to fade. It thereby redefined itself as "the document company."[9] But, the failure of Xerox to substantially profit from most of the innovation developed by Xerox PARC (Palo Alto Research Center) likewise illustrates the difficulty of making the transition. Xerox understood the necessity of making the architectural transition, and invested in it, for many years before being organizationally capable of actually making the transition.[10]

Creating Innovative Teams

Clearly the personalities of members of any team, particularly an innovative architecting team, must be compatible. A series of USC Research Reports[11] by Jonathan Losk, Tom Pieronek, Kenneth Cureton, and Norman P. Geis, based on the Myers-Briggs Type Indicator (MBTI), strongly suggest that the preferred personality type for architecting team membership is NT.[12] That is, members should tend toward systematic and strategic analysis in solving problems. As Cureton summarizes, "Systems architects are made and not born, but some people are more equal than others in terms of natural ability for the systems architecting process, and MBTI seems to be an effective measure of such natural ability. No single personality type appears to be the 'perfect' systems architect, but the INTP personality type often possesses many of the necessary skills."

Their work also shows the need for later including an ENTP (extroversion, intuition, thinking, perceiving), a "field marshal" or deputy project manager, not only to add some practicality to the philosophical bent of the INTPs (introversion, intuition, thinking, perceiving), but to help the architecting team work smoothly with the teams responsible for building the system.

Creating *innovative* teams is not easy, even if the members work well together. The start-up company, having little choice, depends on good fortune in its recruiting of charter members. The established company, to put it bluntly, has to be willing to change how it is organized and staffed from the top down based almost solely on the conclusions of a presumably innovative team of "outsiders," albeit individuals chartered to be such. The charter is a critical element, not so much in defining new directions as in defining freedoms, rights of access, constraints, responsibilities, and prerogatives for the team. For example, can the team go outside the company for ideas, membership, and such options as corporate acquisition? To whom does the team respond and report — and to whom does it not? Obviously, the architecting team better be well designed and managed. Remember, if the team does not succeed in presenting a new and accepted architecture, the company may well fail.

One of the more arguable statements about architecting is the one by Frederick P. Brooks Jr. and Robert Spinrad that *the best architectures are the product of a single mind*. For modest-sized projects, that statement is reasonable enough. As projects get larger and larger, it remains true but in somewhat different form. The complexity and work load of creating large, multidisciplinary, technology-driven architectures would overwhelm any individual. The observation of a single mind is most easily accommodated by a simple but subtle change from "a single mind" to "a team of a single mind." Some would say "of a single vision" composed of ideas, purposes, concepts, presumptions, and priorities. It is also critical to understand the difference between composing multidisciplinary teams and how teams form decisions. The key to a coherent architecture is coherent decision making. Majority votes by large committees are practically the worst-case scenario for gaining coherence of decision making over a long series of related complex decisions.

One architect put the issue succinctly. When asked about the role of multidisciplinary teams, he said: "Multi-disciplinary teams covering all stakeholders and major subsystem areas are critical to effective space architecting, and I love using them. As long as I get to make all of the decisions." His point was simple — good architecting requires diversity of view but unity of decision.

In the simplest case, the single vision would be that of the chief architect and the team would work to it. For practical as well as team cohesiveness reasons, the single vision needs to be a shared one. In no system is that more important than in the entrepreneurially motivated one. There will always be a tension between the more thoughtful architect and the more action-oriented entrepreneur. Fortunately, achieving balance and compromise of their natural inclinations works in the system's favor.

An important corollary of the shared vision is that the architecting team, and not just the chief architect, must be seen as creative, communicative, respected, and of a single mind about the system-to-be. Only then can the team be credible in fulfilling its responsibilities to the entrepreneur, the builder, the system, and its many stakeholders. Internal power struggles, basic disagreements on system purpose and values, and advocacies of special interests can only be damaging to that credibility.

As Ben Bauermeister, Harry Hillaker, Archie Mills, Bob Spinrad,[13] and other friends have stressed in conversations with the authors, innovative teams need to be cultural in form, diverse in nature, and almost obsessive in dedication.

By cultural is meant a team characterized by informal creativity, easy interpersonal relationships, trust and respect, all characteristics necessary for team efficiency, exchange of ideas, and personal identification with a shared vision. To identify with a vision, they must deeply believe in it and in their chief. The members must acknowledge and follow the lead of their chief or the team disintegrates.

Diversity in specialization is to be expected; it is one of the reasons for forming a team. Equally important, a balanced diversity of style and programmatic experience is necessary to assure open-mindedness, to spark creative thinking in others, and to enliven personal interrelationships. It is necessary, too, to avoid the "groupthink" of nearly identical members with the same background, interests, personal style, and devotion to past architectures and programs. Indeed, team diversity is one of the better protections against the second-product risks mentioned earlier.

Consequently, an increasingly accepted guideline is that, to be truly innovative and competitive in today's world: *The team that created and built a presently successful product is often the best one for its evolution — but seldom for creating its replacement.*

A major challenge for the architect, whether as an individual or as the leader of a small architecting team, is to maintain dedication and momentum not only within the team but also within the managerial structure essential for its support. The vision will need to be continually restated as new participants and stakeholders arrive on the scene — engineers, managers active and displaced, producers, users, and new clients. Even more difficult, it will have to be transformed as the system proceeds from a dream to a concrete entity, to a profit maker, and finally to a quality production. Cultural collegiality will have to give way to the primacy of the bottom line and finally to the necessarily bureaucratic discipline of production. Yet the integrity of the vision must never be lost or the system will die.

The role of organizations in architectures, and the architecture of organizations, is taken up at much greater length by one of the present authors.[14]

Architecting "Revolutionary" Systems

A distinction to be made at this point is between architecting in precedented, or evolutionary, environments, and architecting unprecedented systems. Whether we call such systems "revolutionary," "disruptive," or "unprecedented" seems more a matter of fashion. What is important is that the system stands apart from all that came before it, and that is great change of businesses or militaries operate. One of the most notable features of Rechtin (1991)[15] was an examination of the architectural history of clearly successful and unprecedented systems. A central observation is that all such systems have a clearly identifiable architect or small architect team. They were not conceived by the consensus of a committee. Their basic choices reflect a unified and coherent vision of one individual or a very small group. Further reflection, and study by students, has only reinforced this basic conclusion, while also filling in some of the more subtle details.

Unprecedented systems have been both purpose driven and technology driven. In the purpose-driven case, the architect has sometimes been part of the developer's organization and sometimes not. In the technology-driven case, the architect is almost always in the developer's organization. This should be expected as technology-driven systems typically come from intimate knowledge of emerging technology, and someone's vision of where it can be applied to advantage.[16] This person is typically not a current user but is rather a technology developer. It is this case that is the concern of this section.

The architect has a lead technical role. But this role cannot be properly expressed in the absence of good project management. Thus, the pattern of a strong duo, project manager and system architect, is also characteristic of successful systems. In systems of significant complexity, it is very difficult to combine the two roles. A project manager is typically besieged by short-term problems. The median due date of things on the project manager's desk is probably yesterday. In this environment of immediate problems, it is unlikely that a person will be able to devote the serious time to longer-term thinking and broad communicating that are essential to good architecting.

The most important lesson in revolutionary systems, at least those not inextricably tied to a single mission, is that success is commonly not found where the original concept thought it would be. The Macintosh computer was a success because of desktop publishing, not what the market assumed in its original rollout (which was as a personal information appliance). Indeed, desktop publishing did not exist as a significant market when the Macintosh was introduced.* This pattern of new systems becoming successful because of new applications has been common enough in the computer industry to have acquired a nickname, "the killer app(lication)." Taken narrowly, a "killer app" is an application so valuable that it drives the sales of a particular computer platform. Taken more broadly, a "killer app" is any new system usage so valuable that, by itself, it drives the dissemination of the system.

One approach to unprecedented systems is to seek the killer application that can drive the success of a system. A recent noncomputer example that illustrates the need, and the difficulty, is the search for a killer application for reusable space launch vehicles. Proponents believe that there is a stable economic equilibrium with launch costs an order of magnitude lower, and flight rates around an order of magnitude higher, than current. But, if flight rates increase and space payload costs remain the same, then total spending on space systems will have to be far higher (roughly an order of magnitude, counting only the payload costs). For there to be a justification for high flight rate launch, there has to be an application that will realistically exploit it. That is, some application must attract sufficient new money to drive up payload mass.

Various proposals have been floated, including large constellations of communication satellites, space power generation, and space tourism. If the cost of robotic payloads was reduced at the same time, their flight rate might increase without total spending going up so much. But the only

---

* Though the concept was anticipated, as witnessed by the original business plan being composed in publishable form on prototype Macintosh systems.

clear way of doing that is to move to much larger-scale serial production of space hardware to take advantage of learning curve cost reductions.[17] This clearly indicates a radical change to the architecture not only of launch, but to satellite design, satellite operations, and probably to space manufacturing companies as well. And all these changes need to take place synchronously for the happy consequence of lowered cost to result. So far, this line of reasoning has not produced success. Launches remain expensive, and the most efficient course appears to be greater reliability and greater functionality per pound of payload, which has the effect of driving the launch rate down and making a high-rate/low-cost launch approach even more difficult.

Sometimes such synchronized changes do occur. The semiconductor industry has experienced decades of 40% annual growth because such synchronized changes have become ingrained in the structure of the computer industry. As the production and design technology improve, the total production base (in transistor quantity and revenue) goes up. Lowered unit costs result in increased consumption of electronics even larger than the simple scale up of each production generation. The resulting revenue increases are sufficient to keep the process going, and coordinated behavior in the production equipment supplier, design system supplier, and consumer electronic producers smoothes the process sufficiently for it to run stably for decades.

In summary, the successful architect exploits what the market demonstrates as the killer application, assuming he or she can predetermine it. The successful innovator exploits the first-to-market position to take advantage of the market's demonstration of what it really wants faster than the second-to-market player does. The successful follower beats the first-to-market by being able to exploit the market's demonstration more quickly. Each is making a consistent choice of both strategy and architecture (in a technical sense). We explore this issue in depth in Chapter 12.

Systems Architecting and Basic Research

One other relationship should be established, that between architects and those engaged in basic research and technology development. Each group can further the interests of the other. The architect can learn without conflict of interest. The researcher is more likely to become aware of potential sponsors and users.

New technologies enable new architectures, though not singly or by themselves. Consider solid-state electronics, fiber optics, software languages, and molecular resonance imaging for starters. And innovative architectures provide the rationale for underwriting research, often at a very basic level. Yet, though both innovative architecting and basic research explore the unknown and unprecedented, there seems to be little early contact between their respective architects and researchers. The architectures of intelligent machines, the chaotic aerodynamics of active surfaces, the sociology of intelligent transportation systems, and the resolution of conflict in multimedia networks are examples of presumably common interests. Universities might well provide a natural meeting place for seminars, consulting, and the creation and exchange of tools and techniques.

New architectures, driven by perceived purposes, sponsor more basic research and technology development than is generally acknowledged. Indeed, support for targeted basic research undoubtedly exceeds that motivated by scientific inquiry. Examples abound in communications systems that sponsor coding theory, weapons systems that sponsor materials science and electromagnetics, aircraft that sponsor fluid mechanics, and space systems that sponsor the fields of knowledge acquisition and understanding.

It is therefore very much in the mutual interest of professionals in research and development (R&D) and systems architecting to know each other well. Architects gain new options. Researchers gain well-motivated support. Enough said.

Heuristics for Architecting Technology-Driven Systems

General

- An insight is worth a thousand market surveys.
- Success is defined by the customer, not by the architect.
- In architecting a new program, all the serious mistakes are made in the first day.
- The most dangerous assumptions are the unstated ones.
- The choice between products may well depend upon which set of drawbacks the users can handle best.
- As time to delivery decreases, the threat to user utility increases.
- If you think your design is perfect, it is only because you have not shown it to someone else.
- If you do not understand the existing system, you cannot be sure you are building a better one.
- Do the hard parts first.
- Watch out for domain-specific systems. They may become traps instead of useful system niches, especially in an era of rapidly developing technology.
- The team that created and built a presently successful product is often the best one for its evolution — but seldom for creating its replacement. (It may be locked into unstated assumptions that no longer hold.)

Specialized

From Morris and Ferguson[5]:

- Good products are not enough. (Their features need to be owned.)
- Implementations matter. (They help establish architectural control.)
- Successful architectures are proprietary, but open. (Maintain control over the key standards, protocols, etc., that characterize them but make them available to others who can expand the market to everyone's gain.)

  From Chapters 2 and 3:
- Use open architectures. You will need them once the market starts to respond.

Conclusion

Technology-driven, builder-architected systems, with their greater uncertainty of customer acceptance, encounter greater architectural risks than those that are purpose driven. Risks can be reduced by the careful inclusion of options, the structuring of their innovative teams, and the application of heuristics found useful elsewhere. At the same time, they have lessons to teach in the control of critical system features and the response to competition enabled by new technologies.

Exercises

1. The architect can have one of three relationships to the builder and client. The architect can be a third party, can be the builder, or can be the client. What are the advantages and disadvantages of each relationship? For what types of system is one of the three relationships necessary?
2. In a system familiar to you, discuss how the architecture can allow for options to respond to changes in client demands. Discuss the pros and cons of product versus product-line architecture as strategies in responding to the need for options. Find examples among systems familiar to you.
3. Architects must be employed by builders in commercially marketed systems because many customers are unwilling to sponsor long-term development; they purchase systems after evaluating the finished product according to their then-perceived needs. But placing the architect in the builder's organization will tend to dilute the independence needed by the architect. What organizational approaches can help to maintain independence while also meeting the needs of the builder organization?

4. The most difficult type of technology-driven system is one that does not address any existing market. Examine the history of both successful and failed systems of this type. What lessons can be extracted from them?

Notes and References

1. Majchrzak, Ann, *The Human Side of Automation*. San Francisco: Jossey-Bass, 1988, pp. 95–102. The challenge: 50%–75% of the attempts at introducing advanced manufacturing technology into U.S. factories are unsuccessful, primarily due to lack of human resource planning and management.
2. Automobile Club of Southern California speaker, Los Angeles, August 1995.
3. The Goldstone California planetary radar system of the early 1960s tied together antenna, transmitter, receiver, signal coding, navigation, and communications research and development programs. All had to be completed and integrated into a working radar at the fixed dates when target planets were closest to the Earth and well in advance of commitment for system support of communication and navigation for as-yet-not-designed spacecraft to be exploring the solar system. Similar research coordination can be found in NASA aircraft, applications software, and other rapid prototyping efforts.
4. Hewlett-Packard marketers have long tested customer acceptance of proposed instrument functions by presenting just the front panel and asking for comment. Both parties understood that what was behind the panel had yet to be developed, but could be. A more recent adaptation of that technique is the presentation of "reference architectures."
5. Morris, Charles R., and Charles H. Ferguson, How Architecture Wins Technology Wars, *Harvard Business Review*, March–April 1993; and its notably supportive follow-up review, Will Architecture Win the Technology Wars? in the same publication, May–June 1993 — a seminal article on the subject.
6. Morris and Ferguson (see Note 5).
7. Rechtin, E., *Systems Architecting, Creating and Building Complex Systems*. Englewood Cliffs, NJ: Prentice Hall, 1991, p. 301. Note that throughout the rest of the chapter, this reference will be referred to as Rechtin 1991.
8. It is important to distinguish between a product-line architecture and a product in that line. It is the first of these that is the subject here. Hewlett-Packard is justly famous for the continuing creation of innovative products to the point where half or less of their products on the market were there 5 years earlier. Their product *lines* and their architectures, however, are much longer lived. Even so, these, too, were replaced in due course, the most famous of which was the replacement of "dumb" instruments with smart ones.
9. Described by Robert Spinrad in a 1992 lecture to a USC systems architecting class. Along with Hewlett-Packard, Xerox demonstrates that it is possible to make major architectural, or strategic, change without completely dismantling the organization. But the timing and nature of the change is critical. Again, a product architecture and the organization that produces it must match, not only as they are created but as they decline. For either of the two to be too far ahead or behind the other can be fatal.
10. Hiltzik, M. A., *Dealers of Lightning*. New York: Harper Business Books, 1999.

11. Geis, Norman P., *Profiles of Systems Architecting Teams*. USC Research Report, June 1993, extends and summarizes the earlier work of Losk, Pieronek, and Cureton.
12. It is beyond the scope of this book to show how to select individuals for teams, only to mention that widely used tools are available for doing so. For further information on MBTI and the meaning of its terms, see Briggs Myers, Isabel, and Mary H. McCaulley, *Manual: A Guide to the Development and Use of the Myers-Briggs Type Indicator*. Palo Alto, CA: Consulting Psychologists Press, 1989.
13. Benjamin Bauermeister is an entrepreneur and architect in his own software company. He sees three successive motivations as product development proceeds: product, profit, and production corresponding roughly to cultural, bottom line, and bureaucratic as successive organization forms. Harry Hillaker is the retired architect of the Air Force F-16 fighter and a regular lecturer in systems architecting. Archie W. Mills is a middle-level manager at Rockwell International and the author of an unpublished study of the practices of several Japanese and American companies over the first decades of the 20th century in aircraft and electronics systems. Bob Spinrad is a XEROX executive and former director of XEROX Palo Alto Research Center (PARC).
14. Rechtin, E., *Systems Architecting of Organizations, Why Eagles Can't Swim*. Boca Raton, FL: CRC Press, 1999.
15. Rechtin 1991.
16. In contrast, and for reasons of demonstrated avoidance of conflict of interest, independent architects rarely suggest the use of undeveloped technology. Independent architects, interested in solving the client's purposes, are inherently conservative.
17. This is not the first time these arguments have happened. See Rechtin, E., A Short History of Shuttle Economics, NRC Paper, April, 1983, which demonstrated this case in detail.

Case Study 2: Mass and Lean Production

Introduction

Today, mass production is pervasive. Everything from cars to electronics is made in quantities of hundreds of thousands to millions. From the perspective of 100 years ago, products of extraordinary complexity are made in huge numbers. The story of mass production is significantly an architectural story. It is also a story of the interaction of architectures, in this case the interaction and synergy between the architectures of system-products and systems that built those products. The revolution that took place in production was dependent on changes in how the produced systems were designed, and design changes had synergistic effects with production. The characteristics and structures of the surrounding human systems were also critical to the story, notions that we will take up in later chapters.

This case study is a high-level survey that emphasizes the sweep of changes over time instead of details, and the nature of architectural decision making in mass production. We start by reviewing the history of mass production, from architecture perspective, focusing on the auto industry. We cover from the era of auto production as a cottage industry, through the seminal development of mass production by the Ford Motor Corporation, to the era of competition from other U.S. manufacturers, and end of the development with the Toyota Production System (TPS).

An Architectural History of Mass Production

The auto industry is hardly the only example of mass production, but it is usually considered as prototypical. The innovations in production at Ford, and later Toyota, substantially define the basic structures of modern mass production. The Ford system of production became the model for industry after industry, and the concepts filtered into society at large. The Toyota Production System is the prototype for Lean Production, now likewise a fundamental paradigm for organization in multiple industries, increasingly including service industries.

In the sections following, we cover major blocks of time and consider how decisions about basic organizing structure of production were synergistic (or antagonistic) with how systems were designed. For convenience, refer to Figure CS2.1 for the sequencing and relationship of events.

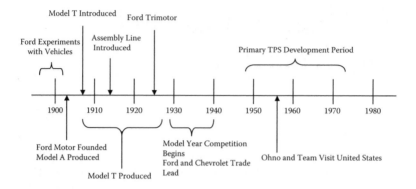

Figure CS2.1 Key events in the architecture of automobile mass production.

Cottage Industry (1890s to 1910s)

As auto production began in the 1890s, it was a classic cottage industry. Small groups of workers assembled each vehicle in a shop. The process involved bringing in a stream of parts (or machining them locally) and assembling them as a small team in one place. When the vehicle was complete, it was driven or otherwise moved away.

Automobiles built this way were very expensive. Of course, high prices and the small market went hand-in-hand. Because the vehicles were expensive, they were a luxury item with a very narrow customer base. Because the market was small, economies of scale were impossible and so prices were high.

Henry Ford was very aware of the problem, and was personally convinced that the way forward was in lower prices and larger production. He developed a conviction that high-quality automobiles could be, and should be, produced at cost low enough for average people to afford. The "car for the masses" would revolutionize society. He clashed repeatedly with his business partners over this, as they were convinced higher profits could be realized by concentrating on more expensive, high-margin vehicles. Over the short run, they were almost certainly right. Over the long run, the situation in automobiles was analogous in some ways to the situation in commercial aircraft just before the introduction of the DC-3 discussed in "Case Study 1." The introduction of a new system would create a qualitative change in the structure of the market (and drive structural, architectural change in both production and systems).

Birth of Mass Production (1908–1913)

Ford's dream of a car for the masses was realized with the famous Model T. The Model T was introduced in 1908, and was eventually produced in numbers vastly greater than any car previously. For the first few years it was produced at the Ford Piquette Avenue plant, Detroit, Michigan. In the terms of this book, we would say that Henry Ford was the sponsor of the mass production system,[1] whose architecture would become a decades-long invariant. The architecting was done by a very small group, with leading credit probably best given to Charles Sorenson,[2] although several others played key roles. Sorenson had primary responsibility for the production system, with several others individually having leadership in other basic elements of the Ford production system architecture. According to Sorenson, the first experiments in the production line took place at the Piquette Avenue plant in mid-1908 on the Model N, an immediate predecessor to the Model T.

The Model N had been introduced as an incomplete prototype at the 1906 Detroit auto show. It was not disclosed that the show car was incomplete, and so the announced price of $500 was a sensation and generated terrific demand. The Model N demonstrated the latent demand for a solid, low-cost car. The Model T, with its superior engineering for production, was able to exploit that demand.

As Sorenson recounts,[3] he and a small team spent Sundays during the summer of 1908 experimenting on the production floor of the Piquette Avenue plant. They laid out the parts required for a car from one end of the long narrow building floor to the other. They mounted a frame on skids, and then dragged the skid down the floor, stopping along the way to add the parts that had been preplaced.

As an amusing aside, and as a wonderful indication of how obvious things go unnoticed when great innovations are made, Sorenson points out why the assembly line model was not actually used in production until 5 years later, in 1913. The main problem was that at the Piquette Avenue plant, the assembly floor was the third floor of the building, the top floor. In retrospect, this is laughable. Why put the place where you need to move all the production parts to and from three floors up off the ground? But in the early 1900s, this did not seem so obvious. When you make only a few cars, why put that messy operation on the ground floor, which has the nicer space for the staff (including sales)?

Once the Model T was introduced, and demand immediately exploded beyond the capacity of the Piquette Avenue plant, Ford built an all new plant at Highland Park, Michigan, where the assembly line was brought to fruition in 1913. As we shall see in a later section of this case study, there is more to the structure of the Ford system than the assembly line, and those other structural elements play at least as important a role.

Competition from New Quarters (1920s to 1930s)

The Model T and its production system were based on a simple, virtuous cycle. Lowering costs allowed prices to be lowered, which increased sales and production, which enabled greater economies of scale, which lowered costs. Ford's pursuit of the Model T was driven by an innate belief in the value of a car for the masses. The vision was eventually overturned by an alternative vision spawned by market forces.

By the mid 1920s, Chevrolet was rapidly catching up to Ford in production numbers. They were catching up primarily by making better-looking, more-exciting cars, and marketing looks and excitement. Although the Model T was a very solid car, a new era had begun, based on market penetration of automobiles being large enough so that people began to see them as partially fashion-driven goods. When market penetration for automobiles became high, the purely utilitarian aspect of automobile ownership began to be replaced by automobiles as status symbols. When status played an important role, it quickly became the case that status was no longer conveyed simply by having a car, but by the car one had.

Model T production was shut down in 1927. Over the next decade, competition between Ford and its competitors (most famously General Motors, also Plymouth and Chrysler) moved to the model-year change system. Different models were produced for different market segments, and those models were regularly changed in external style and engineering features. The changes were synchronized with marketing campaigns to drive demand. Economies of scale in mass production were still of great importance, but the scale was not unlimited. The Model T had tested the outer envelope of focusing purely on cost reduction through scale, and was displaced by a more complex mixture of engineering, production, and marketing.

The Toyota Production System (1940s to 1980s)

The development of the Toyota Production System (TPS)[4] can be said to have revolutionized manufacturing as did Ford's mass production system. Although the revolution was slower and less dramatic, it was in some ways more surprising as it occurred in an industry already apparently mature. By the 1950s, the automobile business appeared mature. Cars were much improved, but their architecture had changed little in decades, and the architecture of production likewise changed little. The revolution of the TPS has no dramatic moments like the assembly experiments at Piquette Avenue. The TPS revolution was a revolution by evolution, a case where incrementally changed, accreted steadily enough and long enough, it takes on a qualitatively different flavor.

The TPS did not outwardly change the architecture of either cars or production. Both cars and factories built in accordance with TPS appear much the same as did cars and factories before. But, the improvements in quality and cost brought Toyota from a nonentity in the business to a neck-and-neck contender for largest auto manufacturer. This displacement of multiple, dominant, profitable firms is very unusual.

The architecture of the TPS is The Toyota Way (see below). Thus, the TPS is a sociotechnical system, and its architecture is likewise more social than technical. The most important elements are the shared principles and the means of their application.

Metaphor or Vision Changes

At each of the stages, the story is captured by a metaphor or basic vision. It is hard to know exactly how important the conceptual vision is, but the testimony of the people directly involved indicates that the coherent vision, the thing they could aim at, was an inspiration and guide, and they gave it great weight. Sorenson reports repeatedly that Henry Ford was devoted to his vision of cars for the masses, and his reluctance to recognize that it had run out of force caused great difficulty when it finally became obvious to everybody except Henry Ford that the time of the Model T was past.

Craftsmen

Early automobiles were craftsmanly products, like bespoke suits. They were made by individual craftsmen and possessing one was a mark of status. Being made by individual craftsmen, they carried the marks of those craftsmen (sometimes good, sometimes bad). Like nearly all craftsmanly products, these cars were very expensive.

The craftsmanly approach to cars is still not quite dead. A few cars, naturally very expensive and basically toys for adults, are built by individual teams of craftsman. The individual attention is a selling point, even if it objectively probably yields poorer quality than the best cars made in lean factories.

A Car for the Masses, or If We Build It, It Will Sell

Henry Ford's most famous quote is probably "The customer can have any color he wants, as long as it is black." Black was apparently chosen mostly because the high-quality black paint of the time was the fastest drying and thus allowed the production line to operate more efficiently. The paradigm for Ford operations from the introduction of the Model T to the mid-1920s was that the only real problem was making more Model Ts, cheaper. If they could be made, they could be sold or so the belief ran. This was the virtuous cycle of economies of scale and cost reductions. For roughly 15 years, this was an effective strategy and reflected the (temporary) correctness of Henry Ford's basic vision.

Cars as Fashion

By the mid 1920s, cars were no longer a rarity in the United States. There were enough reliable cars around that a used car market had begun. As Chevrolet and others introduced frequent style and model changes, they brought a fashion sensibility to automobiles. Henry Ford's simple vision of cheap transportation for the masses gave way to affordable status and transportation for the masses, and eventually a whole hierarchy of desire and status much like other mature product areas.

The Supermarket Metaphor

In Taiichi Ohno's book on the Toyota Production System, he makes a striking observation about his inspiration for the TPS.[5] He says that when he toured the United States in 1956 to see the Ford and General Motors factories, he was more impressed by supermarkets. He adopted a supermarket metaphor for the organization of production. The idea was that the consumer (who in a production system is also a supplier to a later phase) can reach into the supermarket and get exactly what he or she needs, and the act of the consumer taking it "pulls" a replacement onto the shelf. In contrast to Henry Ford's paradigm of pushing automobiles out, knowing they would be sold, the TPS model is to produce and deliver just what is sold, and refill just what is taken. Ohno writes that the supermarket metaphor had been in use since the late 1940s, but his trip to the United States solidified his commitment to the metaphor.

The Toyota Way

Beyond the supermarket metaphor, Toyota promulgates a larger philosophy known as "The Toyota Way." The Toyota Way[6] could be thought of as a metaphor or vision in the large, composed as it is of 14 principles that themselves are reasonably complex. The Toyota Way defines an overall approach to doing a production-oriented business in general, and is not restricted to automobiles. It does not have a distinct end point (as Ford's vision did); rather one of the principles is to embrace a sense of urgency for continuous improvement, regardless of current business conditions. The Toyota Way is, by design, a more embracing philosophy than single vision.

Elements of the Architecture of the Ford Production System

The architecture of Ford mass production was not just the assembly line, or the River Rouge factory (Dearborn, Michigan), or the Model T. The architecture of the enterprise as a whole, the architecture that brought mass production its power, had three major components: the production line, distributed production with synergistic system design, and management processes.

The Assembly Line

By far the most famous element of the mass production enterprise is the assembly line. As noted above, the experiments in fixing assembly stations and moving the vehicle down the factory floor began with the Model N in the Piquette Avenue plant. The physical constraints of the plant prevented full implementation until the Highland Park plant was built to produce the Model T.

The assembly line also led to a variety of other possibilities for efficiencies. Once the basic notion of configuring the flow to optimize material handling was present, the full power of engineering and statistics could be brought to bear to further improve the process. Moreover, assembly production should be (and eventually was) synergistic with design. Automobiles eventually were designed to be easy to assemble within the Ford enterprise, and the enterprise adjusted itself to what it was possible to design.

Enterprise Distribution

The assembly line was just one of the major innovations that enabled mass production. As production volumes grew larger and larger the problem of factory scaling began to appear. There are upper limits to the practical size of a factory. Eventually, the major constraint is transportation. A factory in the Detroit area (or anywhere else) simply cannot bring arbitrarily large quantities of raw materials and parts and cannot move out arbitrarily large quantities of product. Eventually transportation capacity runs out.

So, when it is necessary to build more factories in geographically distributed locations, how do we divide up the production tasks? The solution eventually arrived at in automobiles is to divide production along vehicle subsystem lines. So, engines are made in one location, chassis in another, bodies in still another, and all are brought together in assembly plants. The assembly plants can be located relatively close to major markets, and the others can be distributed based on what areas are favorable to the particular manufacturing task.

This division on vehicle subsystem lines is, or can be, synergistic with vehicle design. Design should be synergistic both with the detailed problem of assembly and the larger problem of how the production enterprise is distributed. For example, tight tolerance processes should be inside subsystems, and the interfaces between them should be less demanding. The subsystems should be designed in ways that facilitate testing and quality control at the point of production. Over time, the production processes, vehicle designs, and supplier networks coevolved.

Management Processes

The assembly line, distribution of plants, and vehicle subsystems are all obviously physical structures. But the history also identifies certain management processes and the synergistic changes they drove as fundamental structural elements (that is, architectural elements) in the development of mass production.

Quality Assurance for Distributed Production

Consider how quality assurance and quality control changes when production becomes distributed. If all production steps are under the same roof, when a problem appears, an engineer can simply walk from one part of the factory to another to understand the source of the problem. When the engine, frame, and transmission factories are in different parts of the United States, and the year is 1920, moving among the factories to straighten out problems was a serious burden.

Part of the success of mass production was the development of new quality assurance and control techniques to manage these problems. Similarly, new supplier management techniques were introduced. Many of the techniques like just-in-time production and negotiated learning curves that are considered very modern techniques were known to Ford and his architects. In Ford's time, the sophistication level was much lower, and the technology did not allow optimization in the ways that it is possible today, but the concepts were already known.

Moving to the TPS era, as quality control improved, it eventually became possible to make architectural-level changes to the assembly process. For example, when very high-quality levels are attained, testing and inspection processes can be greatly reduced and simplified. If the defect rate is low enough, it is no longer economic to conduct multistep inspection and testing processes. With an extremely low defect rate, testing and inspection can be pushed to the final, full system level.

Devotion to Component-Level Simplification

Ford and his architects were devoted to component-level simplification. They continually looked for ways to simplify the production of individual components and to simplify major subsystems. A major method was to cast larger and more complicated iron assemblies. This eventually resulted in the single piece casting of the V8 engine block used in the most successful

Ford immediately prior to World War II. That basic engine block casting design and technique was used for decades afterward.

The movement to larger and more complex castings is a fine example of the *Simplify* heuristic at work. A dictate to "simplify" sounds good, but how does one actually apply it? The application must be in the architect's current context. In the case of Ford and Sorenson, castings that were very complex to develop were ultimately "simple" because of the simplification they brought to the assembly process. Making the castings was only complex up to the point it was fully understood. Once it was understood, it could be carried out very consistently and allowed for great simplifications in downstream assembly.

Social Contract

On the labor relations front, Henry Ford is both famous and infamous. He is famous for introducing much higher wages, specifically targeting his wages to allow all of his workers to be able to realistically afford one of the cars they were building. This was consistent with Ford's overall vision of cars for the masses. After all, what masses could he be building cars for if not the masses that he himself employed? Henry Ford is also infamous for some of his other labor practices, such as his intrusions into the private lives of his workers. The architects of the TPS were well aware of both sides of Ford's labor relations and believed that the architecture of the production system must be reconciled with a stable social contract with the workers.

All systems of productivity improvement must reconcile the improvements that are in the interests of owners with the interests of the workers. If each improvement simply leads to higher worker production quotas and job losses, it is hardly likely that workers will be enthusiastic participants in the improvement process. In the rapid growth days of Ford, when wages were doubled over those otherwise prevailing, Ford workers had obvious reasons for believing their own interests were aligned with Ford's. Toyota faced the same difficulty, but under worse circumstances in the early years as growth was not so spectacular. However, it is probably notable that the Toyota Production System was extensively developed during the 1960s when Japan had an extremely high economic growth rate.

Conclusion

Ford and Toyota are the two classic examples of mass production. Both have recognizable architectural histories and easily identified architects. Both created changes that have rippled into fields well beyond their own. Ford was able to pioneer mass production of systems as complex as the automobile. The architecture of the Ford production system was sociotechnical, but with a heavy emphasis on the technical. We can see directly the technical innovations that made it work and that defined its essential structures (the assembly line, distributed production, new management techniques).

The TPS architectural success was smaller in that it did not create a new industry, but TPS succeeded against a backdrop of established and strong competitors. The development of the TPS is also an example of where incremental change, sufficiently accumulated, can eventually become revolutionary. The architecture of the TPS is much more socio than technical. In its embodiment in the Toyota Way, it is described essentially as philosophy, albeit an operative philosophy, one directly usable in practical decision making.

Notes and References

1. This is also the position taken by Charles Sorenson. Sorenson, C., *My Forty Years with Ford*. Detroit, MI: Wayne State University Press, 2006 (first published in 1956).
2. The story of Ford here draws very heavily on the perspectives of Sorenson, given his association with the entire story and his role as, effectively, the architect of the production system.
3. Sorenson, C., *My Forty Years with Ford*. Detroit, MI: Wayne State University Press, 2006 (first published in 1956), p. 118.
4. Ohno, T., *Toyota Production System: Beyond Large Scale Production*. Florence, KY: Productivity Press, 1968.
5. Ohno, T., *Toyota Production System: Beyond Large Scale Production*. Florence, KY: Productivity Press, 1968, p. 26.
6. See, for example, Liker, J., *The Toyota Way: 14 Management Principles from the World's Greatest Manufacturer*. New York: McGraw-Hill Professional, 2004.

chapter 4

Manufacturing Systems

Introduction: The Manufacturing Domain

Although manufacturing is often treated as if it were but one step in the development of a product, it is also a major system in itself. It has its own architecture.[1] It has a system function that its elements cannot perform by themselves — making other things with machines. And it has an acquisition waterfall for its construction quite comparable to those of its products. Moreover, the architecture of the manufacturing system and the architecture of the system of interest must relate to each other. More broadly, both exist within the structure of the development program, which should be chosen consciously and deliberately to yield the desired properties for the client.

From an architectural point of view, manufacturing has long been a quiet field. Such changes as were required were largely a matter of continual, measurable, incremental improvement — a step at a time on a stable architectural base. Though companies came and went, it took decades to see a major change in its members. The percentage of sales devoted to research and advanced development for manufacturing, per se, was small. The need was to make the classical manufacturing architecture more effective — that is, to evolve and engineer it.

Beginning two decades or so ago, the world that manufacturing had supported for almost a century changed — and at a global scale. Driven by political change in China and other countries, and by new technologies in global communications, transportation, sources, markets, and finance, global manufacturing became practical and then, shortly thereafter, dominant. It quickly became clear that qualitative changes were required in manufacturing architectures if global competition were to be met. In the order of conception, the architectural innovations were ultraquality,[2] dynamic manufacturing,[3] lean production,[4] and "flexible manufacturing."* The results to date, demonstrated first by the Japanese and now spreading globally, have been greatly increased profits and market share, and sharply decreased inventory and time-to-market. Each of these innovations will be presented in turn.

Even so, rapid change is still underway. As seen on the manufacturing floor, manufacturing research as such has yet to have a widespread effect.

---

* Producing different products on demand on the same manufacturing line.

Real-time software is still a work-in-progress. Trend instrumentation, self-diagnosis, and self-correction, particularly for ultraquality systems, are far from commonplace. So far, the most sensitive tool for ultraquality is the failure of the product being manufactured.

Manufacturing in Context

Before discussing architectural innovations in manufacturing, we need to place manufacturing in context. At some point, a system needs to be built or it is of little interest. The building is "manufacturing." But, there are several distinct scenarios we should consider.

Full Development Followed by Serial Production

This applies to and is common in situations where we build tens to millions of copies of a system after producing one or more complete prototypes. The prototypes, which may themselves be the end of a series of intermediate prototypes, are essentially identical to the system to be manufactured. The testing conducted on the prototypes is commonly referred to as "qualification" testing and is to show that the system to be built is fully suitable (in function, environmental suitability, and all other respects) for end use. It shows that the system to be manufactured meets the purposes of the client in operational use. Because the prototypes are not themselves to be delivered to customer use, they can be tested very strongly, indeed destructively if desired and warranted.

There are several strategies by which we work up a series of prototypes to result in the representative manufactured system. The most common is usually referred to as breadboard-to-brassboard. In this strategy, each prototype contains the full functionality intended for the final system but is not packaged in an operationally representative way. The first development cycle, the breadboard, may exist just as open units in a lab interconnected and discrete subsystems tested individually. A subsequent phase may be packaged into a surrogate platform not yet light or strong enough for final use. The development sequence culminates in the manufacturing of representative prototypes.

Incremental Development and Release

A contrasting strategy is to develop a series of prototypes where each is fully operationally suitable but contains less than the desired level of functionality. This is common in software-intensive systems. In software systems, the cost of manufacturing and delivery is quite low, nearly zero when software is electronically delivered. Thus, the cost impediment of frequent re-release does not exist as in systems where most of the value is in hardware.

An incremental development and release strategy facilitates an evolutionary approach to client desires. Instead of needing to get everything right at the beginning, the developer can experiment with suppositions as to what the client really wants. The client's learning process using the system can be fed back into subsequent releases. A major issue in a frequent release strategy is that test and certification costs are re-incurred each time a release cycle is completed. If the release cycles are frequent (best for learning feedback), the cost of test and certification will rise quickly. The process can be cost efficient only when the costs of test and certification can be driven down, usually by automation. In some sense, the process of testing and certification for release takes the place of serial production in the example of the serial production strategy above.

"Protoflight" Development and Manufacturing

In this strategy, which is common in one-of-a-kind items like spacecraft, the developmental unit is also the delivered manufactured unit. That is, rather than delivering a completed prototype to be manufactured, we deliver the completed prototype to be used (launched, in the case of a satellite). The primary advantage for the protoflight approach is cost. Obviously, when only a singular item needs to be delivered, the cost of manufacturing it is minimized by making only one.

The protoflight test quandary is a mirror image of the test quandary in the serial production case. In serial production we can freely test the prototypes as thoroughly as we like, including destructively. But, we must be concerned about whether or not the prototype units fully represent the manufactured units. Usually, if the production run is large enough, we will take units off of the serial production line and test them as thoroughly as the prototypes were tested. In the protoflight case, we know that the prototype and the delivered system are identical (because they are the same unit), but we risk damaging the system during test. Tests can change the state of the system, perhaps invisibly, and test processes are always vulnerable to accidents. We cannot test in certain ways because we cannot afford test-induced damage to the flight system. We must also continuously trade the risk of not revealing a defect because of lack of testing with the risk of creating defects through testing. The satellite business in particular is full of stories of protoflight systems that were damaged through accidents in testing (for example, over-limit vibration testing, a weather satellite tipped off of its test stand).

In each of these cases, there is a relationship between the system architecture, the architecture of the program that builds the system, the test strategy, and the architecture of any systems used for testing. When we choose an overall program architecture, we induce constraints on how we can test the resulting system. The architecture of the system-of-interest will determine the sorts of test approaches that can be supported. That likewise affects what sorts of systems we can build for conducting tests. Each of these issues cannot be considered and resolved in isolation. In mature situations there may be widely accepted solutions and established architecture breakdowns. In immature situations there may be great leverage in innovative breakdowns.

> *Example: DARPA Grand Challenge* — The U.S. Defense Advanced Research Projects Agency sponsored a Grand Challenge[5] race between autonomous ground vehicles. The competing teams all used the protoflight approach; they built, tested, and raced a single vehicle. Because the single vehicle had to be used for testing as well as racing, there were fundamental architectural choices that arbitrated between these needs. As examples, if more time was devoted to building a mechanically more complex vehicle, the amount of time available to use the vehicle in software testing would be reduced. Was superior mechanical performance worth less software testing time? Any test instrumentation needed to be built into the vehicle so it could be used in field experiments. But, that same test instrumentation would need to be carried in the race, or removed at the last minute (generating its own risks). Where should the trade-off in enhanced testing versus less system burden lie? A vehicle optimized for autonomous operation would not be drivable by a human, but a vehicle that can be alternately human or computer driven leads to much simpler field test operations. Is the loss of performance with retaining human drivability worth the lessened burden in field test operations? As a matter of historical record, different teams participating in the Grand Challenge events took distinctly different approaches along this spectrum, but the most successful teams took relatively similar approaches (simple, production-vehicle-based mechanical system available very early in the development cycle; extensive test instrumentation; and human drivability retained).

Architectural Innovations in Manufacturing

Ultraquality Systems

At the risk of oversimplification, a common perception of quality is that quality costs money — that is, quality is a trade-off against cost and profit. Not coincidentally, there is an accounting category called "cost of quality." A telling illustration of this perception is the "DeSoto Story." As the story goes, a young engineer at a DeSoto automobile manufacturing plant went to his boss with a bright idea on how to make the DeSoto a more reliable automobile. The boss's reply: "Forget it, kid. If it were more reliable it would last more years and we would sell fewer of them. It's called planned obsolescence." DeSoto no longer is in business, but the perception remains in the minds of many manufacturers of many products.

The difficulty with this perception is partly traceable to the two different aspects of quality. The first is quality associated with features like leather seats and air conditioning. Yes, those features cost money, but the buyer perceives them as value added and the seller almost always *makes* money on them. The other aspect of quality is absence of defects. As it has now been shown, absence of defects *also* makes money, and for both seller and buyer, through reductions in inventory, warranty costs, repairs, documentation, testing, and time to market — *provided that* the level of product quality is high enough* and the whole development and production process is architected at that high level.

To understand why absence of defects makes money, imagine a faultless process that produces a product with defects so rare that it is impractical to measure them; that is, none are anticipated within the lifetime of the product. Testing can be reduced to the minimum required to certify system-level performance of the first unit. Delays and their costs can be limited to those encountered during development; if and when later defects occur, they can be promptly diagnosed and permanently eliminated. "Spares" inventory, detailed parts histories, and statistical quality control can be almost nonexistent. First-pass manufacturing yield can be almost 100% instead of today's highly disruptive 20% to 70%. Service in the field is little more than replacing failed units, free.

The only practical measurement of ultraquality would then be an end system-level test of the product. Attempting to measure defects at any subsystem level would be a waste of time and money — defects would have to be too rare to determine with high confidence. Redundancy and fault-tolerant designs would be unnecessary. Indeed, they would be impractical because, without an expectation of a specific failure (which then should be fixed), protection against rare and unspecified defects is

* Roughly less than 1%/year rate of failure at the system level regardless of system size. The failure rate for subsystems or elements clearly must be much less.

not cost effective. Thus, the level of quality achieved could affect the most appropriate architecture of the manufacturing system.

To some readers, this ultraquality level may appear to be hopelessly unrealistic. Suffice it to say that it has been approached for all practical purposes. For decades, no properly built unmanned satellite or spacecraft failed because of a defect known before launch (any defect would have been fixed beforehand). Microprocessors with millions of active elements, sold in the millions, now outlast the computers for which they were built. Like the satellites, they become technically and financially obsolete long before they fail. Television sets are produced with a production line yield of over 99%, far better than the 50% yield of a decade ago, with a major improvement in cost, productivity, and profit.

As a further example, the readers should note that consumer electronic products today are commonly unrepairable, and if defective within a warranty period are simply replaced. This approach carries multiple benefits. If a system does not need to be repaired, the supplier need not maintain a repair and supply network, and can sweep away all the costs associated with one. If repairs are not necessary, the unit can be designed without repair access or diagnostics, which commonly saves space and money, and allows the use of manufacturing techniques (such as sealing) that themselves improve reliability.

Today's challenge, then, is to achieve and maintain such ultraquality levels even as systems become more complex. Techniques have been developed that certainly help.* More recently, two more techniques have been demonstrated that are particularly applicable to manufacturing.

The first is: *Everyone in the production line is both a customer and a supplier, a* customer for the preceding worker and a supplier for the next. Its effect is to place quality assurance where it is most needed, at the source.

The second is: *The Five Why's,* a diagnostic procedure for finding the basic cause of a defect or discrepancy. Why did this occur? Then why did that, in turn, occur? Then, why *that?* and so on until the offending causes are discovered and eliminated.

To these techniques can be added a relatively new understanding: *Some of the worst failures are system failures* — that is, they come from the interaction of subsystem deficiencies which of themselves do not produce an end system failure, but together can and do. Four catastrophic civil

* Rechtin 1991, Chapter 8, pp. 160–187. One technique mentioned there — fault tolerance through redundancy — has proved to be less desirable than proponents had hoped. Because fault-tolerant designs "hide" single faults by working around them, they accumulate until an overall system failure occurs. Diagnostics then become very much more difficult. Symptoms are intertwined. Certification of complete repair cannot be guaranteed because successful-but-partial operation again hides undetected (tolerated) faults. The problem is most evident in servicing modern, microprocessor-rich, automobile controls. The heuristic still holds, *Fault avoidance is preferable to fault tolerance in system design.*

space system failures were of this kind: Apollo 1, Apollo 13, Challenger, and the Hubble Telescope. For Tom Clancy buffs, just such a failure almost caused World War III in his *Debt of Honor.* In all these cases, had any one of the deficiencies not occurred, the near-catastrophic end result could not have occurred. That is, though each deficiency was necessary, none were sufficient for end failure. As an admonition to future failure review boards, until a diagnosis is made that indicates that the set of presumed causes is both necessary and sufficient — and that no other such set exists — the discovery-and-fix process is incomplete and ultraquality is not assured.

Successful ultraquality has indeed been achieved, but there is a price that must be paid. Should ultraquality *not* be produced at any point in the whole production process, the process may collapse. Therefore, when something does go wrong, it must be fixed immediately; there are no cushions of inventory, built-in holds, full-time expertise, or planned work-arounds. Because strikes and boycotts can have instantaneous effects, employee, customer, and management understanding and satisfaction are essential. Pride in work and dedication to a common cause can be of special advantage, as has been seen in the accomplishments of the zero defect programs of World War II, the American Apollo lunar landing program, and the Japanese drive to world-class products.

In a sense, ultraquality-built systems are fine-tuned to near-perfection with all the risks thereof. Just how much of a cushion or insurance policy is needed for a particular system is an important value judgment that the architect must obtain from the client, the earlier the better. That judgment has strong consequences in the architecture of the manufacturing system. Clearly, then, ultraquality architectures are very different from the statistical quality assurance architectures of only a few years ago.*

Most important for what follows, it is unlikely that either lean production or flexible manufacturing can be made competitive at much less than ultraquality levels.

Dynamic Manufacturing Systems

The second architectural change in manufacturing systems is from comparatively static configurations to dynamic, virtually real-time, ones. Two basic architectural concepts now become much more important. The first concerns intersecting waterfalls, and the second, feedback systems.

---

* One of the authors, a veteran of the space business, visited two different manufacturing plants and correctly predicted the plant yields (acceptance versus start rates) by simply looking at the floors and at the titles (not content) and locations of a few performance charts on the walls. In the ultraquality case, the floors were painted white; the charts featured days-since-last-defect instead of running average defect rates; and the charts were placed at the exits of each work area. Details, but indicative.

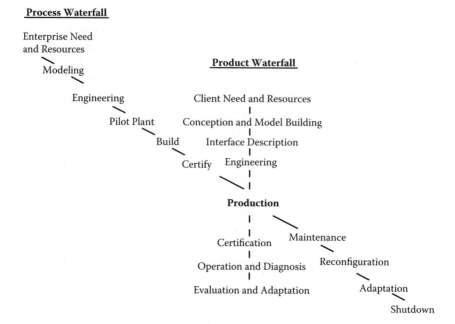

Figure 4.1 The intersecting process and product waterfalls.

Intersecting Waterfalls

The development of a manufacturing system can be represented as a separate waterfall, distinct from, and intersecting with, that of the product it makes. Figure 4.1 depicts the two waterfalls, the process (manufacturing) diagonally and the product vertically, intersecting at the time and point of production. The manufacturing one is typically longer in time, often decades, and contains more steps than the relatively shorter product sequence (months to years) and may end quite differently (the plant is shutdown and demolished). Sketching the product development and the manufacturing process as two intersecting waterfalls helps emphasize the fact that manufacturing has its own steps, time scales, needs, and priorities distinct from those of the product waterfall. It also implies the problems its systems architect will face in maintaining system integrity, in committing well ahead to manufacture products not yet designed, and in adjusting to comparatively abrupt changes in product mix and type. A notably serious problem is managing the introduction of new technologies, safely and profitably, into an inherently high-inertia operation.

There are other differences. Manufacturing certification must begin well before product certification or the former will preclude the latter; in any case, the two must interact. The product equivalent of plant demolition, not shown in the figure, is recycling, both now matters of national law in Europe. Like certification, demolition is important to plan early, given its collateral human costs in the manufacturing sector. The effects of environmental regulations, labor contracts, redistribution of usable resources, retraining, right-sizing of management, and continuing support of the customers are only a few of the manufacturing issues to be resolved — and well before the profits are exhausted.

Theoretically if not explicitly, these intersecting waterfalls have existed since the beginning of mass production. But not until recently have they been perceived as having equal status, particularly in the United States. Belatedly, that perception is changing, driven in large part by the establishment of global manufacturing — clearly not the same system as a wholly owned shop in one's own backyard. The change is magnified by the widespread use of sophisticated software in manufacturing, a boon in managing inventory but a costly burden in reactive process control.[6] Predictably, software for manufacturing process and control, more than any element of manufacturing, will determine the practicality of flexible manufacturing. As a point in proof, Hayes, Wheelright, and Clark[7] point out that a change in the architecture of [even] a mass production plant, particularly in the software for process control, can make dramatic changes in the capabilities of the plant without changing either the machinery or layout.

The development of manufacturing system software adds yet another production track. The natural development process for software generally follows a spiral,* certainly not a conventional waterfall, cycling over and over through functions, form, code (building), and test. The software spiral shown in Figure 4.2 is typical. It is partially shaded to indicate that one cycle has been completed with at least one more to go before final test and use. One reason for such a departure from the conventional waterfall is that software, as such, requires almost no manufacturing, making the waterfall model of little use as a descriptor.† The new challenge is to synchronize the stepped waterfalls and the repeating spiral processes of software-intensive systems. One of the most efficient techniques is through the use of stable intermediate forms,[8] combining software and hardware into partial but working precursors to the final system. Their important feature is that they are *stable* configurations; that is, they are reproducible, well-documented, progressively refined baselines — in other words, they are identifiable architectural waypoints and must be treated as such. They

* See also Chapter 2.
† Efforts to represent the manufacturing and product processes as spirals have been comparably unsuccessful. Given the need to order long-lead-time items, to "cut metal" at some point, and to write off the cost of multiple rapid prototypes, the waterfall is the depiction of choice.

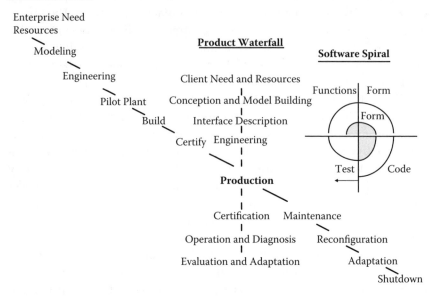

Figure 4.2 Product, process, and software system tracks.

can also act as operationally useful configurations,* built-in "holds" allowing lagging elements to catch up, or parts of strategies for risk reduction as suggested in Chapter 3.

The Spiral-to-Circle Model

Visualizing the synchronization technique for the intersecting waterfalls and spirals of Figure 4.2 can be made simpler by modifying the spiral so that it remains from time to time in stable concentric circles on the four-quadrant process diagram. Figure 4.3 shows a typical development from a starting point in the function quadrant cycling through all quadrants three times — typical of the conceptualization phase — to the first intermediate form. There the development may stay for a while, changing only slightly, until new functions call for the second form, say an operational prototype. In Air Force procurement, that might be a "fly-before-buy" form. In space systems, it might be a budget-constrained "operational prototype" that is actually flown. In one program, it turned

---

* Many Commanders of the Air Force Space and Missiles Division have insisted that all prototypes and interim configurations have at least some operational utility, if only to help increase the acceptance in the field once the final configuration is delivered. In practice, field tests of interim configurations almost always clarify if not reorder prior value judgments of what is most important to the end user in what the system can offer.

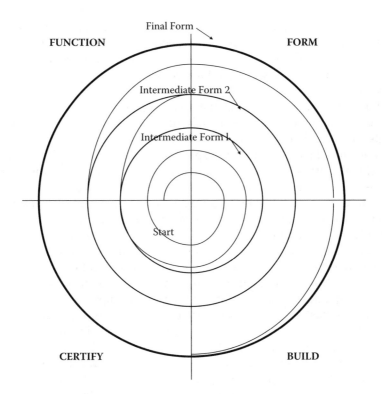

FUNCTION   Final Form   **FORM**

Intermediate Form 2

Intermediate Form 1

Start

CERTIFY   **BUILD**

Figure 4.3 The spiral-to-circle model.

out to be the *only* system flown. But, to continue, the final form is gained, in this illustration, by way of a complete, four-quadrant cycle.

The spiral-to-circle model can show other histories, for example, a failure to spiral to the next form, with a retreat to the preceding one, possibly with less ambitious goals, or a transition to a still greater circle in a continuing evolution, or an abandonment of these particular forms with a restart near the origin.

Synchronization can also be helped by recognizing that cycling also goes on in the multistep waterfall model, except that it is depicted as feedback from one phase to one or more preceding ones. It would be quite equivalent to software quadrant spiraling if all waterfall feedback returned to the beginning of the waterfall — that is, to the system's initial purposes and functions, and from there down the waterfall again. If truly major changes are called for, the impact can be costly, of course, in the short run. The impact in the long run might be cost effective, but few hardware managers are willing to invest.

The circle-to-spiral model for software-intensive systems in effect contains both the expanding-function concept of software and the stepwise character of the waterfall. It also helps understand what and when hardware and software functions are needed in order to satisfy requirements by the *other* part of the system. For example, a stable intermediate software form of software should arrive when a stable, working form of hardware arrives that needs that software, and vice versa.

It is important to recognize that this model, with its cross-project synchronizations requirement, is notably different from models of procurements in which hardware and software developments can be relatively independent of each other. In the spiral-to-circle model, the intermediate forms, both software and hardware, must be relatively unchanging and bug-free. A software routine that never quite settles down or that depends upon the user to find its flaws is a threat, not a help, in software-intensive systems procurement. A hardware element that is intended to be replaced with a "better and faster" one later is hardly better. Too many subsequent decisions may unknowingly rest on what may turn out to be anomalous or misunderstood behavior of such elements in system test.

To close this section, although this model may be relatively new, the process that it describes is not. Stable intermediate forms, blocks (I, II, and so forth), or "test articles" as they are called, are built into many system contracts and perform as intended. Yet there remains a serious gap between most hardware and software developers in their understanding of each other and their joint venture. As the expression goes, "These guys just don't talk to each other." The modified spiral model should help both partners bridge that gap, to accept the reasons both for cycling and for steps, and to recognize that neither acquisition process can succeed without the success of the other.

There should be no illusion that the new challenge will be easy to meet. Intermediate software forms will have to enable hardware phases at specified milestones — not just satisfy separate software engineering needs. The forms must be stable, capable of holding at that point indefinitely, and practical as a stopping point in the acquisition process if necessary. Intermediate hardware architectures must have sufficient flexibility to accommodate changes in the software — as well as in the hardware. And finally, the architects and managers will have a continuing challenge in resynchronizing the several processes so that they neither fall behind nor get too far ahead. Well-architected intermediate stable forms and milestones will be essential.

Concurrent Engineering

To return to Figure 4.2, this intersecting waterfall model also helps identify the source of some of the inherent problems in concurrent (simultaneous, parallel) engineering — in which product designers and manufacturing engineers work together to create a well-built product. Concurrent engineering in practice, however, has proven to be more than modifying designs for manufacturability. However defined, it is confronted with a fundamental problem, evident from Figure 4.2 — namely, coordinating the two intersecting waterfalls and the spirals, each with different time scales, priorities, hardware, software, and profit-and-loss criteria. Because each track is relatively independent of the others, the incentives for each are to optimize locally even if doing so results in an impact on another track or on the end product. After all, it is a human and organizational objective to solve one's own problems, to have authority reasonably commensurate with responsibilities, and to be successful in one's own right. Unfortunately, this objective forces even minor technical disagreements to higher, enterprise management where other considerations than just system performance come into play.

> *A typical example*: A communications spacecraft design was proceeding concurrently in engineering and manufacturing until the question came up of the spacecraft antenna size. The communications engineering department believed that a 14-foot diameter was needed; the manufacturing department insisted that 10 feet was the practical limit. The difference in system performance was a factor of two in communications capability and revenue. The reason for the limit, it turned out, was that the manufacturing department had a first-rate subcontractor with all the equipment needed to build an excellent antenna, but no larger than 10 feet. To go larger would cause a measurable manufacturing cost overrun. The manufacturing manager was adamant about staying within his budget, having taken severe criticism for an overrun in the previous project. In any case, the added communications revenue gain was far larger than the cost of re-equipping the subcontractor. Lacking a systems architect, the departments had little choice but to escalate the argument to a higher level where the larger antenna was eventually chosen and the manufacturing budget increased slightly. The design proceeded normally until software engineers wanted to add more memory well after manufacturing had invested heavily in the original computer hardware design. The argument escalated, valuable time was lost, department prerogatives were again at stake, and so it went.

The example is not uncommon. A useful management improvement would have been to set up a trusted, architect-led team to keep balancing the system as a whole within broad top management guidelines of cost, performance, risk, and schedule.

If so established, the architectural team's membership should include a corporate-level (or "enterprise") architect, the product architect, the manufacturing architect, and a few specialists in system-critical elements, and no more.[9] Such a structure does exist implicitly in some major companies, though seldom with the formal charter, role, and responsibilities of systems architecting.

Feedback Systems

Manufacturers have long used feedback to better respond to change. Feedback from the customer has been, and is, used directly to maintain manufacturing quality and indirectly to accommodate changes in design. Comparably important are paths from sales to manufacturing and from manufacturing to engineering.

The presence or absence of feedback paths, and their time constants, is something that can be deliberately controlled through the architecture of the program and organization that envelop a system of interest. Consider space exploration systems. An exploration organization can choose to pursue large, multimission systems that take a long time, or many more smaller, more rapidly turned over programs. Because the payload fraction of a spacecraft is generally higher as the spacecraft gets bigger, larger, multimission spacecraft are generally more cost efficient. But, because they take much longer, the time constant on which the things learned on one mission can be fed back into the next is longer. The organization has fewer opportunities to incorporate their learning into future mission design. In a very mature mission area where needs change slowly, this might be a fair trade-off. In an immature mission area where each new payload reveals new questions and new preferences, a faster feedback loop yields dramatically different characteristics. Moreover, the pace of feedback affects the people in the organization. They, likewise, learn (and are held accountable) primarily when each full mission feedback loop closes. An organization with a fast feedback loop (but not too fast) is a rapidly learning organization.

What is new in manufacturing is that the pace has changed. Multiyear is now yearly, yearly is now monthly, monthly is now daily, and daily — especially for ultraquality systems — has become hourly if not sooner. What was a temporary slowdown is now a serious delay. What used to affect only adjacent sectors can now affect the whole. What used to be the province of planners is now a matter of real-time operations.

Consequently, accustomed delays in making design changes, correcting supply problems, responding to customer complaints, introducing new products, reacting to competitors' actions, and the like were no longer acceptable. The partner to ultraquality in achieving global competitiveness was to counter the delays by anticipating them, in other words, using anticipatory feedback in as close to real time as practical. The most dramatic industrial example to date has been in lean production,[10] in which feedback to suppliers, coupled with ultraquality techniques, cut the costs of inventory in half and resulted in across-the-board competitive advantage in virtually all business parameters. More recently, criteria for certification, or those of its predecessor, quality assurance, are routinely fed back to conceptual design and engineering — one more recognition that *quality must be designed in, not tested in.*

A key factor in the design of any real-time feedback system is loop delay, the time it takes for a change to affect the system "loop" as a whole. In a feedback system, delay is countered by anticipation based on anticipated events (like a failure) or on a trend derived from the integration of past information. The design of the anticipation, or "correction," mechanism, usually the feedback paths, is crucial. The system as a whole can go sluggish on the one hand or oscillatory on the other. Symptoms are inventory chaos, unscheduled overtime, share price volatility, exasperated sales forces, frustrated suppliers, and, worst of all, departing long-time customers. Design questions are as follows: What is measured? How is it processed? Where is it sent? And, of course, to what purpose?

Properly designed feedback control systems determine transient and steady-state performance, reduce delays and resonances, alleviate nonlinearities in the production process, help control product quality, and minimize inventory. By way of explanation, in nonlinear systems, two otherwise independent input changes interact with each other to produce effects different from the sum of the effects of each separately. Understandably, the end results can be confusing if not catastrophic. An example is a negotiated reduction in wages followed immediately by an increase in executive wages. The combination results in a strike; either alone would not.

A second key parameter, the resonance time constant, is a measure of the frequency at which the system or several of its elements tries to oscillate or cycle. Resonances are created in almost every feedback system. The more feedback paths there are, the more resonances. The business cycle, related to inventory cycling, is one such resonance. Resonances, internal and external, can interact to the point of violent, nonlinear oscillation and destruction, particularly if they have the same or related resonant frequencies. Consequently, a warning: *Avoid creating the same resonance time constant in more than one location in a [production] system.*

Delay and resonance times, separately and together, are subject to design. In manufacturing systems, the factors that determine these parameters include inventory size, inventory replacement capacity, information acquisition and processing times, fabrication times, supplier capacity, and management response times. All can have a strong individual and collective influence on such overall system time responses as time to market, material and information flow rates, inventory replacement rate, model change rate, and employee turnover rate. Few, if any, of these factors can be chosen or designed independently of the rest, especially in complex feedback systems.

Fortunately, there are powerful tools for feedback system design. They include linear transform theory, transient analysis, discrete event mathematics, fuzzy thinking, and some selected nonlinear and time-variant design methods. The properties of at least simple linear systems designed by these techniques can be simulated and adjusted easily. A certain intuition can be developed based upon long experience with them. For example,

- *Behavior with feedback can be very different from behavior without it.*

   Positive example: Provide inventory managers with timely sales information and drastically reduce inventory costs. *Negative example*: Ignore customer feedback and drown in unused inventory.
- *Feedback works. However, the design of the feedback path is critical. Indeed, in the case of strong feedback, its design can be more important than that of the forward path.*

   Positive example: Customer feedback needs to be supplemented by anticipatory projections of economic trends and of competitor's responses to one's own strategies and actions to avoid delays and surprises.

   Negative examples: If feedback signals are "out of step" or of the wrong polarity, the system will oscillate, if not go completely out of control. Response that is too little, too late is often worse than no response at all.
- *Strong feedback can compensate for many known vagaries, even nonlinearities in the forward path, but it does so "at the margin."*

   Example: Production lines can be very precisely controlled around their operating points; that is, once a desired operating point is reached, tight control will maintain it, but off that point or on the way to or from it (e.g., start up, synchronization,[11] and shut down), off-optimum behavior is likely. *Example*: Just-in-time response works well for steady flow and constant volume. It can oscillate if flow is intermittent and volume is small.
- *Feedback systems will inherently resist unplanned or unanticipated change, whether internal or external.*

Satisfactory responses to anticipated changes, however, can usually be assured. *In any case, the response will last at least one time constant (cycle time) of the system.* These properties provide stability against disruption. On the other hand, abrupt mandates, however necessary, will be resisted and the end results may be considerably different in magnitude and timing from what advocates of the change anticipated. *Example*: Social systems, incentive programs, and political systems notoriously "readjust" to their own advantage when change is mandated. The resultant system behavior is usually less than, later than, and modified from, that anticipated.

- *To make a major change in performance of a presently stable system is likely to require a number of changes in the overall system design.*

    Examples: The change from mass production to lean production to flexible production[12] and the use of robots and high technology.

Not all systems are linear, however. As a warning against overdependence on linear-derived intuition, typical characteristics of nonlinear systems are as follows:

- *In general, no two systems of different nonlinearity behave in exactly the same way.*
- *Introducing changes into a nonlinear system will produce different (and probably unexpected) results if they are introduced separately than if they are introduced together.*
- *Introducing even very small changes in input magnitude can produce very different consequences even though all components and processes are deterministic.*

    Example: Chaotic behavior (noiselike but with underlying structure) close to system limits is such a phenomenon. *Example*: When the phone system is saturated with calls and goes chaotic, the planned strategy is to cut off all calls to a particular sector (e.g., California after an earthquake) or revert back to the simplest mode possible (first come, first serve). Sophisticated routing is simply abandoned — it is part of the problem. *Example*: When a computer abruptly becomes erratic as it runs out of memory, the simplest and usually successful technique is to turn it off and start it up again (reboot), hoping that not too much material has been lost.

- *Noise and extraneous signals irreversibly intermix with and alter normal, intended ones, generally with deleterious results.*

    Example: Modification of feedback and control signals is equivalent to modifying system behavior — that is, changing its transient and steady-state behavior. Nonlinear systems are therefore particularly vulnerable to purposeful opposition (jamming, disinformation, overloading).

- *Creating nonlinear systems is of higher risk than creating well-understood, linear ones.*

   The risk is less that the nonlinear systems will fail under carefully framed conditions than that they will behave strangely otherwise. *Example*: In the ultraquality spacecraft business, there is an heuristic: *If you cannot analyze it, do not build it* — an admonition against unnecessarily nonlinear feedback systems.

The two most common approaches to nonlinearity are, first, when nonlinearities are both unavoidable and undesirable, minimize their effect on end-system behavior through feedback and tight control of operating parameters over a limited operating region. Second, when nonlinearity can improve performance as in discrete and fuzzy control systems, be sure to model and simulate performance *outside* the normal operating range to avoid "nonintuitive" behavior.

The architectural and analytic difficulties faced by modern manufacturing feedback systems are that they are neither simple nor necessarily linear. They are unavoidably complex, probably contain some nonlinearities (limiters and fixed time delays), are multiply interconnected, and are subject to sometimes drastic external disturbances, not the least of which are sudden design changes and shifts in public perception. Their architectures must therefore be robust and yet flexible. Though inherently complex, they must be simple enough to understand and modify at the system level without too many unexpected consequences. In short, they are likely to be prime candidates for the heuristic and modeling techniques of systems architecting.

Feedback can be thought of at levels beyond the individual system and the manufacturing enterprise in normal operation. At the strategic level, we configure our enterprises with some level of feedback based on achieving, or failing to achieve success. Some measures of success are tied into enterprise-level feedback behavior.

Consider the strategic problem of an enterprise with a scientific research purpose that conducts missions of varying duration, from a few years to a decade.* There is a feedback loop, at both a scientific and enterprise-programmatic level, from mission to mission. The scientific discoveries on one mission will affect the scientific questions posed on the next mission. When unexpected things are found on one mission, like clear evidence of water on Mars, it deeply affects the enterprise's preference for what to look for on subsequent missions. Likewise, the success or

---

* The example is meant to apply, conceptually, to many different public and private research enterprises that can make trade-offs between the duration and complexities of their missions.

failure of various systems on a mission will affect their potential use on subsequent missions.

The duration of a mission is, in effect, a feedback time constant. The shorter that time constant, the more rapidly scientific discoveries and mission results are fed back into future missions. If the major goal of the enterprise is to produce unexpected scientific discoveries, then a shorter time constant may be an effective trade-off for less single mission cost-effectiveness. Reprising the maligned slogan "Faster, Better, Cheaper," it could be that faster is better at the enterprise level, even if it is not better at a single mission level. But note, such a conclusion is dependent on the overall objectives of the enterprise being subject to change from feedback. If the overall enterprise objectives are stable, and more like stewardship, a shorter time constant would not be a good trade-off.

Similar effects can be imagined at the management level of the enterprise. If the mission time constant is short enough, it will last no more than one person's normal assignment period. Program managers, architects, systems engineers, principal investigators, and others will serve for the full duration of a mission. Instead of end-to-end mission success or failure being fed back over several different leaders (as commonly happens with decade-long programs), accountability for success or failure is attached directly to those leaders. The effects on personnel policies and organizational learning should be obvious.

Lean Production

One of the most influential books on manufacturing of the last decade was the 1990 bestseller, *The Machine That Changed the World: The Story of Lean Production.*[13] Although the focus of this extensive survey and study was on automobile production in Japan, the United States, and Europe, its conclusions are applicable to manufacturing systems in general, particularly the concepts of quality and feedback. A 1994 follow-up book, *Comeback, The Fall and Rise of the American Automobile Industry,*[14] tells the story of the American response and the lessons learned from it, though calling it a "comeback" may have been premature. The story of lean production systems is by no means neat and orderly. Although the principles can be traced back to 1960, effective implementation took decades. Lean production certainly did not emerge full blown. Ideas and developments came from many sources, some prematurely. Credits were sometimes misattributed. Many contributors were very old by the time their ideas were widely understood and applied. Quality was sometimes seen as an end result instead of as a prerequisite for any change. The remarkable fact that virtually *every* critical parameter improved by at least 20%, if not 50%,[15] does not seem to have been anticipated. Then, within a few years,

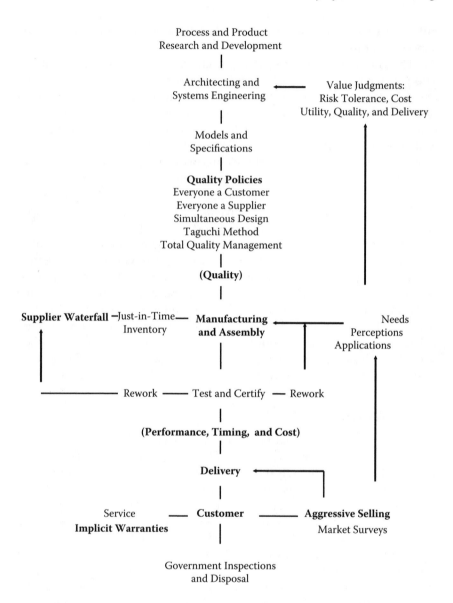

Figure 4.4 An architect's sketch of lean production.

everything worked. But when others attempted to adopt the process, they often failed. Why?

One way to answer such questions is to diagram the lean production process from an architectural perspective. Figure 4.4 is an architect's sketch of the lean production waterfall derived from the texts of the just-mentioned books, highlighting (boldfacing) its nonclassical features and strengthening its classical ones.* The most apparent feature is the number and strength of its feedback paths. Two are especially characteristic of lean production: the supplier waterfall loop and the customer-sales-delivery loop. Next evident is the quality policies box, crucial not only to high quality but to the profitable and proper operation of later steps, just-in-time inventory, rework, and implicit warranties. Quality policies are active elements in the sequence of steps, are a step through which all subsequent orders and specifications must pass, and are as affected by its feedback input as any other step; that is, policies must change with time, circumstance, technology, and product change and process imperatives.

Research and development (R&D) is not "in the loop" but instead is treated as one supplier of possibilities, among many, including one's competitors' R&D. As described in the 1990 study, R&D is not a driver, though it would not be surprising if its future role were different. Strong customer feedback, in contrast, is very much within the loop, making the loop responsive to customer needs at several key points. Manufacturing feedback to suppliers is also a short path, in contrast with the stand-off posture of much U.S. procurement.

The success of lean production has induced mass producers to copy many of its features, not always successfully. Several reasons for lack of success are apparent from the figure. If the policy box does not implement ultraquality, little can be expected from changes further downstream regardless of how much they ought to be able to contribute. Just-in-time (JIT) inventory is an example. Low-quality supply mandates a cushion of inventory roughly proportional to the defect rate; shifting that inventory from the manufacturer back to the supplier without a simultaneous quality upgrade simply increases transportation and financing costs. To decrease inventory, decrease the defect rate, then apply the coordination principles of JIT, not before.

Another reason for limited success in converting piecemeal to lean production is that any well-operated feedback system, including those used in classical mass production, will resist changes in the forward (waterfall) path. The "loop" will attempt to self-correct. And, it will take at least one loop time constant before all the effects can be seen or even be known. To illustrate, if supply inventory is reduced, what is the effect on sales and service inventory? If customer feedback to the manufacturing line is

---

* Strictly speaking, though the authors of the lean production books did not mention it, an architect's perspective should also include the intersecting product waterfalls and software spirals. Interestingly, because it seems to be true for all successful systems, it is possible to find where and by whom systems architecting was performed. Two of the more famous automotive production architects were Taiichi Ohno, the pioneer of the Toyota Motor Company Production System, and Yoshiki Yamasaki, head of automobile production at Mazda.

aggressively sought, as it is in Japan, what is the effect on time-to-market for new product designs?

A serious question raised in both books is how to convert mass production systems into lean production systems. It is not, as the name "lean" might imply, a mass production system with the "fat" of inventory, middle management, screening, and documentation taken out. It is to recognize lean production as a different architecture based on different priorities and interrelationships. How then to begin the conversion? What is both necessary and sufficient? What can be retained?

The place to begin conversion, given the nature of feedback systems, is in the quality policies step. In lean production, quality is not a production result determined postproduction and posttest; it is a prerequisite policy imperative. Indeed, Japanese innovators experienced years of frustration when total quality management (TQM), JIT, and the Taguchi methods at first seemed to do very little. The level of quality essential for these methods to work had not yet been reached. When it was, the whole system virtually snapped into place with results that became famous. Even more important for other companies, unless their quality levels are high enough, even though all the foregoing methods are in place, the famous results will not — and cannot — happen.

Conversely, at an only slightly lower level of quality, lean systems sporadically face at least temporary collapse.[16] As a speculation, there appears to be a direct correlation between how close to the cliff of collapse the system operates and the competitive advantage it enjoys. Backing off from the cliff would seem to decrease its competitive edge, yet getting too close risks imminent collapse — line shutdown, transportation jam-up, short-fuse customer anger, and collateral damage to suppliers and customers for whom the product is an element of a still larger system production.

To summarize, lean production is an ultraquality, dynamic feedback system inherently susceptible to any reduction in quality. It depends upon well-designed, multiple feedback. Given ultraquality standards, lean production arguably is less complex, simpler, and more efficient than mass production. And, by its very nature, it is extraordinarily, fiercely, competitive.

Flexible Manufacturing

Flexible manufacturing is the capability of sequentially making more than one product on the same production line. In its most common present-day form, it customizes products for individual customers, more or less on demand, by assembling different modules (options) on a common base (platform). Individually tailored automobiles, for example, have been coming down production lines for years. But with one out of three or even one out of ten units having to be sent back or taken out of a production stream, flexible manufacturing in the past has been costly in inventory, complex in operation, and high-priced per option compared to all-of-a-kind production.

What changed are customer demands and expectations, especially in consumer products. Largely because of technological innovation, more capability for less cost now controls purchase rate rather than wearout and increasing defect rate — an interesting epilogue for the DeSoto story!* One consequence of the change is more models per year with fewer units per model, the higher costs per unit being offset by use of techniques such as preplanned product improvement, standardization of interfaces and protocols, and lean production methods.

A natural architecture for the flexible manufacturing of complex products would be an extension of lean production with its imperatives — additional feedback paths and ultraquality-produced simplicity — and an imperative all its own, human-like information command and control.

At its core, flexible manufacturing involves the real-time interaction of a production waterfall with multiple product waterfalls. Lacking an architectural change from lean production, however, the resultant multiple information flows could overwhelm conventional control systems. The problem is less that of gathering data than of condensing it. That suggests that flexible manufacturing will need judgmental, multiple-path control analogous to that of an air traffic controller in the new "free flight" regime. Whether the resultant architecture will be fuzzy, associative, neural, heuristic, or largely human, is arguable.

To simplify the flexible manufacturing problem to something more manageable, most companies today would limit the flexibility to a product *line* that is forward and backward compatible, uses similar modules (with many modules identical), keeps to the same manufacturing standards, and is planned to be in business long enough to write off the cost of the facility out of product-line profits. In brief, production would be limited to products having a single basic architecture, for example, producing either Macintosh computers, Hitachi TV sets, or Motorola cellular telephones, but not all three on demand on the same production line.

Even that configuration is complex architecturally. To illustrate: A central issue in product line design is where in the family of products, from low-end to high-end, to optimize. Too high in the series, and the low end is needlessly costly. Too low, and the high end adds too little value. A related issue arises in the companion manufacturing system. Too much

---

* Parenthetically, the Japanese countered the automobile obsolescence problem by quadrennial government inspections so rigorous that it was often less expensive to turn a car in and purchase a new one than to bring the old one up to government standards (Womack et al., 1990, p. 62).

capability, and its overhead is too high; too little, and it sacrifices profit to specialty suppliers.

Another extension from lean production to flexible manufacturing is much closer coordination between the design and development of the product line and the design and development of its manufacturing system. Failure to achieve this coordination, as illustrated by the problems of introducing robots into manufacturing, can be warehouses of unopened crates of robots and in-work products that cannot be completed as designed. Clearly: *The product and its manufacturing system must match.* At the elementary level, this means that the system must be composed of subsystems that distribute cleanly over the manufacturing enterprise. More specifically, their time constants, transient responses, and product-to-machine interfaces must match, recognizing that any manufacturing constraint means a product constraint, and vice versa. At a more sophisticated level, the elements of the process, like quality measurement and control, must be matched across the product-system and manufacturing-system boundaries.

As suggested earlier, the key technological innovation is likely to be the architecture of the joint information system.[17] In that connection, one of the greatest determinates of the information system's size, speed, and resolution is the quality of the end product and the yield of its manufacturing process — that is, their defect rates. The higher these defect rates, the greater the size, cost, complexity, and precision of the information system that will be needed to find and eliminate them quickly.

Another strong determinate of information system capacity is piece part count, another factor dependent on the match of product and manufacturing techniques. Mechanical engineers have known this for years: whenever possible, replace a complicated assembly of parts with a single, specialized piece. Nowhere is the advantage of piece part reduction as evident as in the continuing substitution of more and more high-capacity microprocessors for their lower-capacity predecessors. Remarkably, this substitution, for approximately the same cost, also decreases the defect rate per computational operation. It appears to be an inevitable consequence of the different parts of Moore's law. As technology allows more and more transistors per unit area, the cost of the fabrication plant likewise rises. The rise in cost of the fabrication plant drives the market toward increasing standardization of parts (to spread large capital costs over many units). Increasing standardization means greater regularization, and higher quality to achieve economic throughput in an expensive fabrication plant. The end-user value added can then come only from software (a topic we take up later).

And, especially for product lines, the fewer different parts from model to model, the better, as long as that commonality does not decrease system capability unduly. Once again, there is a close correlation between reduced defect rate, reduced information processing, reduced inventory, and reduced complexity — all by design.

Looking further into the future, an extension of the lean production architecture is not the only possibility for flexible manufacturing. It is possible that flexible manufacturing could take a quite different architectural turn based on a different set of priorities. Is ultraquality production really necessary for simple, low-cost, limited-warranty products made largely from commercial, off-the-shelf (COTS) units (for example, microprocessors and flat screens)? Or is the manufacturing equivalent of parallel processors (pipelines) the answer? Should some flexible manufacturing hark back to the principles of special, handcrafted products or one-of-a-kind planetary spacecraft? The answers should be known in less than a decade, considering the profit to be made in finding them.

Heuristics for Architecting Manufacturing Systems

- The product and its manufacturing system must match. (In many ways.)
- Keep it simple. (Ultraquality helps.)
- Partition for near-autonomy. (A trade-off with feedback.)
- In partitioning a manufacturing system, choose the elements so that they minimize the complexity of material and information flow. (Savagian, Peter J., 1990, USC)
- Watch out for critical mis-fits. (Between intersecting waterfalls.)
- In making a change in the manufacturing process, how you make it is often more important than the change itself. (Management policy.)
- When implementing a change, keep some elements constant to provide an anchor point for people to cling to. (Schmidt, Jeffrey H., 1993, USC) (A trade-off when a new architecture is needed.)
- Install a machine that even an idiot can use and pretty soon everyone working for you is an idiot. (Olivieri, J. M., 1991, USC) (An unexpected consequence of mass production Taylorism — see next heuristic.)
- Everyone a customer, everyone a supplier.
- To reduce unwanted nonlinear behavior, linearize!
- If you cannot analyze it, do not build it.
- Avoid creating the same resonance time constant in more than one location in a [production] system.
- The five why's. (A technique for finding basic causes, and one used by every inquisitive child to learn about the world at large.)

Conclusion

Modern manufacturing can be portrayed as an ultraquality, dynamic feedback system intersecting with that of the product waterfall. The manufacturing systems architect's added tasks, beyond those of all systems architects, include (1) maintaining connections to the product waterfall and the software spiral necessary for coordinated developments, (2) assuring quality levels high enough to avoid manufacturing system collapse or oscillation, (3) determining and helping control the system parameters for stable and timely performance, and (4) looking farther into the future than do most product-line architects.

Exercises

1. Manufacturing systems are complex systems that need to be architected. If the manufacturing line is software intensive, and repeated software upgrades are planned, how can certification of software changes be managed?
2. The feedback lags or resonances of a manufacturing system of a commercial product interact with the dynamics of market demand. Give examples of problems arising from this interaction and possible methods for alleviating them.
3. Examine the following hypothesis: Increasing quality levels in manufacturing enable architectural changes in the manufacturing system that greatly increase productivity but may make the system increasingly sensitive to external disruption. For a research exercise, use case studies or a simplified quantitative model.
4. Does manufacturing systems architecture differ in mass production systems (thousands of units) and very low-quantity production systems (fewer than ten produced systems)? If so, how and why?
5. Current flexible manufacturing systems usually build very small lot sizes from a single product line in response to real-time customer demand; for example, an automobile production line that builds each car to order. Consider two alternative architectures for organizing such a system, one centralized and one decentralized. The first would use close centralized control, centralized production scheduling, and resource planning. The other would use fully distributed control based on disseminating customer/supplier relationships to each work cell; that is, each job and each work cell interact individually through an auction for services. What would be advantages and disadvantages of both approaches? How would the architecture of the supporting information systems (extending to sales and customer support) have to differ in the two cases?

Notes and References

1. Hayes, Robert H., S. C. Wheelwright, and K. B. Clark, *Dynamic Manufacturing, Creating the Learning Organization.* New York: The Free Press, 1988, Chapter 7: The Architecture of Manufacturing: Material and Information Flows, p. 185,

defines a manufacturing architecture as including its hardware, material, and information flows, their coordination and managerial philosophy. This textbook is highly recommended, especially Chapters 7, 8, and 9.

2. Rechtin, E., *Systems Architecting, Creating and Building Complex Systems.* Englewood Cliffs, NJ: Prentice Hall, 1991, Chapter 8, pp. 160–187. (Note that throughout the rest of this chapter, this reference will be referred to as Rechtin 1991.) Ultraquality: Quality (absence of defects) so high as to be impractical to prove by measure with confidence.

3. Hayes, Robert H., S. C. Wheelwright, and K. B. Clark, *Dynamic Manufacturing, Creating the Learning Organization.* New York: The Free Press, 1988.

4. Womack, James P., Daniel T. Jones, and Daniel Roos, *The Machine That Changed the World, The Story of Lean Production.* New York: Harper Collins, 1990 (paperback 1991), p. 4.

5. See www.grandchallenge.org.

6. See also Hayes et al., 1988, Chapter 8, Controlling and Improving the Manufacturing Process.

7. Hayes et al., 1988, Chapter 7, p. 185.

8. See Brooks, Frederick P., Jr., No Silver Bullet, Essence and Accidents of Software Engineering, *Computer*, pp. 10–19, April 1987, especially p. 18; Simon, Herbert, *The Sciences of the Artificial*, 2nd ed. Cambridge, MA: MIT Press, 1987, pp. 200–209.

9. For the equivalent in a surgical team, see Brooks' *Mythical Man-Month*.

10. See Womak et al., 1990.

11. See Lindsey, William C., *Synchronization Systems in Communication and Control.* Englewood Cliffs, NJ: Prentice Hall, 1972.

12. Hayes et al., 1988, p. 204.

13. Womack et al., 1990.

14. For an update, see Ingrassia, Paul, and Joseph B. White, *Comeback, The Fall and Rise of the American Automobile Industry.* New York: Simon & Schuster, 1994.

15. Womack et al., 1994, title page.

16. Womack et al., 1990, p. 62.

17. Hayes et al., 1988, Chapter 7, p. 185.

Case Study 3: Intelligent Transportation Systems

Introduction

Intelligent transport systems (ITSs)[1] are members of a class of systems in which humans and their behavior are inextricably part of the system. They are also systems whose architectures are distributed, in both a logical and physical sense, and are equally distributed in their development, procurement, and management. The key characteristics of such systems, which will jointly help define the concept of a collaborative system in Chapter 7, include (1) the lack of a single client with ownership and developmental responsibility for the system, (2) considerable uncertainty in system purposes and a recognition that purposes will evolve in unknown directions over its lifetime, and (3) the need for extensive voluntary cooperation in their deployment and use. This last point, creation through voluntary cooperation and interaction, will be the central insight of Chapter 7. In Chapter 5, we deal more generally with the concept of sociotechnical systems, where humans and their behaviors are inside rather than outside the system boundary.

ITS concepts have been around for several decades but started getting serious attention in the 1990s. ITS in general refers to transportation-related guidance, control, and information systems. These systems use computer and information technology to address transportation functions at the level of individual vehicles, roadways, and large transportation networks. The motivator for developing such systems is the belief that they will improve transport network flow, improve safety, reduce environmental impact, and be large commercial market opportunities. Many believe that the transport improvements gained through the application of information technology promise to be cheaper and less environmentally damaging than further expansion of the physical transport infrastructure. Over the long term, ITS could evolve into automated highways where vehicles are automatically controlled for even larger gains in system performance.

At the time of the writing of this third edition, a variety of intelligent transport services are already commonly available. Many vehicles have built-in electronic navigators using Global Positioning System (GPS) and digital map databases. The systems provide route planning and real-time route guidance. Position monitoring systems are in fairly wide use in commercial vehicle fleets. Real-time traffic conditions are available at various Web sites. Several different online services provide free maps of virtually every city, and nonurban areas in industrialized countries, with route planning services. Many metro areas use intelligent traffic control methods in managing their stoplights, road admittance systems, and demand lanes.

In most areas, what does not exist today is the interconnection of these various services and the centralized exploitation of both information and controls. This is striking because many of the early concepts and proponents emphasized the role and value of centralized control. Some steps toward centralized systems have been made in a few cities, and there is continuing interest in further integration, although perhaps somewhat less than in the first run of enthusiasm in the 1990s. The actual experience points out that the split between public and private control and responsibility is an architectural choice. When certain services are to be provided voluntarily by market means, that is a choice on the overall structure of the system. When certain services are reserved to government control, likewise that is a choice on the overall structure of the system. To understand the architecture of social systems, one element is division among public and private means.

ITS Concepts

Possible ITSs have been extensively described in the literature.[2] The most common decomposition of ITS services is into five core services and automated highways, which is considered somewhat farther out. The five core services as have been usually defined, with an indication of how they are provided today, are as follows:

Advanced Traveler Information Services (ATIS)
ATIS is the provision of accurate, real-time information on transportation options and conditions to a traveler anytime, anywhere. For example:

1. Computer-assisted route planning to a street location anywhere in the country. This service could be coupled with traffic prediction and multimode transport information. Extensive capabilities in this category are available from multiple Web sites (although not necessarily tied to traffic conditions) and through in-vehicle GPS navigators.
2. Computer-assisted route guidance to a street location anywhere in the country, again possibly coupled to real-time traffic information and predictions. Absent the traffic conditions component, this is the main selling feature of GPS in-vehicle navigators.
3. Access to full public transportation schedules in a distant city before leaving for that city. Again, this is available today, at least in some jurisdictions.

4. Broadcast of real-time and predictive traffic conditions on the major roads of an urban area. Today this is available with moderate fidelity on Web sites and through radio.
5. Emergency situation and location reporting, manual and automated. Various private services, some tied to particular auto manufacturers, now provide this.

Advanced Traffic Management Systems (ATMS)

The intent of ATMS is to improve the carrying capacity and flow of the road network by integrating traffic sensors, remotely operated traffic signals, real-time monitoring and prediction, and dissemination of route information. The service components of ATMS include sensing traffic conditions in real time over wide areas, real-time prediction of traffic conditions, and remotely controlling traffic signals and signage from central control centers to optimize road network conditions. A long-term concept in ATMS is coupling traffic management with route selection in individual vehicles.

ATMS exist today, although their penetration has been less than many of the advocates hoped. Coupling of traffic management with individual route selection is almost nonexistent, wide area prediction is limited, but wide area real-time monitoring is common. Some jurisdictions make use of considerable centralized control, including additional mechanisms not previously listed, such as demand pricing.

Advanced Vehicle Control Systems (AVCS)

AVCS covers driver assistance systems within a single vehicle. This is an area of continuing interest with some roll-out in production vehicles, mostly in high-end private automobiles. Some examples include:

1. Partially automated braking systems. Today, antilock brake automation is common, with additional levels of automation rare.
2. Automated driver assistance in distance following or lateral lane keeping. A few high-end vehicles have limited capabilities here.
3. Obstacle warning and avoidance. Backup sensors are common in larger vehicles today, and some limited stability enhancement in emergency avoidance maneuvers has been done.
4. Vision enhancement in reduced visibility conditions. Again, a few systems are available.

Commercial Vehicle Operations (CVO)

CVO deals with the automation of regulatory functions and record keeping, especially in interstate travel. The goal is reduction of time and expense due to regulatory requirements in road transport. Although the roll-out to public infrastructure has been limited, there has been extensive usage in private fleets. Some examples of proposed CVO capabilities include:

1. Weigh-in-motion for trucks.
2. Electronic license/tag/permit checking and record keeping.
3. Hazardous cargo monitoring (coupled with navigation and position reporting).
4. Position monitoring and reporting for fleet management.

Advanced Public Transport (APT)

The goal of APT is performance improvements in public transport through application of intelligent vehicle and highway system (IVHS) technology. Some examples include:

1. Real-time monitoring of bus, subway, or train position coupled with waiting area displays and vehicle dispatch. These systems have proved popular and reasonably effective where deployed.
2. Electronic fare paying systems to improve stopping times and allow time-sensitive pricing. Many systems have moved to smart card and related electronic payment systems.

ITS Sociotechnical Issues

An ITS is unquestionably a sociotechnical system, in the sense that humans and their behavior are irreducible components. People decide whether or not to use route planning and guidance systems. When given route advice, they choose to use it or ignore it. People choose to purchase (or not) various components of an ITS. At the political level, people make joint decisions through their government whether or not to make infrastructural investments. So, any discussion of the architecture of an ITS must include people, and architecting of an ITS must incorporate the sociotechnical nature of the system. The heuristics of sociotechnical systems are the primary focus of Chapter 5, with Chapter 7 taking up the somewhat narrower, more specialized case of collaborative systems. To illustrate, consider how the issues have been resolved, in practice, for the ITSs that now exist and are emerging.

Who Is the Client for an Architect?

Borrowing a phrase,[3] an ITS will be a system no one owns. ITS planning, at least in the United States, assumes that purchase, deployment, and use will be the result of distributed decisions by governments and consumers.[4] In consequence, the integrity of any ITS architecture must be maintained through some similarly distributed means. When a single client (agency, company, or individual) commissions a system, the integrity of that system can be maintained by an architect hired by the client. When no client with that power exists, no architect with the power can exist either. This complicates the architecting problem. Lacking the power to directly establish and maintain the architecture, the architect must find indirect means to do so.

In systems architecting, it is common for the actual users to be different from the system sponsor. When this occurs, the architect must be conscious of the possibility that the preferences and needs of the ultimate users are different from those of the sponsors, or as perceived by the sponsors. The system might seem perfectly satisfactory to the sponsors, and yet be unacceptable to the users. If the system is rejected by the users, the sponsor is unlikely to perceive the system as successful.

This situation is even more extreme in the case of an ITS. To date, there has been little centralized architecting of ITSs, at least of those elements deployed and widely used in the United States. The U.S. Department of Transportation (DOT) sponsored rather extensive ITS architecture studies. But, the U.S. DOT has only limited authority to direct or mandate transportation developments in the United States. Execution is up to states, metropolitan traffic authorities, cities, and individuals. By analogy, an architect for an ITS is more in the position of an urban planner than a civil architect. The urban planner has a great deal of influence, but relatively little power. For an urban planner to be effective requires considerable political skill, and sponsors who understand the limits of their own paper. Through city governments and zoning boards, urban planners can possess a "no" authority, that is the ability to say "no" to nonconformant plans. However, their ability to say "yes" is much less, and happens only if the government employing the urban planner is willing to commit funds.

Public or Private?

In the first wave of excitement over ITSs, many of the concepts seemed to assume a dominantly public infrastructure. But, in practice today, many, perhaps most, of the interesting traveler information services are private. The in-vehicle GPS navigators are privately produced and purchased. The online map services are private and advertising supported. Because the architecture of an ITS is not currently centrally directed by government (at least in the United States), it is probably not a surprise that the private side has been the side that has most extensively been developed.

Even if more centralized architecting had been done, the result might well have been the same, although there might have been niche areas where centralized decision making could take hold. As an elaboration, consider the problem of social collaboration and fully coupled control. In order that any of the fully coupled control schemes can be effectively used, the driver compliance rate must be high. Why will drivers willingly comply with centralized route guidance that is knowingly being computed with the benefit of the whole in mind? Will people just game the system?

There is a useful heuristic from Rechtin (1991)[5]:

> In introducing technological and social change, *how* you do it is often more important than *what* you do.

> If social cooperation is required, the way in which a system is implemented and introduced must be an integral part of its architecture.

Using this heuristic requires identifying those architectural characteristics that lead to cooperative acceptance and use. In the ITS case, what system characteristics are most likely to foster public cooperation and acceptance? The answer will not be identical for all countries and cultures. We suggest for the United States that ITS general acceptance will be greater for those services that are privately and voluntarily contracted for.

The deployment mode for an architectural element should be, in order of preference:

1. Private development and purchase.
2. Private development and purchase subject to governmental guidelines or standards.
3. Private development and purchase subject to government mandating.
4. Government-financed development and private purchase.
5. Government mandating of deployment with direct finance.

A consequence of using these criteria is that ITS services should be partitioned to support and encourage private development of particular packages. Such packaging requires groupings that provide income streams a private firm can capture and defensible markets. This criterion suggests that the technical architecture should support such decentralization in development and deployment even where centralization would be more "efficient," say on a life-cycle cost basis.

Although the issue of public–private partitioning is not so prominent in early writings on ITSs, as noted it has played an important role in actual development. The greatest ferment in ITSs has been in private systems, and private systems have avoided problems of perception of invasion of privacy and monitoring. Of course, the reality is that private monitoring is also monitoring, but this only reemphasizes that perceptions may matter more than facts.

Facts and Perceptions

Continuing on that same line of facts versus perceptions, consider how ITS-like sociotechnical systems are judged as successes or failures. How would we know if ITSs were successful or not? A traditional systems engineering approach would immediately appeal to measures of effectiveness, probably measures like average speed on the roads, road throughput, life-cycle cost, incident rate, and various other measures easy to imagine and cite. But, do the actual stakeholders of ITSs (government authorities and the traveling public) perceive those as measures of success? Another heuristic: *Success is in the eye of the beholder,* suggests not. Consider the following thought experiment:

> *Scenario 1:* It is 15 years later than the present. Intelligent transportation systems are widely deployed in most major urban areas and very heavily used. Most urban areas have five or more competing, private traffic information service providers. There is extremely active, competitive development of supporting communication, display, and algorithmic systems. Market penetration of rich services is above 85%. But, traffic in major urban areas is much worse than current. Most measures of effectiveness (for example, average travel time, average speed) have decayed.
>
> *Scenario 2:* Again, it is 15 years beyond the present. ITSs are likewise widely deployed and widely used. Now the various measures are substantially improved. But, deployment and compliance have come only from vigorous mandates and enforcement. In many jurisdictions, the mandates have been tossed out by popular vote. Effectiveness is demonstrably highest in those jurisdictions with the strictest enforcement and the least responsiveness to popular demand.

Although the reader may consider the scenarios fanciful and unrealistic, that is not the point. The point is to ask, seriously, whether a system should be judged by whether it gives the users what they want, regardless of whether the architect thinks what they want makes any sense. The classical paradigm says what the sponsors want is what matters, not what the architect thinks makes sense. In a sociotechnical and a collaborative system, where voluntary interaction is essential to system operation, what the users think they want is more important than what the sponsor wants. What we have now is a situation more like scenario 1, because the dominant deployments are private.

Although the exact scenarios are not reality, the underlying observation about success not necessarily coming from the original objectives has been proved in practice. In several cases where ITSs have been found successful in use, a major source of satisfaction has been through the impact on travel time variance and predictability, and not on average speed. That is, the most valued impact has been how it enables users to accurately predict how long a trip will take, and making travel times more consistent, rather than making the average time shorter.

Architecture as Shared Invariants

One way of envisioning an architecture of something as complex as an ITS is by looking at the shared invariants. For an ITS this means looking for the things shared across many or all users, and that do not change much with time. In the sense of shared invariants, some of the things that could be an ITS architecture include the following:

- Shared positioning services (GPS).
- Map data, specifically the network of roads and their positions relative to GPS locations.
- How digital traffic messages are encoded. How do you digitally indicate that the flow at a given point on a given road has some value, in a system-independent way.
- Mobile communication networks over which traffic information flows (networks that may not belong to ITS).

Much less of this kind of invariant definition has been done than could have been done. In practice it has been hard to get centralized attention to elements that are supportive behind the scenes but are not delivering services that are benefits directly.

Dominance of Economics

Finally, a theme of sociotechnical systems that is strongly evident in ITS is the role of, or the dominance of, economics. Today, what we have in an ITS has largely been driven by what makes a profitable business, in many cases rather independently of doing anything about travel. The online map services, for example, are widely used and popular. In the course of a few short years, people have gone from mostly using paper maps to where people commonly pass around map service printouts for directional instructions. Many Web sites are now modified to simply link to one of the main map services whenever a location must be provided.

But, what drives the map services? Because they are free and rather modest appendages to larger Internet firms, the answer is mainly advertising. A popular Web service attracts users, a large user base brings advertising dollars. Map services are a particularly fine advertising target because the nature of the search strongly suggests what the user is looking for, hence assisting in advertising targeting. A user can opt out of advertising by using a purchased, disk-based version instead, but usage rates have proven that most users are not reluctant to abandon a small slice of privacy about their location searches in order to have continuously updated, online map information. It works because the economics works.

The broader lesson to consider in sociotechnical systems is that business is usually a part of them. "Stable forms" includes the notion of economically stable, not just technically stable. In Chapter 13 in Part IV, we will discuss politicotechnical systems. In a politicotechnical system, stability is likewise critical, but there it comes mostly from the nature of the constituency instead of the business model.

Notes and References

1. This case study is drawn heavily from a previous work of one of the authors, Maier, M. W., On Architecting and Intelligent Transport Systems, Joint Issue, *IEEE Transactions on Aerospace and Electronic Systems/System Engineering*, AES33:2, pp. 610–625, April 1997.
2. Strategic Plan for Intelligent Vehicle-Highway Systems in the United States, IVHS America, Report No. IVHS-AMER-92-3, Intelligent Vehicle-Highway Society of America, Washington, DC, 1992.
3. Schuman, R., Developing an Architecture That No One Owns: The U.S. Approach to System Architecture, in *Proceedings of the First World Congress on Applications of Transport Telematics and Intelligent Vehicle-Highway Systems*, Paris, France, 1994.
4. Federal Highway Administration, Intelligent Vehicle Highway Systems: A Public Private Partnership, Washington, DC, 1991.
5. Rechtin, E., *Systems Architecting*. New York: Prentice Hall, 1991, p. 315.

chapter 5

Social Systems

Introduction: Defining Sociotechnical Systems

> *Social:* Concerning groups of people or the general public.
> *Technical:* Based on physical sciences and their application.
> *Sociotechnical Systems:* Technical works involving significant social participation, interests, and concerns.

Sociotechnical systems are technical works involving the participation of groups of people in ways that significantly affect the architectures and design of those works. Historically, the most conspicuous have been the large civil works — monuments, cathedrals, urban developments, dams, and roadways among them. Lessons learned from their construction provide the basis for much of civil (and systems) architecting and its theory.[1]

More recently, others, physically quite different, have become much more sociotechnical in nature than might have been contemplated at their inception — ballistic missile defense, air travel, information networks, welfare, and health delivery, for example. Few can even be conceived, much less built, without major social participation, planned or not. Experiences with them have generated a number of strategies and heuristics of importance to architects in general. Several are presented here. Among them are three heuristics: the four who's, economic value, and the tension between perceptions and facts. In the interests of informed use, as with all heuristics, it is important to understand the context within which they evolved, the sociotechnical domain. Then, at the end of the chapter are a number of general heuristics of applicability to sociotechnical systems in particular.

Public Participation

At the highest level of social participation, members of the public directly use — and may own a part of — the system's facilities. At an intermediate level, they are provided a personal service, usually by a public or private utility. Most important, individuals and not the utilities — the architect's clients — are the end users. Examples are highways, communication and information circuits, general aviation traffic control, and public power. Public cooperation and personal responsibility are required for effective operation. That is, users are expected to follow established rules with a minimum of policing or control. Drivers and pilots follow the rules of the road. Communicators respect the twin rights of free access and privacy.

At the highest level of participation, participating individuals choose and collectively own a major fraction of the system structure — cars, trucks, aircraft, computers, telephones, electrical and plumbing hardware, and so on. In a sense, the public "rents" the rest of the facilities through access charges, fees for use, and taxes. Reaction to a facility failure or a price change tends to be local in scope, quick and focused. The public's voice is heard through specialized groups such as automobile clubs for highways, retired persons associations for health delivery, professional societies for power and communications services, and the like. Market forces are used to considerable advantage through adverse publicity in the media, boycotts, and resistance to stock and bond issues on the one hand and through enthusiastic acceptance on the other. Recent examples of major architectural changes strongly supported by the public are superhighways, satellite communications, entertainment cable networks, jet aircraft transportation, health maintenance organizations, and a slow shift from polluting fossil fuels to alternative sources of energy.

Systems of this sort are best described as collaborative systems, systems that operate only through the partially voluntary initiative of their components. This collaboration is an important subject in its own right, because the Internet, the World Wide Web, and open source software are collaborative assemblages. We address this topic in detail in Chapter 7.

At the other extreme of social participation are social systems used solely by the government, directly or through sponsorship, for broad social purposes delegated to it by the public; for example, National Aeronautics and Space Administration (NASA) and U.S. Department of Defense (DoD) systems for exploration and defense, Social Security, and Medicare management systems for public health and welfare, research capabilities for national well-being, and police systems for public safety. The public pays for these services and systems only indirectly, through general taxation. The architect's client and end user is the government. The public's connection with the design, construction, and operation of these systems is sharply limited. Its value judgments are made almost solely through the *political* process, the subject of Chapter 10. They might best be characterized as "politicotechnical."

The phrase "system-of-systems" is now commonly used in the systems engineering literature, although not with a consistent definition. Because the term system-of-systems is ambiguous on its face (is any system whose subsystems are complex enough to be regarded as systems a system-of-systems?) we prefer the terms we use here. In many writings, sociotechnical systems and systems-of-systems are conflated. In others, collaborative systems and systems-of-systems are conflated. For the purposes of this chapter, we will establish the foregoing definition for sociotechnical systems and will explore the notion of collaborative systems at some length.

The Foundations of Sociotechnical Systems Architecting

The foundations of sociotechnical systems architecting are much the same as for all systems: a systems approach, purpose orientation, modeling, certification, and insight. Social system quality, however, is less a foundation than a case-by-case trade-off; that is, the quality desired depends on the system to be provided. In nuclear power generation, modern manufacturing, and manned space flight, ultraquality is an imperative. But in public health, pollution control, and safety, the level of acceptable quality is only one of many economic, social, political, and technical factors to be accommodated.

But if sociotechnical systems architecting loses one foundation, ultraquality, it gains another — a direct and immediate response to the public's needs and perceptions. Responding to public perceptions is particularly difficult, even for an experienced architect. The public's interests are unavoidably diverse and often incompatible. The groups with the strongest interests change with time, sometimes reversing themselves based on a single incident. Three Mile Island was such an incident for nuclear power utilities. Pictures of the Earth from a lunar-bound Apollo spacecraft suddenly focused public attention and support for global environmental management. An election of a senator from Pennsylvania avalanched into widespread public concern over health insurance and Medicare systems.

The Separation of Client and User

In most sociotechnical systems, the client, the buyer of the architect's services, is not the user. This fact can present a serious ethical, as well as technical, problem to the architect: how should conflicts between the preferences, if not the imperatives, of the utility agency and those of the public (as perceived by the architect) be treated when preferences strongly affect system design.

It is not a new dilemma. State governments have partly resolved the problem by licensing architects and setting standards for systems that affect the health and safety of the public. Buildings, bridges, and public power systems come to mind as systems that affect public safety. Information systems are already on the horizon. The issuing or denial of licenses is one way of making sure that public interest comes first in the architect's mind. The setting of standards provides the architect some counterarguments against overreaching demands by the client. But these policies do not treat such conflicts as that of the degree of traffic control desired by a manager of an intelligent transportation system (ITS) as compared with that of the individual user/driver, or the degree of governmental regulation of the Internet to assure a balance of access, privacy, security, profit making, and system efficiency.

One of the ways of alleviating some of these tensions is through economics. Economics has important insights, as economics is fundamentally the study of social constructs. In addition, markets have evolved a variety of mechanisms, such as market segmentation, that effectively deal with essential problems that arise from the nature of sociotechnical, social, and collaborative systems architecting.

Socioeconomic Insights

Social economists bring two special insights to sociotechnical systems. The first insight, which might be called *the four who's*, asks four questions that need to be answered *as a self-consistent set* if the system is to succeed economically — namely: *Who benefits? Who pays? Who provides? And, as appropriate, Who loses?*

> *Example*: The answers to these questions of the Bell Telephone System were: (1) the beneficiaries were the callers and those who received the calls; (2) the callers paid based on usage because they initiated the calls and could be billed for them; (3) the provider was a monopoly whose services and charges were regulated by public agencies for public purposes; and (4) the losers were those who wished to use the telephone facilities for services not offered or to sell equipment not authorized for connection to those facilities. The telephone monopoly was incentivized to carry out widely available basic research because it generated more and better service at less cost, a result the regulatory agencies desired. International and national security agreements were facilitated by having a single point of contact, the Bell System, for all such matters. Standards were maintained and the financial strategy was long term, typically 30 years. The system was dismantled when the losers evoked antitrust laws, creating a new set of losers, complex billing, standards problems, and a loss of research. Arguably, it enabled the Internet sooner than otherwise. Subsequently, separate satellite and cable services were established, further dismantling what had been a single service system. The dismantlement also may

have assisted in allowing the rapid rollout of wireless cellular telephone systems. In other countries, some of the most successful wireless cellular rollouts have occurred where wireless companies were the only allowed alternative to a government-sponsored telephone monopoly.

Example: The answers to the four who's for the privatized Landsat System, a satellite-based optical-infrared surveillance service, were as follows: (1) the beneficiaries were those individuals and organizations who could intermittently use high-altitude photographs of the Earth; (2) because the *value* to the user of an individual photograph was unrelated to its cost (just as is the case with weather satellite TV), the individual users could not be billed effectively; (3) the provider was a private, profit-making organization that understandably demanded a cost-plus-fixed-fee contract from the government as a surrogate customer; and (4) when the government balked at this result of privatization, the Landsat system faced collapse. Research had been crippled, a French government-sponsored service (SPOT) had acquired appreciable market share, and legitimate customers faced loss of service. Privatization was reversed and the government again became the provider.

Example: Serious debates over the nature of their public health systems are underway in many countries, triggered in large part by the technological advances of the last few decades. These advances have made it possible for humanity to live longer and in better health, but the investments in those gains are considerable. The answers to the four who's are at the crux of the debate. Who benefits — everyone equally at all levels of health? Who pays — regardless of personal health or based on need and ability to pay? Who provides — and determines cost to the user? Who loses — anyone out of work or above some risk level, and who determines who loses?

Regardless of how the reader feels about any of these systems, there is no argument that the answers are matters of great social interest and concern. At some point, if there are to be public services at all, the questions must be answered and decisions made. But who makes them and on what basis? Who and where is "the architect" in each case? How and where is the architecture created? How is the public interest expressed and furthered?

The second economics insight is comparably powerful: *In any resource-limited situation, the true value of a given service or product is determined by what a buyer is willing to give up to obtain it.* Notice that the subject here is *value,* not price or cost.

> *Example*: The public telephone network provides a good example of the difference between cost and value. The cost of a telephone call can be accurately calculated as a function of time, distance, routing (satellite, cable, cellular, landline, and so forth), location (urban or rural), bandwidth, facility depreciation, and so on. But the value depends on content, urgency, priority, personal circumstance, and message type, among other things. As an exercise, try varying these parameters and then estimating what a caller might be willing to pay (give up in basic needs or luxuries). What is then a fair allocation of costs among all users? Should a sick, remote, poor caller have to pay the full cost of remote TV health service, for example? Should a business that can pass costs on to its customers receive volume discounts for long-distance calling via satellite? Should home TV be pay-per-view for everyone? Who should decide on the answers?

These two socioeconomic heuristics, used together, can alleviate the inherent tensions among the stakeholders by providing the basis for compromise and consensus among them. The complainants are likely to be those whose payments are perceived as disproportionate to the benefits they receive. The advocates, to secure approval of the system as a whole, must give up or pay for something of sufficient value to the complainants that they agree to compromise. Both need to be made to walk in the other's shoes for a while. And therein can be the basis of an economically viable solution.

The Interaction between the Public and Private Sectors

A third factor in sociotechnical systems architecting is the strong interplay between the public and private sectors, particularly in the advanced democracies where the two sectors are comparable in size, capability, and influence — but differ markedly in how the general public expresses its preferences.

By the middle of the 1990s, the historic boundaries between public and private sectors in communications, health delivery, welfare services, electric power distribution, and environmental control were in a state of flux. This chapter is not the place to debate the pros and cons. Suffice it to say, the imperatives, interests, and answers to the economists' questions are sharply different in the two sectors.[2] The architect is well advised to understand the imperatives of both sectors prior to suggesting architectures that must accommodate them. For example, the private sector must make a profit to survive; the public sector does not and treats profits as necessary evils. The public sector must follow the rules; the private sector sees rules and regulations as constraints and deterrents to efficiency. Generally speaking, the private sector does best in providing well-specified things at specified times. The public sector does best at providing services within the resources provided.

Because of these differences, one of the better tools for relieving the natural tension between the sectors is to change the boundaries between them in such negotiable areas as taxation, regulation, services provided, subsidies, billing, and employment. Because perceived values in each of these areas are different in the two sectors and under different circumstances, possibilities can exist where each sector perceives a net gain. The architect's role is to help discover the possibilities, achieve balance through compromise on preferences, and assure a good fit across boundaries.

Architecting a system, such as a public health system, that involves both the public and private sectors can be extraordinarily difficult, particularly if agreement does not exist on a mutually trusted architect, on the answers to the economist's questions, or on the social value of the system relative to that of other socially desirable projects. The problem is only exacerbated by the fundamental difficulties of diverse preferences. Public-sector projects look for a core of common agreement and an absence of "losers" sufficient to generate a powerful negative constituency. Private ventures look for segments of users (more is often better) with the resources to fund their own needs.

Facts versus Perceptions: An Added Tension

Of all the distinguishing characteristics of social systems, the one that most sharply contrasts them with other systems is the tension between facts and perceptions about system behavior. To illustrate the impact on design, consider the following: Architects are well familiar with the trade-offs between performance, schedule, cost, and risk. These competing factors might be thought of as pulling architecting four different directions as sketched in

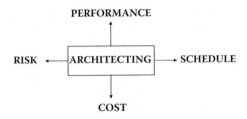

Figure 5.1 Four basic tensions in architecting.

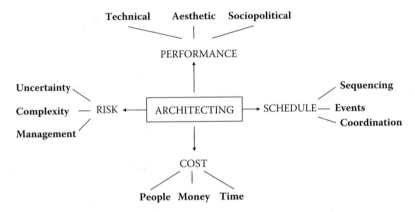

Figure 5.2 Underlying sources of the four tensions.

Figure 5.1 and Figure 5.2 can be thought of as the next echelon or ring — the different sources or components of performance, schedule, cost, and risk. Notice that performance has an aesthetic component as well as technical and sociopolitical sources. Automobiles are a clear example. Automobile styling often is more important than aerodynamics or environmental concerns in their architectural design. Costs also have several components, of which the increased costs to people of cost reduction in money and time are especially apparent during times of technological transition, and so on.

To these well-known tensions must be added another, one that social systems exemplify but which exist to some degree in all complex systems — namely, the tension between perceptions and facts, shown in Figure 5.3. Its sources are shown in Figure 5.4. This added tension may be dismaying to technically trained architects, but it is all too real to those who deal with public issues. Social systems have generated a painful design heuristic: *It is not the facts; it is the perceptions that count.* Some real-world examples include the following:

- It makes little difference what facts nuclear engineers present about the safety of nuclear power plants, their neighbors' perception is

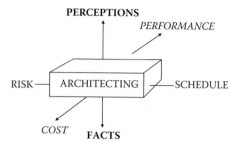

Figure 5.3 Adding facts versus perceptions.

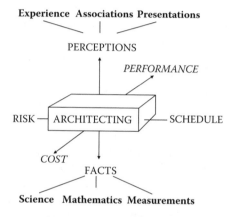

Figure 5.4 Sources of facts and perceptions.

that someday their local plant will blow up. Remember Three Mile Island and Chernobyl? A. M. Weinberg of Oak Ridge Associated Universities suggested perhaps the only antidote: "The engineering task is to design reactors whose safety is so transparent that the skeptical elite is convinced, and through them the general public."[3]

- Airline travel has been made so safe that the most dangerous part of travel can be driving to and from the airport. Yet, every airliner crash is headline news. A serious design concern, therefore, is how many passengers an airliner should carry — 200? 400? 800? — because even though the average accident rate per departure would probably remain the same,[4] more passengers would die at once in the larger planes and a public perception might develop that larger airliners are less safe, facts not withstanding.

- One of the reasons that health insurance is so expensive is that health care is perceived by employees as nearly "free" because almost all its costs are paid for either by the employee's company or the government. The facts are that the costs are either passed on to the consumer, subtracted from wages and salaries, taken as a business deduction against taxes, or paid for by the general taxpayer, or all of the above. As any economist will explain, free goods are overconsumed.

- One of the most profound and unanticipated results of the Apollo flights to the Moon was a picture of the Earth from afar, a beautiful blue, white, brown, and green globe in the blackness of space. We certainly had understood that the Earth was round, but that distant perspective changed our perception of the vulnerability of our home forever, and with it, our actions to preserve and sustain it. Just how valuable was Apollo, then and in our future? Is there an equivalent value today?

Like it or not, the architect must understand that perceptions can be just as real as facts, just as important in defining the system architecture, and just as critical in determining success. As one heuristic states, *The phrase, "I hate it," is direction.*[5] There have even been times when, in retrospect, perceptions were "truer" than facts that changed with observer, circumstance, technology, and better understanding. Some of the most ironic statements begin with, "It can't be done, because the facts are that..."

Alleviating the tension between facts and perceptions can be highly individualistic. Some individuals can be convinced — in *either* direction — by education, some by prototyping or anecdotes, some by A. M. Greenberg's antidote given earlier, some by better packaging or presentation, and some only by the realities of politics. Some individuals will never be convinced, but they might be accepting. In the end, it is a matter of achieving a balance of perceived values. The architect's task is to search out that area of common agreement that can result in a desirable, feasible system.

Looking more broadly, this is just a strengthened version of the basic admonition that an architect must know his or her client and what communicates to that client. It does no good to communicate precise and accurate representations that the client does not understand. Some clients are convinced only by prototypes. Some are convinced by analyses. In any case, the architect must understand what the audience in the domain of interest understands and will accept.

Heuristics for Social Systems

- Success is in the eyes of the beholder (not the architect).
- Do not assume that the original statement of the problem is necessarily the best, or even the right one. (Most customers would agree.)
- In conceptualizing a social system, be sure there are mutually consistent answers to the Four Who's: Who benefits? Who pays? Who supplies (provides)? And, as appropriate, Who loses?

- In any resource-limited situation, the true value of a given service or product is determined by what one is willing to give up to obtain it.
- The choice between the architectures may well depend upon which set of drawbacks the stakeholders can handle best. (Not on which advantages are the most appealing.)
- Particularly for social systems, it is not the facts, it is the perceptions that count. (Try making a survey of public opinion.)
- The phrase, "I hate it." is direction. (Or were you not listening?)
- In social systems, *how* you do something may be more important than *what* you do. (A sometimes bitter lesson for technologists to learn.)
- When implementing a change, keep some elements constant as an anchor point for people to cling to. (At least until there are some new anchors.)
- It is easier to change the technical elements of a social system than the human ones. (Enough said.)

Conclusion

Social systems, in general, place social concerns ahead of technical ones. They exemplify the tension between perception and fact. More than most systems, they require consistent responses to questions of who benefits? who pays? who supplies (provides, builds, and so forth), and, sociologically at least, who loses?

Perhaps more than other complex systems, the design and development of social ones should be amenable to insights and heuristics. Social factors, after all, are notoriously difficult to measure, much less predict. But, like heuristics, they come from experience, from failures as well as successes, and from lessons learned.

Exercises

1. Public utilities are examples of sociotechnical systems. How are the heuristics discussed in this chapter reflected in the regulation, design, and operation of a local utility system?
2. Apply the *four who's* to a sociotechnical system familiar to you. Examples: the Internet, air travel, communication satellites, a social service.
3. Many efforts are underway to build and deploy intelligent transport systems using modern information technologies to improve existing networks and services. Investigate some of the current proposals and apply the *four who's* to the proposal.
4. Pollution and pollution control are examples of a whole class of sociotechnical systems where disjunctions in the *four who's* are common. Discuss how governmental regulatory efforts, both through mandated standards and pollution license auctions, attempt to reconcile

the *four who's*. To what extent have they been successful? How did you judge success?

5. Among the most fundamental problems in architecting a system with many stakeholders is conflicts in purposes and interests. What architectural options might be used to reconcile them?

6. Give an example of the application of the heuristic, In introducing technological change, *how* you do it is often more important than *what* you do.

Notes and References

1. Lang, Jon, *Creating Architectural Theory, The Role of the Behavioral Sciences in Environmental Design.* New York: Van Nostrand Reinhold, 1987.

2. See Rechtin, E., *Systems Architecting, Creating and Building Complex Systems.* Englewood Cliffs, NJ: Prentice Hall, Organizations as Purposeful Systems, pp. 270–274, 1991; Rechtin, E., Why Not More Straight Commercial Buying, *Government Executive,* pp. 46–48, October 1976.

3. Weinberg, Alvin M., Engineering in an Age of Anxiety: The Search for Inherent Safety. *Engineering and Human Welfare NAE 25, Proceedings of the 25th Annual Meeting*, Washington, DC: National Academy of Engineering, 1990.

4. U.S. Airline Safety, Scheduled Commercial Carriers, *The World Almanac® and Book of Facts,* 1994, Funk and Wagnalls Corporation ©1993. According to the National Transportation Safety Board source, the fatal accidents per 100,000 miles have remained at or less than 0.100 from 1977 through 1992 despite a 40% increase in departures, major technological change, and the addition of increasingly larger aircraft to the airline fleets.

5. Gradous, Lori I., University of Southern California, October 18, 1993.

Case Study 4: Hierarchical to Layered Systems

A core concept we will encounter in Chapter 6 is the concept of a layered rather than a hierarchical system. Software is naturally constructed as a layered system rather than in the classic hierarchy, the basic paradigm of systems engineering. As with the other chapters in Part II, we introduce the core concepts with examples taken from life before proceeding with the chapter. The case study in this section differs from some of the others in that it is not drawn from a single, named system. For this case study, it has been more convenient to combine and abstract a number of stories the authors have encountered over time. The individual stories either illustrate only a limited range of issues or are not available to publish with full acknowledgment. Nevertheless, a reader with experience should have little trouble drawing parallels in his or her own personal experiences. The basic stories and issues in making the hierarchical to layered transition are encountered consistently.

Key points to consider in this case study include the following:

- The contrasting logic of layered versus hierarchical construction. In each, what constitutes components, what constitutes connectors between components, and how does each relate to others?
- The technical structure is (or should be) a reflection of business strategy. Choosing a layered architecture is foremost a business, or operational, strategic choice.
- The implementation, and the means of implementation, matter a great deal in whether or not the business strategy is realized. Simply converting a hierarchical architecture to a layered architecture does not embody a coherent business strategy. Implementation of the strategy requires details of the implementation (a repeat of the heuristic of variable technical depth, but in a different guise).

Many of these points are echoed and expanded in later chapters. For example, the relationship between business strategy and architecture is studied in depth in Chapter 12. We introduce many of the key points here that we will return to at greater length in later chapters.

Business Background

Our fictitious case study company, MedInfo, makes a wide variety of medical imaging systems, including conventional x-ray, computed tomography

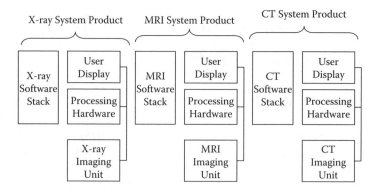

Figure CS4.1 MedInfo's initial situation. Multiple products are structured as stand-alone systems. Their nature as a product line is restricted to marketing and branding.

(CT), magnetic resonance imaging (MRI), and others. The systems are sold to hospitals and clinics, both in the United States and internationally. Wherever one of the MedInfo systems is deployed, it will be integrated into the user's technical infrastructure, at least so far as possible. At the beginning point in this story, the systems are structured as "stovepipes"; that is, each system is designed, manufactured, sold, and operated as its own, stand-alone, system. This is illustrated in Figure CS4.1.

The progression of business has been dominated by steady upgrades to the individual systems and occasional introduction of new imaging systems. The upgrade path is what one would expect, additional user features, lowered cost, greater throughput, enhanced sensitivity or coverage area, and so forth. The management model for the product family is likewise simple and as one would expect. Each of the products has an individual product manager. That manager is responsible end-to-end for that product. The manager leads design efforts, runs development and production, and is responsible for field performance, maintenance, and support. Although each product manager has many subordinates responsible for different aspects, all responsibility for that product ultimately lies with the product manager.

Each system has associated with it a supply chain of subcontractors and other suppliers. The subsystems or components supplied are each defined through specifications and interface control documents, written as needed based on the patterns of interconnecting each system.

Motivation for Change

If MedInfo has a solid record of success with things as they are, what motivation is there for change, especially relatively radical, architectural change? The motivation for change is clearly not incremental improvement. Incremental, steady improvement is clearly possible, and is already being realized, with the current architecture. However, business strategy issues are pushing MedInfo toward restructuring their family of products. The business strategic drivers for change are as follows:

Software cost reduction: MedInfo management has noticed that the fraction of development cost expended on software has risen steadily, and now tops 70%. What used to be hardware-dominated products are now software-dominated products. The shift comes from multiple causes. The first is the continued commodification of hardware. Custom processors have disappeared, and larger and more complex hardware units are available through subcontracting. Second, and related, is that competitive differentiation is increasingly software based. When competitors have access to the same hardware components, it is possible to competitively differentiate only through software. User demands are also increasingly about software capabilities, such as processing algorithms, display forms, user customization, and the ability to support process automation. A major source of user demand, and a source of the movement toward higher value fractions in software, is the need for interconnection and integration.

User demand for interconnection and integration: Users are increasingly dissatisfied with stand-alone systems. A radiologist might need access to five different imaging technologies during a day, and have to report from any or all of them on a hospital network. Few radiologists (much less other types of doctors) are happy with five computers on their desks and with manual file transfer among systems. Users are increasingly demanding both interconnection and integration. A very simple form of integration is collapsing the number of displays and computers on the desk needed to access multiple imaging systems. A form of interconnection and integration is the ability to move data from different imaging systems onto a common reporting platform. A complex form of integration is being able to combine, overlay, and otherwise jointly process images from different systems. The most complex form of integration is where integration leads itself to new products and new concepts of operation by customers. An example here would be integrating medical imaging data into multidisciplinary diagnostic decision support systems.

Rate of product turnover: In MedInfo's world, as in many other product spaces, there is increasing pressure to turn over products. User expectations for product cycles are getting shorter. As MedInfo's competition works to lower development cycle time, MedInfo is forced to match.

Lateral and vertical product space expansion: Finally, the pressure to grow makes it important to continuously challenge the horizontal and vertical boundaries of the product space. If MedInfo machines are going to be integrated into larger medical information systems, then failing to move one's own boundary outward to encompass some of that larger information space leaves one open to being laterally consumed. If the processing and user interface side of a MedInfo imager is subsumed into a shared information system, the information system supplier will want to capture that part of the value stream and push MedInfo back to being a narrower hardware supplier. Integrated system markets can easily become "winner take all" markets, meaning one better try to be the winner.

The Layered Alternative

As MedInfo systems become software dominated (in development cost), and increasingly interconnected, it becomes obvious that different products are sharing a great deal of software. Networking, data storage and transformation, significant processing, and much user interface code are either the same, or easily could be the same. Achieving integration is largely a matter of achieving protocol sharing, which is often most easily accomplished by code sharing. Systems that build with layers that carefully isolate the parts with the greatest potential to change from each other, through relatively invariant layers, generally have the highest rates of effective change. But, hierarchical system decomposition with end-to-end product managers with complete vertical responsibility does not encourage the discovery and management of that shared code.

An alternative architecture for the MedInfo product line that emphasizes horizontal layering is illustrated in Figure CS4.2. In this variant, the actual deployed products may or may not look different than before. If a customer desired a stand-alone system, he will receive a stand-alone system. On the other hand, if a customer desires that the system display on a shared display system, then that can be accommodated. In either case, the same subsystems are present as before, but now those subsystems are drawn from a shared base. There is extensive, designed-in code sharing among the different elements of the family. The shared elements form layers.

In a classic hierarchy, a lower-level element is a "part of" a higher-level element. This is the relationship among the elements of a chair. The legs of the chair, the seat of the chair, and the back of the chair are all parts of the chair. A set of chair legs belongs to exactly one chair (although the design and manufacturing of those legs may be shared across many identical chairs). In a layered system, a lower-layer element provides services to a higher-layer element. The lower element does not belong to the upper; it is

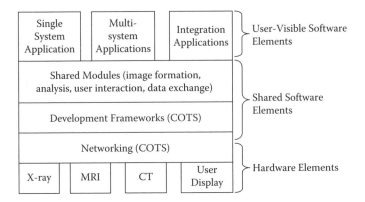

Figure CS4.2 Transformed structure of MedInfo product line. The product line allows "mixing and matching" of hardware elements and assembles end-to-end applications from a large base of shared modules and commercial products.

used by the upper. A layered architecture has a component relationship of the "uses" form instead of the "part-of" form.

The layered model is borrowed originally from communication networks, where it originated as the well-known seven-layer model. The seven-layer model of the ISO Open System Interconnect (OSI) standard is now of largely historic interest, having been replaced by the 5+ layered model of the Internet. In both, the lower four layers (physical, data link, network, and transport) and the top layer (application) are largely the same. What is different is what is in between. The original OSI model defined two specific in-between layers, the session and the presentation. In practice, these are not used. Many Internet applications simply are written directly onto the transport layer. In modern development libraries, and in this case study, the middle area is occupied by various forms of middleware (for example, message servers, .NET, Common Object Request Broker Architecture [CORBA®]).

Various real companies, and our abstracted MedInfo company, have made the transition from a hierarchical to a layered architecture. One can easily find reports on how the transition went that emphasize the following:

1. The success of the transition is critical to realizing the business strategic objectives (those discussed above).
2. The transition was intensely traumatic for staff and management, leading to extensive attrition and financial difficulties while trying to carry it off.

If the benefits to MedInfo of making this transition are clear, what are the sources of pain?

The Pain of the Transition

The first source of pain is in how end-to-end management responsibility changes. In a stovepiped world, the product manager has everything necessary within his or her scope of responsibility. When there are problems, there is no doubt where to go to demand a fix, and one point of decision on how to make the fix. Once the fix is made, the scope of its impact is on the product for which the manager is responsible. In the layered construct, the situation becomes more complex.

In the layered construct, the end-user product is now assembled out of components shared across the product family. It may be delivered, in part, on platforms out of the product manager's responsibility. For example, in the layered construct, it may be that an x-ray imager is delivered by delivering the imaging hardware and software, but that the software and user interface reside entirely on computers shared with other imaging systems, imaging systems that may not be made by MedInfo (in a more highly integrated case). A large part of the imaging and user interface software is shared with other products in the MedInfo family. Being able to do this is a major part of the stated MedInfo business strategy.

When things go wrong, either during development or in deployment, who is responsible for the fix? The product manager no longer has vertical control over the elements that combine to produce a valuable product. If a change is required in the shared code base, that change could conceivably impact all of the other products that use the shared elements. Various companies and government development groups have reported that handling this diffusion of end-to-end responsibility was the most difficult aspect of the change to a layered architecture. It is not practical to make the first convergence point for technical issues across the different products the chief executive officer (CEO). There has to be a point to resolve the issues lower down, but conceivably there is no point of common financial responsibility lower down. It is not hard to institute cross-product or cross-functional engineering teams, but it is likewise not hard to make such teams toothless when all financial accountability resides elsewhere.

Related to management responsibility is how MedInfo must do quality management. In the stovepiped construct, quality can be managed product by product. The quality requirements can come directly from the expectations of the market for each product. But, how do we do quality management for shared infrastructure components? Granted, coming up with measures is no problem for the experienced, but where do the thresholds lie, given that quite different thresholds might apply to different products within the family? If trade-offs for shared components resolve quite differently in different product applications, which trade-off should be selected? And, how do we enforce standards when those standards do not directly relate to delivered customer quality but do have immediate financial consequences (probably bad)?

Again, various companies have similarly reported on the difficulty of these issues and on how they were successfully dealt with. One heuristic that stands out is

> *Subsystem quality requirements that are sufficient when the component is an element of a stovepiped system are unlikely to be sufficient when the component is shared across many systems. Or, the quality requirements on the components of a shared layer are likely to be much more demanding than when those components are not shared.*

More difficult quality requirements may require new quality assessment tools. Some highlights have included the following:

- The transition to a layered, family-of-systems architecture drove the development and adoption of a massively parallel and automated software regression testing system. All unit-level regression tests needed to be run automatically over a weekend (and they were run every weekend).
- All heavily shared libraries were required to be written with assertion statements on all function entrances and exits. All designs with assertions must be formally reviewed before production. All software had to be tested with the assertions compiled. Any calling function that causes an assertion to fail is assumed to be at fault and must correct itself.

A related problem in end-to-end management is in how subcontracting or outsourcing is organized. In the hierarchical construct, subcontracting tends to follow physical lines. A subcontractor delivers a box or a board, and the specification is written at the box or board level. In a layered system, one can likewise imagine doing the subcontracting of a whole layer, or components within a layer. But, new difficulties are introduced, such as

- The specification for a layer typically looks nothing like the specification for a box. Is the expertise available in-house to write specifications and manage subcontracts when the interfaces change in dramatic ways?
- Testing and integration of layered elements presupposes access to the shared programming libraries. How will shared programming environments be managed with a subcontractor? If the prime contractor has selected some overall software framework to facilitate

integration, will all subcontractors buy licenses to the same frame-work? Is that financially feasible? How will the configurations of the separately purchased frameworks be managed to ensure compatibility?

- What happens when a subcontractor supplying a component that cuts across the whole family-of-systems goes out of business, or decides to drop support, or simply releases a poor-quality version?

All of these, leaving aside the detailed technical issues, fall generally under the heading of management culture and skills. It is not as if there are no solutions to these issues, many companies and government depart-ments have encountered them and solved them. The impact on manage-ment culture and practices is most likely when companies frequently report high attrition as a cost of transition. In one case known to the author, the chief operating officer (COO) of a major company responded when asked how his company had successfully managed a stovepipe to layered transition that others had failed at, "We became successful when management attrition reached 50%." Unfortunately, this does not appear to be uncommon.

Results

Solving the problems imposed by changing architectures is possible, but typically quite painful. Is the solution and the pain of arriving at the solution worth the strategic gains? In our composite example, MedInfo answers "yes," but with a certain degree of qualification. The transition from a stovepiped to layered system addresses the business objectives, at least it can when the devilish details are worked out.

- An effective layered architecture can drop the total line of code count across a family of systems. If the total size is dropped, cost and devel-opment time advantages can be expected to follow. However, even where there is a software size savings, the savings can be lost if the newer development environment has higher overhead, is much more expensive, or if access constraints make development more difficult.
- A layered architecture can allow much more complete integration among elements of a product line, when all elements of the line have made the transition. The end point might be very integrated, but it might be a very long march to get to the point at which significant benefits are realized. Management needs to know where the cut-over point is to make a rational decision.
- If the layered architecture effectively isolates areas of change from each other, it can allow for much faster product evolution. The key is good choice of invariants. The invariants must flow from a wise

identification of things that change, and where an invariant structure can isolate change. The Transmission Control Protocol/Internet Protocol (TCP/IP) are an outstanding example.

- The transition is almost invariably very painful. The pain is related much more to the difficulties of the human enterprise than to inherent difficulties in the technologies involved. The new architecture is not more complex than the old one; it is simply different, and many success factors relevant to the old one must be replaced before the new one can be equally as successful.

chapter 6

Software and Information Technology Systems

> Today I am more convinced than ever. Conceptual
> integrity is central to product quality. Having a
> system architect is the most important step toward
> conceptual integrity.

Frederick P. Brooks, Jr.
The Mythical Man-Month after Twenty Years

Introduction: The Status of Software Architecting

Software is rapidly becoming the centerpiece of complex system design, in
the sense that an increasing fraction of system performance and complexity is captured in software, and that software considerations drive overall
system development. Software is increasingly the portion of the system that
enables the unique behavioral characteristics of the system. Competitive
developers of end-user system products find themselves increasingly
developing software, even though the system combines both hardware
and software. The reasons stem from software's ability to create intelligent behavior and quickly to accommodate technical-economic trends in
hardware development. This capability of software is matched against
increasing maturity in many other fields containing complex systems. As
examples, the physical architectures of aircraft have been slowly varying
since 1970, and the physical architectures of spacecraft have been slowly
varying since at least 1990.

Although detailed quantitative data are hard to come by, anecdotal
stories tell a consistent story. A wide variety of companies in different
industries (for example, telecommunications, consumer electronics, industrial controls) have reported a dramatic shift in the relative engineering
efforts devoted to hardware and software.* Where 15 to 20 years ago the
ratio was typically 70% hardware and 30% software, it is now typically

---

* The numbers are anecdotal but reflect private communications to one of the present
  authors from a wide variety of sources.

reversed, 30% hardware and 70% software. And the software fraction is continuing to grow. This should not be surprising, given how the semiconductor industry has changed. Where product developers used to build from relatively simple parts (groups of logic gates), they now use highly integrated microprocessors with most peripheral devices on the chip. The economies of scale in semiconductor design and production have pushed the industry toward integrated solutions where the product developer primarily differentiates through software. Moreover, microcontrollers have become so inexpensive and have such low power consumption that they can be placed in nearly any product, even throwaway products. The microprocessor-based products acquire their functionality by the software that executes on them. The product developer is transformed from a hardware designer to a hardware integrator and software developer. As software development libraries become larger, more capable, and accepted, many of the software developers will be converted to software integrators.

The largest market for software today is usually termed "information technology," which is a term encompassing the larger domain of computers and communications applied to business and public enterprises. We consider both here, as software architecture as a field is becoming a distinct specialty. What is usually called software architecture, at least in the research community, is usually focused on developing original software rather than building information-processing systems through integration of large software and hardware components. Information technology practice is less and less concerned with developing complete original applications and more and more concerned with building systems through integration. What is usually called enterprise architecture, to the extent that it is dealing with the architecture of software, is normally dealing with integrating large, preexisting software applications.

The focus of this chapter is less on the architecting of software (though that is discussed here and in Part III) than it is on the impact of software on system architecting. Software possesses two key attributes that affect architecting. First, well-architected software can be very rapidly evolved. Evolution of deployed software is much more rapid than evolution of deployed hardware, because an installed base of software can be regularly replaced at moderate cost. The cost of "manufacturing" software is essentially zero (although the cost of certifying it for use may be high), and so unlike in hardware systems, regular total replacement is efficient. As a result, annual and even quarterly replacement is common. Annual or more frequent field software upgrades are normal for operating systems, databases, end-user business systems, large-scale engineering tools, and communication and manufacturing systems. This puts a demanding premium on software architectures because they must be explicitly designed to accommodate future changes and to allow repeated certification with those changes.

Second, software is an exceptionally flexible medium. Software can easily be built which embodies many logically complex concepts such as layered languages, rule-driven execution, data relationships, and many others. This flexibility of expression makes software an ideal medium with which to implement system "intelligence." In both the national security and commercial worlds, intelligent systems are far more valuable to the user and far more profitable for the supplier than their simpler predecessors.

In addition, a combination of technical and economic trends favor building systems from standardized computer hardware and system-unique software, especially when computing must be an important element of the system. Building digital hardware at very high integration levels yields enormous benefits in cost per gate but requires comparably large capital investments in design and fabrication systems. These costs are fixed, giving a strong competitive advantage to high production volumes. Achieving high production volumes requires that the parts be general purpose. For a system to reap the benefits of very high integration levels, its developers must either use the standard parts (available to all other developers as well) or be able to justify the very large fixed expense of a custom development. If standard hardware parts are selected, the remaining means to provide unique system functionality is software.

Logically, the same situation applies to writing software. Software production costs are completely dominated by design and test. Actual production is nearly irrelevant. So, there is likewise an incentive to make use of large programming libraries or components and amortize the development costs over many products. In fact, this is already taking place. Even though much of the software engineering community is frustrated with the pace of software reuse, there are many successful examples. One obvious one is operating systems. Very few groups who use operating systems write one from scratch anymore. Either they use an off-the-shelf product from one of the remaining vendors, or they use an open source distribution and customize it for their application. Databases, scripting languages, and Web applications are all examples of successful reuse of large software infrastructures.

The rapid proliferation of open source software is likewise a successful example of wide-scale software reuse. When the source code is completely open and available for modification and redistribution, many groups have built vigorous communities of developers and users. The availability of the source code, and the licensing terms for redistribution, appear to be key to making this form of reuse work, as is the quality of the design.

A consequence of software's growing complexity and central role is recognition of the importance of software architecture and its role in system design. An appreciation of sound architectures and skilled architects is broadly accepted. The soundness of the software architecture will strongly influence the quality of the delivered system and the ability of the developers to further evolve the system. When a system is expected to undergo extensive evolution after deployment, it is usually more important that the system be easily evolvable than that it be exactly correct at first deployment.

Software architecture is frequently discussed, from both academic and industrial perspectives.[1] Within the software architecture community, there is limited consensus on the borders of what constitutes "architecture." Many groups focus on architecture as high-level physical structure, primarily of source code. A distillation of commonly used ideas is that the architecture is the overall structure of a software system in terms of components and interfaces. This definition would include the major software components, their interfaces with each other and the outside world, and the logic of their execution (single threaded, interrupted, multithreaded, combination). To this is often added principles defining the system's design and evolution, an interesting combination of heuristics with structure to define architecture. A software architectural "style" is seen as a generic framework of components or interfaces that defines a class of software structures. The view taken in this book, and in some of the literature,[2] is more expansive than just high-level physical structure, and includes other high-level views of the system: behavior, constraints, and applications as well.

High-level advisory bodies to the Department of Defense are calling for architects of ballistic missile defense, C4I (command, control, communications, computers, and intelligence), global surveillance, defense communications, Internetted weapon systems, and other "systems-of-systems." Formal standards have been developed, defining the role, milestones, and deliverables of system architecting. Many of the ideas and terms of those standards come directly from the software domain, for example, object-oriented, spiral process model, and rapid prototyping. The carryover should not be a surprise; the systems for which architecting is particularly important are behaviorally complex, data intensive, and software rich. Examples of software-centered systems of similar scope are appearing in the civilian world, such as the information superhighway, the Internet, global cellular telephony, health care, manned space flight, and flexible manufacturing operations.

The consequences to software design of this accelerating trend to smarter systems are now becoming apparent. For the same reason that guidance and control specialists became the core of systems leadership in the past, software specialists will become the core in the future. In the systems world, software will change from having a support role (usually after the hardware design is fixed) to becoming the centerpiece of complex systems design and operation. As more of the behavioral complexity of systems is embodied in software, software will become the driver of system configuration. Hardware will be selected for its ability to support software instead of the reverse. This is now common in business information systems and other applications where compatibility with a software legacy is important.

If software is becoming the centerpiece of system development, it is particularly important to reconcile the demands of system and software development. Even if 90% of the system-specific engineering effort is put into software, the system is still the end product. It is the system, not the software inside, the client wishes to acquire. The two worlds share many common roots, but their differing demands have led them in distinctly different directions. Part of the role of systems architecting is to bring them together in an integrated way.

Software engineering is a rich source for integrated models; models that combine, link, and integrate multiple views of a system. Many of the formalisms now used in systems engineering had their roots in software engineering. This chapter discusses the differences between system architecting and software architecting, current directions in software architecting and architecture, and heuristics and guidelines for software. Chapter 10 provides further detail on four integrated software modeling methods, each aimed at the software component of a different type of system.

Software as a System Component

How does the architecture and architecting of software interact with that of the system as a whole? Software has unique properties that influence overall system structure:

1. Software provides a palette of abstractions for creating system behavior. Software is extensible through layered programming to provide abstracted user interfaces and development environments. Through the layering of software, it is possible to directly implement concepts such as relational data, natural language interaction, and logic programming that are far removed from their computational implementation. Software does not have a natural hierarchical structure, at least not one that mirrors the system-subsystem-component hierarchy of hardware.

2. It is economically and technically feasible to use evolutionary delivery for software. If architected to allow it, the software component of a deployed system can be completely replaced on a regular schedule.
3. Software cannot operate independently. Software must always be resident on some hardware system and, hence, must be integrated with some hardware system. The interaction between, and integration with, this underlying hardware system becomes a key element in software-centered system design.

For the moment there are no emerging technologies that are likely to take software's place in implementing behaviorally complex systems. Perhaps some form of biological or nano-agent technology will eventually acquire similar capabilities. In these technologies, behavior is expressed through the emergent properties of chaotically interacting organisms. But the design of such a system can be viewed as a form of logic programming in which the "program" is the set of component construction and interface rules. Then the system, the behavior that emerges from component interaction, is the expression of an implicit program, a highly abstracted form of software.

System architecting adapts to software issues through its models and processes. To take advantage of the rich functionality, there must be models that capture the layered and abstracted nature of complex software. If evolutionary delivery is to be successful, and even just to facilitate successful hardware/software integration, the architecture must reconcile continuously changing software with much less frequently changing hardware.

Software for Modern Systems

Software plays disparate roles in modern systems. Mass market application software, one-of-a-kind business systems, real-time analysis and control software, and human interactive assistants are all software-centered systems, but each is distinct from the other. The software attributes of rich functionality and amenability to evolution match the characteristics of modern systems. These characteristics include the following:

1. Storage of, and semiautonomous and intelligent interpretation of, large volumes of information.
2. Provision of responsive human interfaces that mask the underlying machines and present their operation in metaphor.
3. Semiautonomous adaptation to the behavior of the environment and individual users.
4. Real-time control of hardware at rates beyond human capability with complex functionality.

5. Constructed from mass-produced computing components and unique system software, with the capability to be customized to individual customers.
6. Coevolution of systems with customers as experience with system technology changes perceptions of what is possible.

The marriage of high-level language compilers with general-purpose computers allows behaviorally complex, evolutionary systems to be developed at reasonable cost. Although the engineering costs of a large software system are considerable, they are much less than the costs of developing a pure hardware system of comparable behavioral complexity. Such a pure hardware system could not be evolved without incurring large manufacturing costs on each evolutionary cycle. Hardware-centered systems do evolve, but at a slower pace. They tend to be produced in similar groups for several years, and then make a major jump to new architectures and capabilities. The time of the jump is associated with the availability of new capabilities and the programmatic capability of replacing an existing infrastructure.

Layering of software as a mechanism for developing greater behavioral complexity is exemplified in the continuous emergence of new software languages and in Internet and Web applications being built on top of distributed infrastructures. The trend in programming languages is to move closer and closer to application domains. The progression of language is from machine level (machine and assembly languages) to general-purpose computing (FORTRAN, Pascal, C, C++, Ada) to domain specific (MATLAB, Visual Basic for Applications, dBase, SQL, PERL, and other scripting languages). At each level, the models are closer to the application, and the language components provide more specific abstractions. By using higher and higher level languages, developers are effectively reusing the coding efforts that went into the language's development. Moreover, the new languages provide new computational abstractions or models not immediately apparent in the architecture of the hardware on which the software executes. Consider a logic programming language like PROLOG. A program in PROLOG is more in the nature of hypothesis and theorem proof than arithmetic and logical calculation. But it executes on a general-purpose computer as invisibly as does a C or even FORTRAN program.

Systems, Software, and Process Models

An architectural challenge is to reconcile the integration needs of software and hardware to produce an integrated system. This is both a problem of representation or modeling and of process. Modeling aspects are taken up subsequently in this chapter, and in Part III. On the process side, hardware is best developed with as little iteration in production as possible, but software can (and often should) evolve through much iteration.

Hardware should follow a well-planned design and production cycle to minimize cost, with large-scale production deferred to as close to final delivery as possible (consistent with adequate time for quality assurance). But software cannot be reliably developed without access to the targeted hardware platform for much of its development cycle. Production takes place nearly continuously, with release cycles now often daily in many advanced development organizations.

Software distribution costs are comparatively so low that repeated complete replacement of the installed base is normal practice. When software firms ship their yearly (or more frequent) upgrades, they ship a complete product. Firms commonly "ship" patches and limited updates on the Internet, eliminating even the cost of media distribution. The cycle of planned replacement is so ingrained that some products (for example, software development tools) are distributed as a subscription; a quarterly CD-ROM or Internet download with a new version of the product, application notes, documentation, and prerelease components for early review.

In contrast, the costs of hardware are often dominated by the physical production of the hardware. If the system is mass produced, this will clearly be the case. Even when production volumes are very low, as in unique customized systems, the production cost is often comparable to or higher than the development cost. As a result, it is uneconomic, and hence impractical, to extensively replace a deployed hardware system with a relatively minor modification. Any minor replacement must compete against a full replacement, a replacement with an entirely new system designed to fulfill new or modified purposes.

One important exception to the rule of low deployment costs for software is where the certification costs of new releases are high. For example, one does not casually replace the flight control software of the Space Shuttle any more than one casually replaces an engine. Extensive test and certification procedures are required before a new software release can be used. Certification costs are analogous to manufacturing costs in that they are a cost required to distribute each release but do not contribute to product development.

Waterfalls for Software?

For hardware systems, the process model of choice is a waterfall (in one of its pure or more refined incarnations). The waterfall model development stages and tries to keep iterations local — that is, between adjacent tasks such as requirements and design. Upon reaching production, there is no assumption of iteration, except the large-scale iteration of system assessment and eventual retirement or replacement. This model fits well within the traditional architecting paradigm as described in Chapter 1.

Software can, and sometimes does, use a waterfall model of development. The literature on software development has long embraced the sequential paradigm of requirements, design, coding, test, delivery. But dissatisfaction with the waterfall model for software led to the spiral model and variants. Essentially all successful software systems are iteratively delivered. Application software iterations are expected as a matter of course. Weapon system and manufacturing software is also regularly updated with refined functionality, new capabilities, and fixes to problems. One reason for software iterations is to fix problems discovered in the field. A waterfall model tries to eliminate such problems by doing a very high quality job of the requirements. Indeed, the success of a waterfall development is strongly dependent on the quality of the requirements. But in some systems, the evolvability of software can be exploited to reach the market faster and avoid costly, and possibly fruitless, requirements searches.

> *Example*: Data communication systems have an effective requirement of interoperating with whatever happens to be present in the installed base. Deployed systems from a global range of companies may not fully comply with published standards, even if the standards are complete and precise (which they often are not). Hence, determining the "real" requirements to interoperate is quite difficult. The most economical way to do so may be to deploy to the field and compile real experience. But that, in turn, requires that the systems support the ability to determine the cause of interoperation problems and be economically modifiable once deployed to exploit the knowledge gained.

But, in contrast, a casual attitude toward evolution in systems with safety or mission-critical requirements can be tragic.

> *Example*: The Therac 25 was a software-controlled radiation treatment machine in which software and system failures resulted in six deaths.[3] It was an evolutionary development from a predecessor machine. The evidence suggests that the safety requirements were well understood but that the system and software architectures both failed to maintain the properties. The system architecture was flawed in that all hardware safety interlocks (which had been present in the predecessor model) were removed, leaving software checks as the sole

safety safeguard. The software architecture was
flawed because it did not guarantee the integrity of
treatment commands entered and checked by the
system operator.

One of the most extensively described software development problems is customized business systems. These are corporate systems for accounting, management, and enterprise-specific operations. They are of considerable economic importance, are built in fairly large numbers (though no two are exactly alike), and are developed in an environment relatively free of government restrictions. Popular and widely published development methods have strongly emphasized detailed requirements development followed by semiautomated conversion of the requirements to program code — an application-specific waterfall.

Even though this waterfall is better than ad hoc development, results have been disappointing. In spite of years of experience in developing such business systems, large development projects regularly fail. As Tom DeMarco has noted,[4] "somewhere, today, an accounts payable system development is failing" in spite of the thousands of such systems that have been developed in the past. Part of the reason is the relatively poor state of software engineering compared to other fields. Another reason is failure to make effective use of methods known to be effective. An important reason is the lack of an architectural perspective and the benefits it brings.[5]

The architect's perspective is to explicitly consider implementation, requirements, and long-term client needs in parallel. A requirements-centered approach assumes that a complete capture of documentable requirements can be transformed into a satisfactory design. But existing requirements modeling methods generally fail to capture performance requirements and ill-structured requirements like modifiability, flexibility, and availability. Even where these nonbehavioral requirements are captured, they cannot be transformed into an implementation in any even semiautomated way. And it is the nature of serious technological change that the impact will be unpredictable. As technology changes and experience is gained, what is demanded from systems will change as well.

The spiral model as originally described did not embrace evolution. Its spirals were strictly risk based and designed to lead to a fixed system delivery. Rapid prototyping envisions evolution, but only on a limited scale. Newer spiral model concepts do embrace evolution.[6] Software processes, as implemented, spiral through the waterfall phases but do so in a sequential approach to moving release levels. This modified model was introduced in Chapter 4, in the context of integrating software with manufacturing systems, and it will be further explored below.

Spirals for Hardware?

To use a spiral model for hardware acquisition is equivalent to repeated prototyping. A one-of-a-kind, hardware-intensive system cannot be prototyped in the usual sense. A complete "prototype" is, in fact, a complete system. If it performs inadequately, it is a waste of the complete manufacturing cost of the final system. Each one, from the first article, needs to be produced as though it were the only one. As was discussed previously, under the "protoflight" development strategy, the prototype is the final system. A true prototype for such a one-of-a-kind system must be a limited version or component intended to answer specific developmental questions. We would not "prototype" an aircraft carrier, but we might well prototype individual pieces and build subscale models for testing. The development process for one-of-a-kind systems needs to place strong emphasis on requirements development and attention to detailed purpose throughout the design cycle. Mass-produced systems have greater latitude in prototyping because of the prototype-to-production-cost ratio, but still have less than in software. However, the initial "prototype" units still need to be produced. If they are to be close to the final articles, they need to be produced on a similar manufacturing line. But setting up a complete manufacturing line when the system is only in prototype stage is very expensive. Setting up the manufacturing facilities may be more expensive than developing the system. As a hardware-intensive system, the manufacturing line cannot be easily modified, and leaving it idle while modifying the product to be produced represents a large cost.

Integration: Spirals and Circles

What process model matches the nature of evolutionary, mixed technology, behaviorally complex systems? As was suggested earlier, a spiral and circle framework seems to capture the issues. The system should possess stable configurations (represented as circles) and development should iteratively approach those circles. The stable configurations can be software release levels, architectural frames, or hardware configurations.

This process model matches the accepted software practice of moving through defined release levels, with each release produced in cycles of requirements-design-code-test. Each release level is a stable form that is used while the next release is developed. Three types of evolution can be identified. A software product, like an operating system or

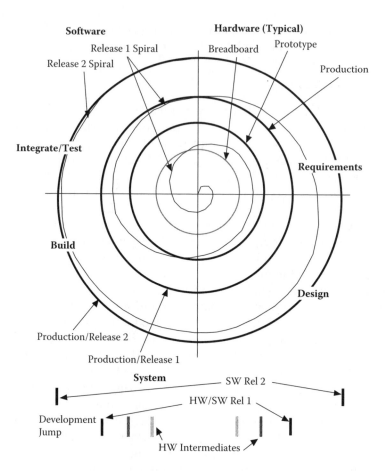

Figure 6.1 A typical arrangement of spirals and circles in a project requiring hardware and software integration. This illustrates the stable intermediate configurations of hardware (typically breadboard and prototype) integrating with a software spiral. Software is developed on the stable intermediate systems.

shrink-wrapped application, has major increments in behavior indicated by changes in the release number, and more minor increments by changes in the number after the "point." Hence, a release 7.2 product would be major version seven, second update. The major releases can be envisioned as circles, with the minor releases cycling into them. On the third level are those changes that result in new systems or re-architected old systems. These are conceptually similar to the major releases but represent even bigger changes. The process with software annotations is illustrated in Figure 6.1. By using a side view, one can envision the major releases as vertical jumps. The evolutionary spiral process moves out to the stable major configurations and then jumps up to the next major change.

In practice, evolution on one release level may proceed concurrently with development of a major change.

> *Example*: The Internet and World Wide Web provide numerous examples of stable intermediate forms promoting evolution. The architecture of the Internet, in the sense of an organizing or unifying structure, is clearly the Internet Protocol (IP), the basic packet switching definition. IP defines how packets are structured and addressed, and how the routing network interacts with the packets. It determines the kinds of services that can be offered on the Internet, and in so doing constrains application construction. As the Internet has undergone unprecedented growth in users, applications, and physical infrastructure, IP has remained stable. As of the writing of this book, the transition from version 4 to version 6 of IP is occurring very slowly. It likely will occur within a few years, but the nature of IPv4 underpins so much of how the Internet operates that transition is necessarily quite slow. The World Wide Web has similarly undergone tremendous growth and evolution on top of a simple set of elements, the Hypertext Transfer Protocol (HTTP) and the Hypertext Markup Language (HTML). Both the Internet and the World Wide Web are classic examples of systems with nonphysical architecture, a topic that becomes central in the discussion of collaborative systems in Chapter 7.

Hardware–software integration adds to the picture. The hardware configurations must also be stable forms but should appear at different points than the software intermediates on the development timeline. Some stable hardware should be available during software development to facilitate that development. A typical development cycle for an integrated hardware–software system illustrates parallel progressions in hardware and software with each reaching different intermediate stable forms. The hardware progression might be breadboard, production prototype, production, then (possibly) field upgrade. The software moves through a development spiral aiming at a release 1.0 for the production hardware. The number of software iterations may be many more than for the hardware. In late development stages, new software versions may be built weekly.[7] Before that there will normally be partial releases that run on the intermediate hardware forms (the breadboards and the production prototypes). Hardware–software codesign research is working toward environments in which developing hardware can be represented faithfully enough so that physical prototypes are unnecessary for early integration. Such tools may become available, but iteration through intermediate hardware development levels is still the norm in practice.

A related problem in designing a process for integration is the proper use of the heuristic: *Do the hard part first.* Because software is evolved or iterated, this heuristic implies that the early iterations should address the most difficult challenges. Unfortunately, honoring the heuristic is often difficult. In practice, the first iterations are often good-looking interface demonstrations or constructs of limited functionality. If interface construction is difficult or user acceptance of the interface is risky or difficult, this may be a good choice. But if operation to time constraints under loaded conditions is the key problem, some other early development strategy should be pursued. In that case, the heuristic suggests gathering realistic experimental data on loading and timing conditions for the key processes of the system. That data can then be used to set realistic requirements for components of the system in its production configuration.

> *Example*: Call distribution systems manage large numbers of phone personnel and incoming lines as in technical support or phone sales operation. By tying the system into sales databases, it is possible to develop sophisticated support systems that ensure that full customer information is available in real time to the phone personnel. To be effective, the integration of data sources and information handling must be customized to each installation and evolve as understanding of what information is needed and available develops. But, because the system is real time and critical to customer contact, it must provide its principal functionality reliably and immediately upon installation.

Thus, an architectural response to the problems of hardware–software integration is to architect both the process and the product. The process is manipulated to allow different segments of development to match themselves to the demands of the implementation technology. The product is designed with interfaces that allow separation of development efforts where the efforts need to proceed on very different paths. How software architecture becomes an element of system architecture, and more details on how this is to be accomplished, are the subjects to come.

The Problem of Hierarchy

A central tenet of classic systems engineering is that all systems can be viewed in hierarchies. A system is composed of subsystems that are composed of small units. A system is also embedded in higher-level systems in which it acts as a component. One person's system is another person's component. A basic strategy is to decompose any system into subsystems, decompose the requirements until they can be allocated to subsystems, carefully specify and control the interfaces among the subsystems, and repeat the process on every subsystem until you reach components you can buy or are the products of disciplinary engineering. Decomposition in design is followed by integration in reverse. First, the lowest-level components are integrated into the next-level subsystems, those subsystems are integrated into larger subsystems, and so on until the entire system is assembled.

Because this logic of decomposition and integration is so central to classical systems engineering, it is difficult for many systems engineers to understand why it often does not match software development very well. To be sure, some software systems are very effectively developed this way. The same logic of decomposition and integration matches applications built in procedural languages (like C or Pascal*) and where the development effort writes all of the application's code. In these software systems, the code begins with a top-level routine, which calls first-level routines, which call second-level routines, and so forth, to primitive routines that do not call others. In a strictly procedural language, the lower-level routines are contained within or encapsulated in the higher-level routines that use them. If the developer organization writes all the code, or uses only relatively low-level programming libraries, the decomposition chain terminates in components much like the hardware decomposition chain terminates. Like in the classical systems engineering paradigm, we can integrate and test the software system in much the same way, testing and integrating from the bottom-up until we reach the topmost module.

As long as the world looks like this, on both the hardware and software sides, we can think of system decompositions as looking like Figure 6.2. This figure illustrates the world, and the position of software, as classical systems engineers would portray it. Software units are contained within the processor units that execute them. Software is properly viewed as a subsystem of the processor unit.

However, if we instead went to the software engineering laboratory of an organization building a modern distributed system and asked the software engineers to describe the system hierarchy, we might get a very

---

* Strictly speaking, C is not a procedural language, and some of what follows does not precisely apply to it. Those knowledgeable in comparative programming languages can consider the details of the procedural versus object-oriented paradigms in the examples to come.

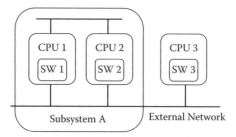

Figure 6.2 System/hardware hierarchy view of a system.

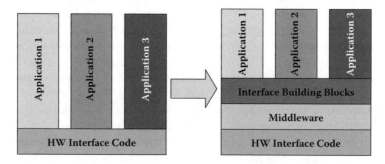

Figure 6.3 Increasingly, thick applications are being replaced by much thinner implementations that rely on thicker shared infrastructure layers. This transition is of high value but introduces problems in quality control and development methods often unfamiliar to groups accustomed to building thick applications.

different story. Much modern software is written using object-oriented abstractions, is built in layers, and makes extensive use of very large software infrastructure objects (like operating systems or databases) that do not look very much like simple components or the calls to a programming library. The transition is illustrated in Figure 6.3, and was discussed in the "Case Study 4" (prior to this chapter). When expanded to the level of interacting bodies of code, the world looks as illustrated in Figure 6.4. Each of these issues (object orientation and layering) creates a software environment that does not look like a hierarchical decomposition of encapsulated parts, and to the extent that a hierarchy exists, it is often quite different from the systems/hardware hierarchy. We consider each of these issues in turn.

Object Orientation

The software community engages in many debates about exactly what "object oriented" should mean, but only the fundamental concepts are important for *systems* architecting. An object is a collection of functions

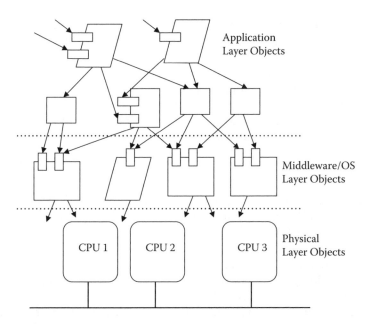

Figure 6.4 Layered software hierarchy view of a system.

(often called methods) and data. Some of the functions are public; that is, they are exposed to other software objects and can be called by them. Depending on the specific language and runtime environment, calling a function may be a literal function call, or it may simply mean sending a message to the target object, which interprets it and takes action. Objects can be "active"; that is, they can run concurrently with other objects. In some software environments, concurrent objects can freely migrate from computer to computer over an intervening network. Often the software developer does not know, and does not want to know, on which machine a particular object is running and does not want to directly control its migration. Concurrent execution of the objects is passed to a distributed operating system, which may control object execution through separately defined policies.

In object-oriented systems, the number of objects existing when the software executes can be indeterminate. An object has a defining "template" (although the word "template" means something slightly different in many object-oriented languages) known as a "class." A class is analogous to a type in procedural programming. So, just as one can declare many variables of type "float," so one can declare many objects corresponding to a given class. In most object-oriented languages, the creation of objects from classes happens at runtime, when the software is executing. If objects are not created until runtime, the number of them can be controlled by external events.

This is a very powerful method of composing a software system. Each object is really a computational machine. It has its own data (potentially a very large amount) and as much of its own program code as the class author decides. This sort of dynamic object-oriented software can essentially manufacture logical machines, in arbitrary numbers, and set them to work on a network, in response to events that happen during program execution. To compare this to classical notions of decomposition, it is as though one could create subsystems on the fly during system operation.

Layered Design

The objects are typically composed in a layered design as is further illustrated in Figure 6.4. Layers are a form of hierarchy, with a critical difference. In a layered system, the lower-level elements (those making up a lower layer) are not contained in the upper-layer elements. The elements of a layer interact to produce a set of services, which are made available to the next higher layer (in a strictly layered system). Objects in the next higher layer can use the services offered by the next lower layer but cannot otherwise access the lower-layer objects. Within a layer, the objects normally treat each other as peers; that is, no object is contained within another object. However, object orientation has the notion of encapsulation. An object has internals, and the internals (functions and data) belong to that object alone, although they can be duplicated in other objects with the same class.

A modern distributed application may be built as a set of interacting, concurrent objects. The objects interact with a lower layer, often called "middleware services." The middleware services are provided by externally supplied software units. Some of the services are part of commercial operating systems; others are individual commercial products. Those middleware components ride on lower layers of network software, supplied as part of the operating system services. In a strong distributed environment, the application programmers, who are writing the objects at the top level, do not know what the network configuration is on which their objects ride. Of course, if there are complex performance requirements, it may be necessary to know and control the network configuration and to program with awareness of its structure. But in many applications, no such knowledge is needed, and the knowledge of the application programmers about what code is actually running ceases when the thread of execution leaves the application and enters the middleware and operating systems.

The hierarchy problem is that at this point the software hierarchy and the hardware hierarchy have become disconnected. To the software architect, the natural structure of the system is layers of concurrent objects, again as illustrated in Figure 6.3 and Figure 6.4. This means the systems and software architects may clash in their partitioning of the system, and inappropriate constraints may be placed on one or the other. Before investigating the issue of reconciliation, we must complete the discussion with the nature of software components.

Large, Autonomous Components

When taking a decompositional approach to design, the designer decomposes until he or she reaches components that can be bought or easily built. In both hardware and software, some of the components are very large. In software, in particular, the design decomposition often results in very large software units, such as operating systems and databases. Both of these are now often millions of lines of programming language code and possess rich functionality. More significantly, they act semiautonomously when used in a system. An operating system is not a collection of functions to be passively called by an application. To be sure, that is one of the services offered by modern operating systems. But modern operating systems manage program memory, schedule program units on processors, and synchronize concurrent objects across multiple processors. An advanced operating system may present unified services that span many individual computers, possibly widely geographically spread.

These large and autonomous components change architecting because the architect is forced to adapt to the components. In principle, of course, the architect and client need not adapt. They can choose to sponsor a from-scratch development instead. But the cost of attempting to replicate the enormous software infrastructure that applications now commonly reuse is prohibitive. So, for example, the market dominance and complexity of very large databases forces us to use commercial products in these applications. The commercial products support particular kinds of data models and do not support others. The architecture must take account of the kinds of data models supported, even when those are not a natural choice for the problem.

Reconciling the Hierarchies

Our challenge is to reconcile the systems and software worlds. Because software is becoming the dominant element, in terms of its cost pacing what can be developed, one might argue for simply adopting software's models and abandoning the classic systems view. This is inappropriate for several reasons. First, the migration of software to object-oriented, layered structures is only partial. Much software is procedurally structured and is likely to remain so for many years to come. The infrastructure for supporting distributed, concurrent, object-oriented applications is not mature. Although leading-edge applications take this path, many others with strong reliability or just predictability requirements will use more traditional structures.

Second, both approaches are fundamentally valid. Figure 6.2 and Figure 6.4 are correct views of the system, they just represent different aspects. No single view can claim primacy. As we move into complex, information-centric systems, we will have to accept the existence of many views, each representing different concerns, and each targeted at a different stakeholder audience. The architect, and eventually systems engineers, will have to be sure the multiple views are consistent and complete with respect to the stakeholder's concerns.

Third, every partitioning has its advantages and drawbacks. Building a system in which each computational unit has its own software confined within it has distinct advantages. In that case, each unit will normally have much greater autonomy (because it has its own software and does not depend on others). That means each unit can be much more easily outsourced or independently developed. Also, the system does not become dependent on the presence of some piece of software infrastructure. Software infrastructure elements (operating systems and middleware) have a poor record for on-schedule delivery and feature completeness. Anybody depending on an advanced feature of an operating system to be delivered more than a year out runs a high risk of being left with nothing when the scheduled delivery date comes by and the operating system vendor has decided to delay the feature to a future version or has simply pushed the delivery cycle out another year.

Nevertheless, the modern approaches have tremendous advantages in many situations. Consider the situation when the units in Figure 6.2 share a great deal of functionality. If separate development teams are assigned to each, the functionality is likely to be independently developed as many times as there are units. Redundant development is likely to be the least of the problems; however, because those independent units probably interact with each other, the test burden has the potential for rising as the square of the number of units. Appropriate code sharing — that is, the use of layered architectures for software — can alleviate both problems.

The Role of Architecture in Software-Centered Systems

In software as in systems, the architect's basic role is the reconciliation of a physical form with the client's needs for function, cost, certification, and technical feasibility. The mindset is the same as described for system architecting in general, though the areas of concentration are different. System architecting heuristics are generally good software heuristics, though they may be refined and specialized. Several examples are given in Chapter 9. In addition, there are heuristics that apply particularly to software. Some of these are mentioned at the end of this chapter.

The architect develops the architecture. Following Brooks' term,[8] the architect is the user's advocate. As envisioned in this book, the architect's responsibility goes beyond the conceptual integrity of the systems as seen by the user, to the conceptual integrity of the system as seen by the builder and other stakeholders. The architect is responsible for both what-the-system-does and well as how-the-system-does-it. But that responsibility extends, on both counts, only as far as is needed to develop a satisfactory and feasible system concept. After all, the sum of both is nearly the whole system, and the architect's role must be limited if an individual or small team is to carry it out. The latter role, of defining the overall implementation structure of the system, is closer to some of the notions of software architecture in recent literature.

The architect's realm is where views and models combine. Where models that integrate disparate views are lacking, the architect can supply the insight. When disparate requirements must interact if satisfaction is to be achieved, the architect's insight can ensure that the right characteristics are considered foremost and that an architecture that can reconcile the disparate requirements is developed. The perspective required is predominantly a system perspective. It is the perspective of looking at the software and its underlying hardware platforms as an integrated whole that delivers value to the client. Its performance as a whole, behavioral and otherwise, is what gives it its value.

Architecting for evolution is also an example of the *greatest leverage is at the interfaces* heuristic. Make a system evolvable by paying attention to the interfaces. In software, interfaces are very diverse. With a hardware emphasis, it is common to think of communication interfaces at the bit, byte, or message level. But in software communication, interfaces can be much richer and capture extensively structured data, flow of control, and application-specific notions. Current work in distributed computing is a good example. The trend in middleware is to find abstractions well above the network socket level that allow flexible composition. Network-portable languages like Java allow each machine to express a common interface for mobile code (the Java virtual machine). The ambition of service-oriented architectures is to provide a rich set of intermediate abstractions to allow end-user development to be further abstracted away from the low-level programming details.

Programming Languages, Models, and Expression

Models are languages. A programming language is a model of a computing machine. Like all languages, they have the power to influence, guide, and restrict our thoughts. Programmers with experience in multiple languages understand that some problems will decompose easily in one language, but only with difficulty in another, an example of fitting the architecture of the solution to that of a prescriptive solution heuristic. The development of programming languages has been the story of moving successively higher in abstraction from computing hardware.

The layering of languages is essential to complex software development because a high-level language is a form of software reuse. Assembly languages masked machine instructions; procedural languages modeled computer instructions in a more language-like prose. Modern languages containing object and strong structuring concepts continue the pattern by providing a richer palette of representation tools for implementing computing constructs. Each statement in FORTRAN, Pascal, or C reuses the compiler writer's machine-level implementation of that construct. Even more important examples are the application-specific languages like mathematical languages or databases. A statement in a mathematical language like MATLAB or Mathematica may invoke a complex algorithm requiring long-term development and deep expertise. A database query language encapsulates complex data storage and indexing code. The current enthusiasm for service-oriented architectures is (or should be) the same phenomena. By assembling abstractions closer to what end users are interested in, while maintaining a low enough level of abstraction to be reusable, we greatly enhance development productivity.

One way of understanding this move up the ladder of abstraction is a famous software productivity heuristic on programmer productivity. A purely programming-oriented statement of the heuristic is

> *Programmers deliver the same number of lines of code per*
> *day regardless of the language they are writing in.*

Hence, to achieve high software productivity, programmers must work in languages that require few lines of code.[9] This heuristic can be used to examine various issues in language and software reuse. The nature of a programming language, and the available tools and libraries, will determine the amount of code needed for a particular application. Obviously, writing machine code from scratch will require the most code. Moving to high-level languages like C or Ada will reduce the amount of original code needed, unless the application is fundamentally one that interacts with the computing hardware at a very low level. Still less original code will be required if the language directly embodies application domain concepts, or, equivalently, application-specific code libraries are available.

Application-specific languages imitate domain language already in use and make it suitable for computing. One of the first and most popular is spreadsheets. The spreadsheet combines a visual abstraction and a computational language suited to a range of modeling tasks in business offices, engineering, and science. An extremely important category is database query languages. Today it would be quite unusual to undertake an application requiring sophisticated database functionality and not use an existing database product and its associated query language. Another more recent category includes mathematical languages. These languages, such as Mathematica, MacSyma, and MatLab, use well-understood mathematical syntax and then process those languages into computer-processable form. They allow the mathematically literate user to describe solutions in a language much closer to the problem than a general-purpose programming language.

Application-specific programming languages are likely to play an increasingly important role in all systems built in reasonably large numbers. The only impediment to use of these abstractions in all systems is the investment required to develop the language and its associated application generator and tools. One-of-a-kind systems will not usually be able to carry the burden of developing a new language along with a new system unless they fit into a class of system for which a "meta-language" exists. Some work along these lines has been done, for example, in command and control systems.[10] As mentioned before, service-oriented architectures are a currently fashionable take on the same theme.

Architectures, "Unifying" Models, and Visions

Architectures in software can be definitions in terms of tasks and modules, language or model constructs, or, at the highest abstraction level, metaphors. Because software is the most flexible and ethereal of media, its architecture, in the sense of a defining structure, can be equally flexible and ethereal.

The most famous example is the original use by Macintosh of the desktop metaphor, a true architecture. To a considerable degree, when the overall human interface guidelines are added, this metaphor defines the nature of the system. It defines the types of information that will be handled and it defines much of the logic or processing. The guidelines force operation to be human centered; that is, the system continuously parses user actions in terms of the effects on objects in the environment. As a result, Macintosh, and now Microsoft Windows, programs are dominated by a main event loop. The foremost structure the programmer must define is the event loop, a loop in which system-defined events are sequentially stripped from a queue, mapped to objects in the environment, and their consequences evaluated and executed.

The power of the metaphor as architecture is twofold. First, the metaphor suggests much that will follow. If the metaphor is a desktop, its components should operate similarly to their familiar physical counterparts. This results in fast and retentive learning "by association" to the underlying metaphor. Second, it provides an easily communicable model for the system that all can use to evaluate system integrity. System integrity is being maintained when the implementation to metaphor is clear.

Directions in Software Architecting

Software architecture and architecting have received considerable recent attention. There have been several special issues of *IEEE Software* magazine devoted to software architecture. Starting with Shaw and Garlan's book,[11] a whole series has appeared. Much of the current work in software architecture focuses on architectural structures and their analysis. Much as the term "architectural style" has definite meaning in civil architecture, usage is attached to style in current software work. In the terminology of this book, work on software architecture styles is attempting to find and classify the high-level forms of software and their application to particular software problems.

Focusing on architecture is a natural progression of software and programming research that has steadily ascended the ladder of abstraction. Work on structured programming led to structured design and to the multitasking and object-oriented models to be described in Chapter 10. The next stage of the progression is to further classify the large-scale structures that appear as software systems become progressively larger and more complex.

Current work in software architecture primarily addresses the product of architecting (the structure or architecture) rather than the process of generating it. The published studies cover topics such as classifying architectures, mapping architectural styles to particularly appropriate applications, and using software frameworks to assemble multiple related software systems. However, newer books are addressing process, and the work on software architecture patterns is effectively work on process, in that it provides directive guidance in forming a software architecture. This book presents some common threads of the architectural process that underlie the generation of architectures in many domains. Once a particular domain is entered, such as software, the architect should make full use of the understood styles, frameworks, or patterns in that domain.

The flavor of current work in software architecture is best captured by reviewing some of its key ideas. These include the classification of architectural styles, patterns and pattern languages in software, and software frameworks.

Architectural Styles

At the most general level, a style is defined by its components, connectors, and constraints. The components are the things from which the software system is composed. The connectors are the interfaces by which the components interact. A style sets the types of components and connectors that will make up the system. The constraints are the requirements that define system behavior. In the current usage, the architecture is the definition in terms of form, which does not explicitly incorporate the constraints. To understand the constraints, one must look to additional views.

As a simple example, consider the structured design models described previously. A pure structured style would have only one component type, the routine, and only one connector type, invocation with explicit data passing. A software system composed using only these components and connectors could be said to be in the structured style. But the notion of style can be extended to include considerations of its application and deviations.

David Garlan and Mary Shaw give this discussion of what constitutes an architectural style*:

> An architectural style, then defines a family of such systems in terms of a pattern of structural organization. More specifically, an architectural style determines the vocabulary of components and connectors that can be used in instances of that style. Additionally, a style might define topological constraints on architectural descriptions (e.g. no cycles). Other constraints — say, having to do with execution semantics — might also be part of the style definition.

> Given this framework, we can understand what a style is by answering the following questions: What is the structural pattern — the components, connectors, and topologies? What is the underlying computational model? What are the essential invariants of the style — its "load bearing walls"? What are some common examples of its use? What are the advantages and disadvantages of using that style? What are some of the common specializations?

Garlan and Shaw have gone on to propose several root styles. As an example, their first style is called "pipe and filter." The pipe and filter style contains one type of component, the filter, and one type of connector, the pipe. Each component inputs and outputs streams of data. All filters can potentially operate incrementally and concurrently. The streams flow through the pipes. Likewise, all stream flows are potentially concurrent. Because each component acts to produce one or more streams from one or

* Garlan, D., and M. Shaw, An Introduction to Software Architecture, Technical Report, School of Computer Science, Carnegie Mellon University, Pittsburgh, PA, p. 6.

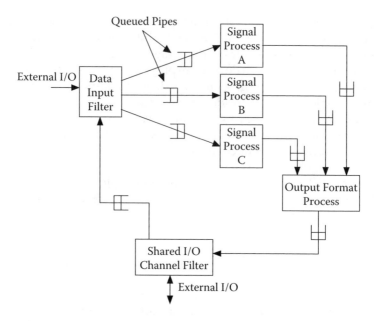

Figure 6.5 A pipe and filter system. Data flow through the system in pipes, which may actually have several types depending on their semantics for queuing, data push or pull, and so forth. Data are processed in filters that read and write pipes.

more streams, it can be thought of as an abstract sort of filter. A pipe and filter system is schematically illustrated in Figure 6.5.

UNIX shell programs and some signal processing systems are common pipe and filter systems. The UNIX shell provides direct pipe and filter abstractions with the filters concurrent UNIX processes and the pipes interprocess communication streams. The pipe and filter abstraction is a natural representation for block-structured signal-processing systems in which concurrent entities perform real-time processing on incoming sampled data streams.

Some other styles proposed include object oriented, event based, layered, and blackboard. An object-oriented architecture is built from components that encapsulate both data and function and which exchange messages. An event-based architecture has as its fundamental structure a loop that receives events (from external interfaces or generated internally), interprets the events in the context of system state, and takes actions based on the combination of event and state. Layered architectures emphasize horizontal partitioning of the system with explicit message passing and function calling between layers. Each layer is responsible for providing a well-defined interface to the layer above. A blackboard architecture is built from a set of concurrent components that interact by reading and writing asynchronously to a common area.

Each style carries its advantages and weaknesses. Each of these styles is a description of an implementation from an implementer's point of view, and specifically from the software implementer's point of view. They are not descriptions from the user's point of view, or even from the point of view of a hardware implementer on the system. A coherent style, at least of the type currently described, gives a conceptual integrity that assists the builder but may be no help to the user. Having a coherent implementation style may help in construction, but it is not likely to yield dramatic improvements in productivity or quality because it does not promise to dramatically cut the size of what must be implemented.

This is reflective of a large fraction of the current software architecture literature. The primary focus is on the structure of the software, not on the structure of the problem that the software is to solve. The architecture description languages being studied are primarily higher-level or more abstracted descriptions of programming language constructs. Where user concerns enter the current discussion is typically through analysis. So, for example, an architecture description language developer may be concerned with how to analyze the security properties of a system description written in the language. This approach might be termed "structuralist." It places the structure of the software first in modeling and attempts to derive all other views from it. There is an intellectual attraction to this approach because the structural model becomes the root. If the notation for structure can be made consistent, then the other views derived from it should retain that consistency. There is no problem of testing consistency across many views written in different modeling languages. The weakness of the approach is that it forces the stakeholders other than the software developers to use an unfamiliar language and trust unfamiliar analyses. In the security example, instead of using standard methods from the security community, those concerned with security must trust the security analysis performed on the architectural language. This approach may grow to be accepted by broad communities of stakeholders, but it is likely to be a difficult sell.

In contrast to the perspective that places structure first in architecture, this book has repeatedly emphasized that only the client's purpose should be first. The architect should not be removed from the purpose or requirements; the architect should be immersed in them. This is a distinction between architecting as described here and as is often taught in software engineering. We do not assume that requirements precede architecture. The development of requirements is part of architecting, not its preconditions.

The ideal style is one that unifies both the user's and builder's views. The mathematical languages mentioned earlier are examples. They structure the system from both a user's and an implementer's point of view. Of course, the internals of the implementation of such a complex software system will contain many layers of abstraction. Almost certainly, new styles and abstractions specific to the demands of implementation in real computers will have to arise internally. When ideal styles are not available, it is still reasonable to seek models or architectural views that unify some set of considerations larger than just the software implementer. For implementation of complex systems, it would be a useful topic of research to find models or styles that encompass a joint hardware–software view.

Architecture through Composition

Patterns, styles, and layered abstraction are inherent parts of software practice. Except for the rare machine-level program, all software is built from layered abstractions. High-level programming languages impose an intellectual model on the computational machine. The nature of that model inevitably influences what kinds of programs (systems) are built on the machine.

The modern trend is to build systems from components at higher and higher levels of abstraction. It is necessary because no other means are available to build very large and complex systems within acceptable time and effort limits. Each high-level library of components imposes its own style and lends itself to certain patterns. The patterns that match the available libraries are encouraged, and it may be very difficult to implement architectures that are not allowed for in the libraries.

> *Example*: Graphical Macintosh and Windows pro-
> grams are almost always centrally organized around
> an event loop and handlers, a type of event-driven
> style. This structure is efficient because the operat-
> ing systems provide a built-in event loop to capture
> user actions such as mouse clicks and key presses.
> However, because neither had multithreading abstrac-
> tions (at least before 1995), a concurrent, interact-
> ing object architecture was difficult to construct.
> Many applications would benefit from a concurrent
> interaction object architecture, but these architec-
> tures were very difficult to implement within the
> constraints of existing libraries. As both systems
> evolved, direct support for multithreaded, concur-
> rent processes has slowly worked its way into all
> aspects of both systems, user interfaces included.

The logical extension is to higher and higher level languages and from libraries to application-specific languages that directly match the nature of the problem they were meant to solve. The modern mathematical software packages are, in effect, very high-level software languages designed to mimic the problem they are meant to solve. The object of the packages is to do technical mathematics. So rather than provide a language into which the scientist or engineer must translate mathematics, the package does the mathematics. This is similar for computer-aided design packages, and indeed for most of the shrink-wrap software industry. These packages surround the computer with a layered abstraction that closely matches the way users are already accustomed to working.

Actually, the relationship between application-specific programming language, software package, and user is more symbiotic. Programmers adapt their programs to the abstractions familiar to the users. But users eventually adapt their abstractions to what is available and relatively easy to implement. The best example is the spreadsheet. The spreadsheet as an abstraction partially existed in paper form as the general ledger. The computer-based abstraction has proven so logical that users have adapted their thinking processes to match the structure of spreadsheets. It should probably be assumed that this type of interactive relationship will accelerate when the first generation of children to grow up with computers reaches adulthood.

Heuristics and Guidelines in Software

The software literature is a rich source for heuristics. Most of those heuristics are specific to the software domain and are often specific to restricted classes of a software-intensive system. The published sets of software heuristics are quite large. The newer edition of Brook's *The Mythical Man-Month: Essays in Software Engineering*[12] includes a new chapter, "The Propositions of the Mythical Man-Month: True or False?" which lists the heuristics proposed in the original work. The new chapters reinforce some of the central heuristics and reject a few others as incorrect.

The heuristics given in *Man-Month* are broad ranging, covering management, design, organization, testing, and other topics. Several other sources give specific design heuristics. The best sources are detailed design methodologies that combine models and heuristics into a complete approach to developing software in a particular category or style. Chapter 10 discusses three of the best documented, ADARTS,* structured design,† and object

---

* The published reference on ADARTS, which is quite thorough, is available through the Software Productivity Consortium, ADARTS Guidebook, SPC-94040-CMC, Version 2.00.13, Vols. 1–2, September 1991. ADARTS is an Ada language-specific method, though its ideas generalize well to other languages. In fact, this has been done, although the resulting heuristics and examples are available only to Software Productivity Consortium members.

† Structured design is covered in many books. The original reference is Yourdon, Edward, and Larry L. Constantine, *Structured Design: Fundamentals of a Discipline of Computer Program and Systems Design*. New York: Yourdon Press, 1979.

oriented.* Even more specific guidelines are available for the actual writing of code. A book by McConnell[13] contains guidelines for all phases and a detailed bibliography.

From this large source set, however, there are a few heuristics that particularly stand out as broadly applicable and as basic drivers for software architecting:

- Choose components so that each can be implemented independently of the internal implementation of all others.
- Programmer productivity in lines of code per day is largely independent of language. For high productivity, use languages as close to the application domain as possible.
- The number of defects remaining undiscovered after a test is proportional to the number of defects found in the test. The constant of proportionality depends on the thoroughness of the test but is rarely less than 0.5.
- Very low rates of delivered defects can be achieved only by very low rates of defect insertion *throughout* software development, and by layered defect discovery — reviews, unit test, system test.
- Software should be grown or evolved, not built.
- The cost of removing a defect from a software system rises exponentially with the number of development phases since the defect was inserted.
- The cost of discovering a defect does not rise. It may be cheaper to discover a requirements defect in customer testing than in any other way, and hence the importance of prototyping.
- Personnel skill dominates all other factors in productivity and quality.
- Do not fix bugs later; fix them now.

As has been discussed, the evolvability of software is one of its most unique attributes. A related heuristic is: *A system will develop and evolve much more rapidly if there are stable intermediate forms than if there are not.* In an environment where wholesale replacement is the norm, what constitutes a stable form? The previous discussion has already talked about releases as stable forms and intermediate hardware configurations. From a different perspective, the stable intermediate forms are the unchanging components of the system architecture. These elements that do not change provide the framework within which the system can evolve. If they are well chosen — that is, if they are conducive to evolution — they will be stable and facilitate

---

* Again, there are many books on object-oriented design, and many controversies about its precise definition and the best heuristics or design rules. The book by Rumbaugh, discussed in Chapter 10, is a good introduction, as is the Unified Modeling Language (UML) documentation and associated books.

further development. A sure sign the architecture has been badly chosen is the need to change it on every major release. The architectural elements involved could be the use of specific data or control structures, internal programming interfaces, or hardware–software interface definitions. Some examples illustrate the impact of architecture on evolution.

> *Example*: The Point-to-Point Protocol (PPP) is a publicly defined protocol for computer networking over serial connections (such as modems). Its goal is to facilitate broad multivendor interoperability and to require as little manual configuration as possible. The heart of the protocol is the need to negotiate the operating parameters of a changing array of layered protocols (for example, physical link parameters, authentication, IP control, AppleTalk control, compression, and many others). The list of protocols is continuously growing in response to user needs and vendor business perceptions. PPP implements negotiation through a basic state machine that is reused in all protocols, coupled with a framework for structuring packets. In a good implementation, a single implementation of the state machine can be "cloned" to handle each protocol, requiring only a modest amount of work to add each new protocol. Moreover, the common format of negotiations facilitates troubleshooting during test and operation. During the protocols development, the state machine and packet structure have been mapped to a wide variety of physical links and a continuously growing list of network and communication support protocols.

> *Example*: In the original Apple Macintosh operating system, the architects decided to not use the feature of their hardware to separate "supervisor" and "user" programs. They also decided to implement a variety of application programming interfaces through access to global variables. These choices were beneficial to the early versions because they improved performance. But these same choices (because of backward compatibility demands) greatly complicated efforts to implement advanced operating system features such as protected memory and preemptive multitasking. In the end, dramatic evolution of the operating system required wholesale replacement, with limited

backward compatibility. Another architectural choice was to define the hardware–software interface through the Macintosh Toolbox and the published Apple programming guidelines. The combination proved to be both flexible and stable. It allowed a long series of dramatic hardware improvements and even a transfer to a new hardware architecture, with few gaps in backward compatibility (at least for those developers who obeyed the guidelines). Even as the old operating system was entirely replaced, the old programming interface survived through minimal modifications allowing recompilation of programs.

Example: The Internet Protocol combined with the related Transmission Control Protocol (TCP/IP) has become the software backbone of the global Internet. Its partitioning of data handling, routing decisions, and flow control has proven to be robust and amenable to evolutionary development. The combination has been able to operate across extremely heterogeneous networks with equipment built by countless vendors. Although there are identifiable architects of the protocol suite, control of protocol development is quite distributed with little central authority. In contrast, the proprietary networking protocols developed and controlled by major vendors performed relatively poorly at scaling to diverse networks. One limitation in the current IP protocol suite that has become clear is the inadequacy of its 32-bit address space. However, the suite was designed from the beginning with the capability to mix protocol versions on a network. As a result, the deployed protocol version has been upgraded several times (and will be again to IPv6).

Exercises

1. Consult one or more of the references for software heuristics. Extract several heuristics and use them to evaluate a software-intensive system.
2. Requirements defects that are delivered to customers are the most costly because of the likelihood they will require extensive rework. But discovering such defects anytime before customer delivery is likewise very costly because only the customers' reaction may make the nature of the defect apparent. One approach to this problem is

prototyping to get early feedback. How can software be designed to allow early prototyping and feedback of the information gained without incurring the large costs associated with extensive rework?

3. Pick three software-intensive systems of widely varying scope, for example, a pen computer-based data-entry system for warehouses, an internetwork communication server, and the flight control software for a manned space vehicle. What are the key determinants of success and failure for each system? As an architect, how would these determinants change your approach to concept formulation and certification?

4. Examine some notably successful or unsuccessful software-intensive systems. To what extent was success or failure due to architectural (conceptual integrity, feasibility of concept, certification) issues and to what extent was it due to other software process issues?

5. Are their styles analogous to those proposed for software that jointly represent hardware and software?

Notes and References

1. *IEEE Software*, Special issue on the The Artistry of Software Architecture, November, 1995. This issue contains a variety of papers on software architecture, including the role of architecture in reuse, comparisons of styles, decompositions by view, and building block approaches.

2. Kruchten, P. B., A 4+1 View Model of Software Architecture, *IEEE Software Magazine*, pp. 42–50, November, 1995.

3. Leveson, N. G., and C. S. Turner, An Investigation of the Therac 25 Accidents, *Computer*, pp. 18–41, July, 1993.

4. DeMarco, T., and T. Lister, *Peopleware: Productive Projects and Teams*. New York: Dorset House, 1987.

5. Lambert, B., Beyond Information Engineering: The Art of Information Systems Architecture, Technical Report, Broughton Systems, Richmond, Virginia, 1994.

6. Software Productivity Consortium, Evolutionary Spiral Process, Vol. 1–3.

7. Maguire, S., *Debugging the Development Process*. Redmond, WA: Microsoft Press, 1994.

8. Brooks, F., *The Mythical Man-Month*. Reading, MA: Addison-Wesley, 1975. This classic book has recently been republished with additional commentary by Brooks on how his observations have held up over time. See p. 256 in the new edition.

9. Data associated with this heuristic can be found in several books. Two sources are Capers Jones, *Programming Productivity*. New York: McGraw-Hill, 1986; Capers Jones, *Applied Software Measurement*, 3rd ed. New York: McGraw-Hill, 2008.

10. Domain Specific Software Architectures, University of Southern California Information Sciences Institute, Technical Report, 1996.

11. Shaw, M., and D. Garlan, *Software Architecture: Perspectives on an Emerging Discipline*, New York: Prentice Hall, 1996.

12. Brooks, F., *The Mythical Man-Month*. Reading, MA: Addison-Wesley, 1975.

13. McConnell, S., *Code Complete: A Practical Handbook of Software Construction*, 2nd ed. Redmond, WA: Microsoft Press, 2004.

Case Study 5: The Global Positioning System

The Global Positioning System (GPS)[1] is one of the great success stories of the late 20th century. Most readers of this book will have had some experience with GPS, either in a car or with a handheld unit. GPS is a large system that has undergone extensive evolution. Early in its history, and in the history of its predecessor programs, it was a centrally controlled system. As it has evolved, and become extremely successful, control has gradually migrated away from the GPS program office. The larger enterprise that is now GPS is no longer controlled wholly by a single program office; indeed it is no longer controlled by the U.S. government. If the European Galileo system eventually flies and is compatible with GPS, GPS will have evolved into a full-fledged collaborative system, the subject of the next chapter.

GPS is a fitting case study to close Part II, looking either forward or backward. Looking forward, to Chapter 7, GPS is partially a collaborative system and partially not. As with most partial cases, the fact that it is not firmly in or out provides a better discussion of just what the border is. Looking backward, to earlier chapters, we can see many of the other points we sought to emphasize. GPS illustrates the principles of architecture through invariants, of the importance of a small architecture team, and of ripe timing with respect to both technology and user and institutional needs.

The History

The moment (actually the weekend) when the GPS architecture was determined can be easily identified — but that moment was a long time coming. The history of satellite navigation leads up to GPS and then extends beyond it. That GPS came about when it did and in the form it did is at least somewhat surprising. It need not have come about. The fact that it did when and how it did, and how successful it has become, illustrates major points in the systems architecting of both conventional and collaborative systems.

The Origins of GPS: The Foundational Programs

Position determination and navigation are fundamental to military operations, with air, sea, or land. For all of military history, extensive technological effort has been expended to improve navigational capabilities. At the beginning of the space era (late 1950s), each of the U.S. military services viewed the navigation problem quite differently.

Inertial Navigation and Its Limits

The U.S. Air Force and Navy were heavily investing in inertial navigation systems. Inertial navigation was particularly well suited to nuclear missiles. It was sufficiently accurate to guide a missile (flight times of 30 minutes or less) with a nuclear warhead (assuming the launch position is accurately known) and was immune to external interference. For the Air Force this was a sufficient solution, as launch points were fixed and their locations known a priori precisely. For the Navy, the initial positioning problem was significant. Inertial navigation systems inherently drift on time scales of days. Thus, an inertial navigation system in a ship or submarine, although very accurate over a day or so, must be "recentered" regularly or its accuracy will continuously degrade. Because ships, and especially ballistic missile submarines, are expected to operate at sea for months at a time, the problem of correcting the drifting inertial units was central.

A naval ship requires global position determination to moderately high accuracy (tens to hundreds of meters) at modest update rates (once to a few times a day). Ships know their altitude, so only two-dimensional positioning is required. In the 1960s these capabilities were provided mainly by land-based radio navigation systems (e.g., LORAN [long-range navigation]), but these systems worked much less well than desired. Strategic aircraft would use global three-dimensional position determination but could operate only with the capabilities available at the time.

Weapon Delivery

In the 1950s, air-delivered weapons were notoriously inaccurate. A hardened, fixed target, like a bridge, might require tens of sorties to destroy. A moving target, like a tank, could only be hit by a pilot flying low and in short-range direct sight. At the time the primary solution of interest was sensor-guided weapons. Extensive work at the time was devoted to developing command, infrared, and radar-guided weapons.

The accuracy requirements for weapon delivery are very challenging. Accuracy must be a few meters at the most, three-dimensional position is required, and the platforms move very rapidly. The belief was that only sensor guidance was suitable to the task.

The Transit Program

Shortly after the launch of Sputnik in 1957, Frank McClure of Johns Hopkins Applied Physics Laboratory (APL) determined that measurements of

Doppler shifts on the Sputnik signal could be used to infer lines of position at the receiver. Given multiple satellites in orbit and precision orbit determination, it would be possible to construct a navigation system. This concept was supported by the Advanced Research Projects Agency (ARPA) and led to a dedicated satellite launch in 1960. By 1965, 23 satellites had been launched and the Transit system had been declared operational.

Transit squarely addressed the Navy navigation problem. Transit provided a two-dimensional position update a few times a day with moderate accuracy, and it did it globally, in all weather, without any land-based infrastructure. From an architectural perspective, Transit was purpose driven, had a clear architect and architecture, and the alignment between the user stakeholders and developers was close. As a result, it was a stable, evolving, and successful system. The witness to the strength of the linkage between the user base and the developers is that Transit satellites were being launched until 1988, and the system operated until 1996, long after GPS became operational.

Transit was important in the history of GPS in several respects:

1. Transit provided a useful service for roughly two decades before GPS became operational. Transit demonstrated the feasibility and utility of satellite navigation.
2. Building and operating Transit forced the resolution of important technical issues. In particular, it led to great improvements in orbit determination, in gravity models, and in computing and predicting signal delays due to propagation through the atmosphere.
3. Transit set a precedent for commercial use. Transit was made available for commercial use in the late 1960s, and the number of commercial use receivers came to far outnumber the number of military use receivers (a harbinger of what was to come in GPS).

TIMATION

A related problem to navigation is the problem of time transfer or clock synchronization. This is the problem of accurately synchronizing clocks in distant locations, or of transferring a precise time measurement in one location to a distant location. This was a natural Navy concern as it was the foundation of the revolution in navigation in the 18th century when chronometers allowed the determination of longitude.

At the Naval Research Laboratory (NRL) in the 1960s, Roger Easton made two key realizations. He first realized that satellite clocks with appropriate signal transmission could be used to allow precision time transfer. He then realized that simultaneous time transfer from multiple, synchronized clocks was equivalent to position determination (either two or three dimensions, depending on the number of satellites). These realizations were translated into a concept known as TIMATION.

TIMATION was envisioned as a navigation system that would begin as an improved version of Transit and evolve into a full-blown system not unlike what GPS is today. The early versions would use small constellations of low Earth-orbiting satellites (like Transit). This is relatively simple to build, but provides only intermittent access (a few times a day, like Transit). Over time, the constellation would grow in size and move to higher orbits until global continuous coverage was achieved.

621B

At the same time, the U.S. Air Force was studying satellite navigation through the Space and Missile Center (SMC) and its research center The Aerospace Corporation. The project was known as 621B. The Aerospace Corporation president, Ivan Getting, had conceived of important elements of the satellite navigation concept in the 1950s[2] and strongly advocated for it as the Aerospace Corporation president. The Air Force concentrated on longer-term, more ambitious goals consonant with Air Force mission needs. The Air Force concept was based on three-dimensional, high-precision, global position determination and was also concerned with electronic warfare and other military factors. 621B used the notion of simultaneous measurement of three delays to three known satellite locations, like Roger Easton's concept but with two key differences. First, 621B performed measurement (or "pseudoranging") with the then-new scheme of digital pseudorandom coded signals. Second, 621B was originally based on high-accuracy clocks in each receiver.

The first difference, the signal, was technically very aggressive for the time but led to key advantages. At the time, processing digital signals several Megahertz (MHz) wide was very challenging (though it became trivial as the microchip revolution proceeded in the 1980s). However, the digitally coded signal had significant jam and interference resistance, and largely solved the frequency coordination problem. With pseudorandom coded signals, all transmitters in the system could operate at the same frequency and rely on the code processing to separate them.

The Origin of GPS

By the early 1970s, Transit was a stable program with a satisfied user base, and the larger vision of satellite navigation was not proceeding. The differing stakeholder groups were engaged in bureaucratic warfare and making little headway. But, in a remarkably short interval, an architecturally sweet compromise would be found among them all and converted into a successful development program.

Parkinson and Currie

The person most often associated with the success of the GPS is Bradford Parkinson.[3] Parkinson arrived in Los Angeles, California, in late 1972 to run the 621B program and quickly became a believer in the merits of global satellite navigation. Through happy, though accidental, circumstances, he was able to spend unusual amounts of time with the Director of Defense Research and Engineering (DDR&E) Malcolm Currie. Parkinson convinced Currie of the merits of satellite navigation, and Currie was convinced that Parkinson was the right person to lead the effort. Parkinson was tasked to form a joint proposal to be presented the Defense Systems Acquisition Review Council (DSARC). The proposal he presented was essentially the 621B concept, and the DSARC immediately rejected it.

The Fateful Weekend

Over Labor Day weekend 1973, Parkinson assembled a small team of experts from each of the programs and areas and closeted them away from interference. Over the long weekend, they reached consensus on a revised concept. The revised concept combined features from the predecessor programs. The fundamental features of the revised concept are shown in Table CS5.1.

For those knowledgeable about GPS, the features should look familiar. They are virtually identical to those of today's operational GPS system. The revised concept was again presented to the DSARC, after Currie had assured Parkinson that a true joint program would have strong support from his level, and was approved.

As a result of DSARC approval, a joint program office was formed (the NAVSTAR program office), with Parkinson as the head. They were able to begin development very rapidly by incorporating major elements of the preexisting programs, particularly TIMATION. Critical space elements of

Table CS5.1 Features of the Revised 1973 NAVSTAR (to become GPS) Concept

Concept of operations	Measure pseudoranges to four (or more) satellites and process to compute both the master time and three-dimensional position. All position computations occur in the receivers that operate entirely passively.
Constellation	Twenty-one to twenty-four satellites in inclined half-geosynchronous orbits.
Source of time	Atomic clocks on satellites updated from the ground. Receiver time computed from multiple satellites simultaneously with position.
Signal	Pseudorandom code at L-band (1,200 MHz). Two codes: one narrow and unencrypted, one wide and encrypted.

the NAVSTAR-GPS concept were taken directly from TIMATION, and so the in-development TIMATION hardware was very useful for prototype demonstrations.

The Long Road to Revolution

Of course, the formation of the GPS concept is far from the end of the story. Although the essential architecture, in terms of its basic structure in signals, processing distribution, and constellation, was set in 1973 and has remained largely invariant, there was a very long road to successful development and an operational revolution. During the long road to operations, a key transition was taking place, the transition from a centrally controlled or "monolithic" system to a collaborative system in which there was no central authority over the whole. We discuss these two key points in the following sections.

The Timeline to Operation

The key events in the timeline to operations (and the present day) are shown in Figure CS5.1.

There were more than 10 years of serious activity, including the deployment of an operational system, before the architecture of GPS was set. After the architecture was set, development continued for roughly 20 years before GPS became fully operational. Although it took 20 years to full operation, GPS delivered real utility much earlier. Not only did it deliver real utility, it had already broken out of the confines of the Joint

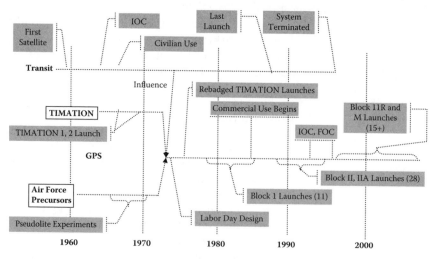

Figure CS5.1 Timeline for development of Global Positioning System (GPS).

Program Office and had a large commercial and multiagency component. By the early 1990s there was even an international component, albeit an unintended one, in the Soviet Union's GLONASS system.

Commercial Markets and the Gulf War

The Transit system set a precedent for allowing civilian use of a military satellite navigation system. This precedent was repeated with GPS in the 1980s. By 1984, in the wake of the shoot-down of the KAL 007 airliner, it became policy to allow the free use of the C/A (coarse/acquisition) coded signal. The information necessary to build a C/A code receiver was made freely available to private industry. Because the C/A code has a narrower bandwidth than the military signal and is only transmitted on one frequency, the accuracy achieved is considerably less than with the military signal. In the 1980s and 1990s, accuracy was further degraded through deliberate introduction of clock noise (known as "selective availability").

The microelectronics evolution of the 1980s enabled commercial development of GPS chipsets. Those chipsets in turn led to low-cost commercial receivers. As the commercial market expanded, receivers dropped very quickly in size and cost. Commercial firms also began developing innovative software applications and even transmitter infrastructure systems.

The first Gulf War in 1991 gave a major impetus to GPS development. The use of receivers by ground troops in the war and GPS support to widely televised precision air strikes led to considerable publicity for GPS. The satellite constellation was sufficiently mature to provide substantial capability, although initial operational capability (IOC) had not been declared. The very public demonstration of the effectiveness of GPS in supporting guided weapons led to further interest in new guidance types. The continuing receiver cost reduction, and a politically driven requirement for weapons for stealth aircraft, led to the Joint Direct Attack Munition (JDAM) concept, a highly successful approach to precision weapons where a low-cost GPS receiver and guidance unit are mated to legacy "dumb" bombs.

Revolution in the Second Generation

The GPS revolution did not come with its deployment in its intended mission and in its original context. The original slogan of the program office was "Five bombs in the same hole." Although that capability has long ago been achieved, it itself has not been as valuable as newly conceived applications. It was really in the second generation, the generation after GPS reached full operational capability, that the revolution began, with applications and markets well outside those in the original architectural concept.

Ubiquitous GPS

Between the achievement of GPS full operational capability and the present day, the number of GPS applications and the number of receivers have exploded. GPS went from a specialized navigation device to something that could be included almost as an afterthought in other devices (for example, cell phones for E911 service). Certain application areas transitioned to deep dependence on GPS. Among them are surveying and time synchronization in power transmission and telecommunications networks.

The ubiquity of GPS was made possible by receiver costs being driven down by the Moore's law advance in digital electronics (which depended on GPS having a digital signal) and the development of new applications.

GPS policy frequently lagged GPS application. One reason was simply the innovation-driven expansion of applications, enabled by cheap receivers, which occurred much more quickly than policy could adapt. The other reason was that GPS had escaped control by the GPS program office and had even significantly escaped the control of the U.S. government. GPS had morphed from a large and complicated, but reasonably conventional, system to a collaborative system, one not under the centralized control of any single entity. An example of the policy lag relative to the technology was the period in the 1990s when the U.S. Federal Aviation Administration (FAA) and Coast Guard were deploying GPS enhancement transmitters at the same time the U.S. Air Force was maintaining the accuracy degrading selective availability features. It was not until after 2000 that the United States abjured the use of selective availability and has only recently begun flying satellites with the dual-frequency civilian code transmitters necessary for higher accuracy without terrestrial augmentation.

GPS-Guided Weapons

A massive increase in the number of military receivers came with the development of GPS-guided weapons. As receiver costs dropped, largely because of the availability of commercial GPS chips, the cost of a receiver became less than even a very simple weapon. At this point, it became feasible to attach a GPS-based guidance system to a huge number of previously unguided weapons. The canonical example is a GPS-based guidance unit attached to 500 to 2,000 lb "dumb" bombs, known as the JDAM.

Even though the JDAM is a fine example of lateral exploitation (to use Art Raymond's term from the DC-3 story), the concepts of operation associated with the JDAM are more revolutionary. With a JDAM, especially the lighter-weight 500 lb JDAM, a large high-altitude aircraft, like a B-52 or B-1 bomber, could become a close support aircraft. This concept of operation was invented on-the-fly during the Afghan war in 2001–2002.

Large aircraft could loiter persistently in the battle area waiting for calls from ground troops. To make the concept work, ground troops needed to be able to precisely measure the GPS coordinates of targets (easily achieved with laser rangefinders coupled to GPS receivers) and communicate directly to the aircraft overhead. When the whole concept was in place, it could be rapidly improved by realizing that smaller guided bombs worked as well, or better, than large guided bombs in close support, given the accurate guidance. With smaller bombs, the large aircraft could carry many more, and were not limited by persistence over target, as older dedicated close support aircraft had been. The synergistic effects were large in combining the technology of GPS with changed concepts of operation and repurposed platforms.

Architecture Interpretation

GPS provides us with important lessons applicable to other systems and that relate back to the topics of Part I and Part II. The lessons are: Right idea, right time, right people; Be technically aggressive, but not suicidal; Consensus without compromise; Architecture through invariants; and Revolution through coupled change.

Right Idea, Right Time, Right People

GPS would almost certainly not have happened when it did and how it did without particular people being in the right place at the right time. It was not obvious that the U.S. Navy and Air Force could reach a consensus on a global navigation concept, sell that concept through the acquisition bureaucracy, and then maintain it for more than the decade it took to become firmly established. Without Parkinson in the key position at that time, it is unlikely that the Air Force program would have discovered and adopted key Navy ideas and expanded its scope enough to become an established program. No stakeholder group in the Air Force needed global satellite navigation badly enough to allow the program to survive.

On the Navy side, they had a stable program plan. The TIMATION concept was intended to lead, eventually, to a global, high-precision system. Had the Navy been left alone, would something like GPS eventually have emerged? Obviously, we cannot ever know for sure, the experiment cannot be carried out. But, two factors speak against the Navy's concept ever growing into the GPS system as it exists today. First, there was no Navy stakeholder with a combination of need and resources to grow the system to the level of capability now provided by GPS. Navy needs were well met by more incremental improvements that were more aligned with the limited resources of Navy space programs. Second, the most important Air Force contribution was the signal, the digital pseudorandom coded ranging signal used in the current GPS. This was a technically aggressive choice in 1973 and was unnecessary for the Navy mission (indeed it had drawbacks for the Navy mission). However, the pseudorandom noise (PRN) signal used in GPS provides it with significant jam resistance and considerably eases the problem of frequency management in crowded areas (such as urban and suburban areas of industrialized countries). The signal, and its placement in L-Band, allows high-precision location (from tens of meters to meters accuracy) to be achieved without severe frequency management problems. This has been an important factor in the long-term success of GPS, but was of little relevance to the Navy mission as understood in the 1970s.

Be Technically Aggressive, But Not Suicidal

Parkinson and his team made technically aggressive choices, with wisdom that is obvious in retrospect but was not so obvious in prospect. The most important over the long term was to base GPS ranging on the digital PRN signal. In the 1970s, processing a digital signal with a modulation rate from 1 to 10 MHz was very difficult, requiring many boards of custom hardware. With decades of advance in Moore's law, processing the same signals today is a trivial hardware exercise easily fit into communications chipsets. Even though choosing an all-digital approach was aggressive in the 1970s, it was central to the achievement of cheap receivers in the 1990s. The price/performance curve for digital electronics has moved orders of magnitude in the intervening decade, but the same curve for analog hardware has moved much less. By the 1990s, commercial firms were able to enter the GPS market with receivers in form factors and prices acceptable to a wide consumer base only because most of the processing required was digital.

The choice of half-geosynchronous orbits was also aggressive, but not excessively. The half-geosynchronous orbit allows for global simultaneous visibility to four satellites with a constellation of 25 or so satellites. The exact number depends on the specification for occasional brief outages. Higher orbits reduce the number of satellites required modestly, but considerably increase the satellite weight (because of the higher power required). Lower orbits either incur a large radiation exposure penalty (in the Van Allen belts) or cause the number of satellites to increase enormously (potentially to hundreds), although lower orbits result in smaller and simpler satellites. Building satellites that survive in the half-geosynchronous orbit is more challenging than in low Earth orbit (because of higher radiation levels), but not excessively so.

Finally, the selected architecture of GPS placed precision clocks on the satellites, and not in the receivers. This meant that receivers needed only digital processing, and all sophisticated computation was done on the ground, in receivers. Over the long term, this was very beneficial. It meant that improved processing techniques could be deployed with each new generation of receiver, and receiver generations have been far shorter than spacecraft generation times. However, it also required that atomic clocks operate precisely and reliably on satellites for up to a decade. Again, the previous work by the Navy had proven the possibility, and had explored various design options for precision clocks in orbit, including both crystal and atomic clocks. Although the technology had to be matured by the GPS program, the essential trades had already been made and the data required for those trades had been acquired in well-designed experiments.

In all three cases, Parkinson's team made technically aggressive decisions but did not incur excessive risk. Although processing megabit/second PRN signals was challenging in the 1970s, it had already been demonstrated. The project 621B experiments, and other projects, had accomplished the task several times. Likewise, the space environment at half-geosynchronous was known, and the techniques for surviving in it were known and tested (albeit uncommon and expensive). High-precision clocks had been, and were being, flown in orbit. In all three cases, the trade-offs could be made with concrete knowledge of the issues. That could not have been the case a few years earlier.

Consensus without Compromise

Even though the architecture of GPS reflected a fusion of ideas from many sources, it was not a watered-down compromise. The fusion genuinely took the best aspects of the approaches of several stakeholders and abandoned inferior aspects. There is little evidence of political compromise, that sort that might have insisted that "Air Force equities require that we do this and Navy equities require that we do that." Instead, the elements selected from the different component programs were those that reflected consensus best choices, or cut through consensus to adopt a clear strategic position.

Using the simultaneous position and time determination method from four pseudoranges can be seen as a clear consensus choice. Essentially all parties would now agree, and agreed even then, that it represented the best choice overall. The impact on overall system simplicity is clear. All position determination is done through the signals broadcast by the satellites, no auxiliary terrestrial signal is required, the receiver is nearly all digital, and all serious computation is done in the receiver.

In the case of the choice of the all-digital signal, the consensus is not clear, but the choice reflects a very clear strategic choice. The all-digital, L-band signal was easy to frequency manage, allowed all satellites to share the same frequency, and led to cheap receivers. On the downside, the L-band signal penetrates poorly in buildings and even foliage and requires more difficult antennas, and the all-digital signal was challenging to process in the 1970s. The choice made a clear strategic decision; GPS would ride the improvements in digital electronics. It would exploit technology developments then well underway but still far from ready. Parkinson's team could have a variety of other choices that would have compromised among the players and been easier in the near-term, but would have missed the long-term opportunities.

Architecture as Invariants

GPS is an example of architecture as invariants. Between the origin of the joint program and 2007, the signals were unchanging, and the orbits underwent minimal change. Of these two, the signals were much more significant as an invariant. The constellation had already been morphed by the inclusion of terrestrial transmitters for local accuracy improvement. Only in the last few years has any change begun to appear in the signals. Currently launched satellites add new military signals, known as the M-code, to augment the legacy military codes. A copy of the unencrypted civilian signal will shortly be added at a second frequency. The second signal improves accuracy by allowing direct measurement of ionospheric delay. This capability has been available for military users since the inception of GPS, but has been unavailable to civilian users.

Architecture through invariants is particularly effective when evolution is important to long-term success. In the case of GPS, the invariant signals have allowed decoupled evolution of the constellation and receivers. The receivers have undergone extensive evolution programmatically and technically without that change having to be coupled to change in the satellites. At the current time, receivers are developed dominantly by commercial firms with no formal relationship to the GPS program office.

Revolution through Coupled Change

The greatest impact of GPS came only through coupled change to affiliated systems and concepts of operation. The original slogan was "five bombs in the same hole and cheap receivers." The latter was achieved and then achieved beyond the original expectations. The former was never as important as was originally thought. Instead, the proliferation of cheap receivers enabled a whole range of new applications. Some of the innovations include the following:

- Extremely compact and low-cost receivers could be distributed to individual soldiers. Small units can accurately and continuously determine their position, and reporting of those positions enables new networked operational concepts.

- As receivers became cheap enough, they could be placed on weapons. Instead of guiding the weapon delivery platform to weapon release, GPS now guides the weapon from release to impact.
- The existence of precision-guided weapons at costs little larger than their unguided predecessors has resulted in a radical shift in the dominant operational concept for aerial weapons delivery (at least for the United States), from less than 10% of all aerial weapons being guided to roughly 90% in a period of 15 years.
- As guided weapons became the norm, the operational concept for platforms has shifted. In the Afghan war begun in 2001, the B-1 bomber was used for close air support, by loitering for long periods at high altitude and dropping GPS-guided bombs on targets located by ground troops. Close air support was performed by small aircraft whose pilots delivered weapons visually, and is now more effectively performed by an airplane designed originally for delivering strategic nuclear weapons.
- The practice of surveying has been pervasively impacted by GPS. Surveyors, because they do not need position determination at high update rates, have been able to exploit a wide range of unanticipated processing techniques.
- The ability of GPS to provide very high accuracy, globally referenced time has led to its embedding into electric power and telecommunications control systems.
- GPS is now typically included in cell phones at a marginal cost to support electronic 911 service. The ability to track large numbers of cell phone users will lead to a wide range of new applications.

Revolution through coupled change is exemplified in GPS, but is hardly unique to GPS. The most dramatic impacts of new technologies typically come from uses beyond the originally envisioned application. Those dramatic applications typically involve rethinking the problem being solved and the concept of operation involved. A simple application of a new technology to old concepts of operation is almost never as valuable as what can be realized from creating new concepts of operation.

Conclusion

GPS is an exceptional example of architecture in a revolutionary system. Its original development is a classic example of architecting by a very small team with a tightly defined mission with the challenges of new technology. As it has developed, it has illustrated evolution toward a collaborative system and revolution through changes to concepts of operation. GPS is not quite a collaborative system. It is still run by a single, joint program office. But, many of the factors that drive GPS development are out of the control of that program office. Commercial receiver builders design in whatever features they wish. Several agencies continue to develop and deploy augmentation transmitters. And, most strikingly, international players are beginning to develop competing programs that may contain compatible or interfering signals.

The greatest impact of GPS has been in areas outside the original conception of its use, and that success is a testament to the quality of the architecture. The core architecture of GPS (the signal and the position determination method) has been robust to extensive lateral exploitation. The willingness of the program office and the sponsors to cede control of some segments and applications and allow a collaborative system to form, has been central to long-term success. The multitude of applications is a witness to the basic insight that ubiquitous, global satellite navigation would be tremendously valuable.

Notes and References

1. There is a great deal of literature on the origins and evolution of the Global Positioning System. The author is indebted to his aerospace colleague Glenn Buchan who wrote an exceptionally fine case study from which we have drawn a great deal. The Air Force Institute of Technology has also published an extensive case study of the Global Positioning System, though it focuses more on the program events after the initial concept was formed than before (O'Brien, P., and J. Griffin, Global Positioning System Systems Engineering Case Study, Air Force Center for Systems Engineering, Air Force Institute of Technology, www.afit.edu/cse).
2. As noted later, considerable controversy exists over credit for GPS. Various elements of the eventual concept were conceived of early but were reduced to practice within the Navy, Air Force, and Aerospace programs described here.
3. A fair amount of literature exists on the controversy over who deserves what credit for GPS. One distinction that might clarify the issue is who "invented" GPS versus who was the "architect" or "father" of GPS. It is abundantly clear that Parkinson was in charge at the key period and led the key decisions that formed GPS as it is today. It is equally clear that Easton originally came up with most of the concept, save for the signal. For some perspectives, see Easton, R., Who Invented the Global Positioning System? *The Space Review,* May 2006, retrieved from www.thespacereview.com/article/626/1; Comments on Navstar: Global Positioning System — Ten Years Later, by Easton, R., with replies by Parkinson and Gilbert, *Proceedings of the IEEE,* Vol. 73, Number 1, January 1985.

chapter 7

Collaborative Systems

Introduction: Collaboration as a Category

Most of the systems discussed so far have been the products of deliberate and centrally controlled development efforts. There was an identifiable client or customer (singular or plural), clearly identifiable builders, and users. Client, in the traditional sense, means the person or organization who sponsors the architect, and who has the resources and authority to construct the system of interest. The role of the architect existed, even if it was hard to trace to a particular individual or organization. The system was the result of deliberate value judgment by the client and existed under the control of the client. However, many systems are not under central control, either in their conception, their development, or their operation. The Internet is the canonical example, but many others exist, including electrical power systems, multinational defense systems, joint military operations, and intelligent transportation systems. These systems are all collaborative, in the sense that they are assembled and operate through the voluntary choices of the participants, not through the dictates of an individual client. These systems are built and operated only through a collaborative process.

Some systems are born as collaborative systems and others evolve that way. The Internet was not originally a collaborative system but has long ago passed out of centralized control. Global Positioning System (GPS) was not originally a collaborative system but is already at least partially one and is likely to soon move farther in that direction. Other systems, such as those architected to be multicompany or multigovernment collaborations are, or should be, considered as collaborative systems from the beginning.

A problem in this area is the lack of standard terminology for categories of system. Any system is an assemblage of elements that together possess capabilities not possessed by an element. This is just saying that a system possesses emergent properties, indeed that possessing emergent properties is the defining characteristic of a system. A microwave oven, a laptop computer, and the Internet are all systems; but each can have radically different problems in design and development.

This chapter discusses systems distinguished by the voluntary nature of the systems assembly and operation. Examples of systems in this category include most intelligent transport systems,[1] military C4I and Integrated Battlespace,[2] and partially autonomous flexible manufacturing systems.[3] The arguments here apply to most of what are often referred to as systems-of-systems, a term some readers may prefer. One of the authors (Maier) has discussed the contrast between the concepts elsewhere.[4]

What exactly is a collaborative system? In this chapter, a system is a "collaborative system" when its components

1. Are complex enough to be regarded as systems in their own right, and interact to provide functions not provided by any of the components alone; that is, the components in combination make up a system.
2. The component systems fulfill valid purposes in their own right and continue to operate to fulfill those purposes if disassembled from the overall system.
3. The component systems are managed (at least in part) for their own purposes rather than the purposes of the whole. The component systems are separately acquired and integrated but maintain a continuing operational existence independent of the collaborative system.

A separate issue is how the components come to be combined together. Our interest here is in systems deliberately constructed. Some people are interested in nondeliberate combinations that form recognizable systems. Some refer to these as "organic" systems, and there are a variety of interesting examples in human society. DeMarco presented an example known as the "Bombay Box-Wallah" system,[5] and others have described the operation of an ungoverned, yet organized urban environment.[6]

Misclassification as a "conventional" system versus a collaborative system (or vice versa) leads to serious problems. Especially important is a failure to architect for robust collaboration when direct control is impossible or inadvisable. This can arise when the developers believe they have greater control over the evolution of a collaborative system than they actually do. In believing this, they may fail to ensure that critical properties or elements will be incorporated by failing to provide a mechanism matched to the problem.

As with other domains, collaborative systems have their own heuristics, and familiar heuristics may have new application. To find them for collaborative systems, we look first at important examples and then generalize to find the heuristics. A key point is the heightened importance of interfaces, and the need to see interfaces at many layers. The explosion of the Internet and the World Wide Web is greatly facilitating collaborative system construction, but we find that the "bricks-and-mortar" of Internet-based collaborative systems are not at all physical. The building blocks are communication protocols, often at higher layers in the communications stack that is familiar from past systems.

Collaborative System Examples

Systems built and operated voluntarily are not unusual, even if they seem very different from classical systems engineering practice. Most of the readers of this book will be living in capitalist democracies where social order through distributed decisions is the philosophical core of government and society. Nations differ in the degree to which they choose to centralize versus decentralize decision making, but the fundamental principle of organization is voluntary collaboration. This book is concerned with technological systems, albeit sometimes systems with heavy social or political overtones. So, we take as our examples the systems whose building blocks are primarily technical. The initial examples are the Internet, intelligent transportation systems (for road traffic), and joint air defense systems.

The Internet

When we say "The Internet," we are not referring to the collection of applications that have become so popular (e-mail, World Wide Web, chats, and so forth). We are referring to the underlying communications infrastructure on which the distributed applications run. A picture of the Internet that tried to show all physical communications links active at one time would be a sea of lines with little or no apparent order. But, properly viewed, the Internet has a clear structure. The structure is a set of protocols called TCP/IP (Transmission Control Protocol/Internet Protocol). Their relationship to other protocols commonly encountered in the Internet is shown in Figure 7.1.[7] The TCP/IP suite includes the IP, TCP, and User Datagram Protocol (UDP) protocols in Figure 7.1. Note in Figure 7.1 that all the applications shown ultimately depend on IP. Applications can use only communications services supported by IP. IP, in turn, runs on many link and physical layer protocols. IP is "link friendly" in that it can be made to work on nearly any communications channel. This has made it easy to distribute widely, but prevents much exploitation of the unique features of any particular communication channel.

The TCP/IP family protocols are based on distributed operation and management. All data are encapsulated in packets, which are independently forwarded through the Internet. Routing decisions are made locally at each routing node. Each routing node develops its own estimate of the connection state of the system through the exchange of routing messages (also encapsulated as IP packets). The distributed estimates of connection state are not, and need not, be entirely consistent or complete. Packet forwarding works in the presence of *some* errors in the routing tables (although introduction of bad information can also lead to collapse).

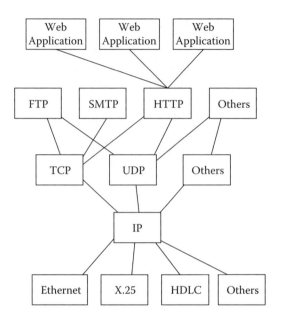

Figure 7.1 Protocol dependencies in the Internet.

The distributed nature of routing information, and the memoryless forwarding, allows the Internet to operate without central control or direction in the classic sense. Of course, control exists, but it is a collaborative, decentralized mechanism based on agreements-in-practice between the most important players. A decentralized development community matches the decentralized nature of control and decentralized architecture itself. There is no central body with coercive power to issue or enforce standards. There is a central body which issues standards, the Internet Engineering Task Force (IETF), but its practices are unlike nearly any other standards body. The IETF approach to standards is, fundamentally, to issue only those which have already been developed and deployed. The IETF acts more in a role of recognizing and promulgating standards than of creating them. Its apparently open structure (almost anybody can go to the IETF and try and form a working group to build standards in a given area) actually has considerable structure, albeit structure defined by customary practices rather than mandates.

The organization accepts nearly any working group that has the backing of a significant subset of participants. The working group can issue "Internet-drafts" with minimal overhead. For a draft to advance to the Internet equivalent of a published standard it must be implemented and deployed by two or more independent organizations. All Internet standards are available for free, and very strong efforts are made to keep them unencumbered by intellectual property. Proprietary elements are usually accepted only as optional extensions to an open standard. But under the surface, the practices of the organization create different forms of virtual mandates. Anybody can try to form a working group, but working code and a willingness to open source it speaks far louder than procedures. Powerful organizations can find their efforts in the IETF stymied by smaller players, if those players are faster and more willing to distribute working implementations, and forge alliances with others who will demonstrate interoperability in working systems.

Distributed operation, distributed development, and distributed management are linked. The Internet can be developed in a collaborative way largely because its operation is collaborative. Because the Internet uses best-effort forwarding and distributed routing, it can easily offer new services, as long as those new services depend only on best effort operation, without changing the underlying protocols. In contrast, services requiring hard network-level guarantees cannot be offered. New services can be implemented and deployed by groups that have no involvement in developing or operating the underlying protocols, but only so long as those new services do not require any new underlying services. So, for example, groups were able to develop and deploy IP-Phone (a voice over the Internet application) without any cooperation from TCP/IP developers or even Internet service providers. However, the IP-Phone application cannot offer any quality of service guarantees, because the protocols it is built on do not offer simultaneous delay and error rate bounding.

In contrast, networks using more centralized control can offer richer building block network services, including quality of service guarantees. However, they are much less able to allow distributed operation. Also, the collaborative environments that have produced telecommunications standards have been much slower moving than the Internet standards bodies. They have not adopted some of the practices of the Internet bodies that have enabled them to move quickly and rapidly capture market share. Of course, some of those practices would threaten the basic structure of the existing standards organizations.

In principle, a decentralized system like the Internet should be less vulnerable to destructive collective phenomena and be able to locally adapt around problems. In practice, both the Internet with its distributed control model and the telephone system with its greater centralization have proven vulnerable to collective phenomena. It turns out that distributed control protocols like TCP/IP are prone to collective phenomena in both transmission and routing.[8] Careful design and selection of parameters has been necessary to avoid network collapse phenomena. One reason is that the Internet uses a "good intentions" model for distributed control, which is vulnerable to nodes that misbehave either accidentally or deliberately. There are algorithms known that are robust against bad intentions faults, but they have not been incorporated into network designs.

The decentralized nature of the system has made it especially difficult to defend against coordinated distributed attacks (for example, distributed denial of service attacks). Centralized protocols often deal more easily with these attacks because they have strong knowledge of where connections originate and can initiate aggressive load-shedding policies under stress.

Wide area telephone blackouts have attracted media attention and shown that the more centralized model is also vulnerable. The argument about decentralized versus centralized fault tolerance has a long history in the electric power industry, and even today has not reached full resolution.

Intelligent Transportation Systems

The goal of most initiatives in intelligent transportation is to improve road traffic conditions through the application of information technology. The subject is broad and cannot be addressed in detail here.[9] We already discussed several aspects of ITSs in "Case Study 3," preceding Chapter 5. We pick out one issue to illustrate how a collaborative system may operate and the architectural challenges in making it happen.

One intelligent transportation concept is called "fully coupled routing and control." In this concept, a large fraction of vehicles are equipped with devices that determine their position and periodically report it to a traffic monitoring center. The device also allows the driver to enter his or her destination when beginning a trip. The traffic center uses the traffic conditions report to maintain a detailed estimate of conditions over a large metropolitan area. When the center gets a destination message, it responds with a recommended route to that destination, given the vehicle's current position. The route could be updated during travel if warranted. The concept is referred to as fully coupled because the route recommendations can be coupled with traditional traffic controls (for example, traffic lights, on-ramp lights, and reversible lanes).

Obviously, the concept brings up a wide array of sociotechnical issues. Many people may object to the lack of privacy inherent in their vehicle periodically reporting its position. Many people may object to entering their destination and having it reported to a traffic control center. Although there are many such issues, we narrow down once again to just one that best illustrates collaborative system principles. The concept works only if:

1. A large fraction of vehicles have, and use, the position reporting device.
2. A large fraction of drivers enter their (actual) destination when beginning a trip.
3. A large fraction of drivers follow the route recommendations they are given.

Under current conditions, vehicles on the roads are mostly privately owned and operated for the benefit of their owners. With respect to the collaborative system conditions, the concept meets it if using the routing system is voluntary. The vehicles continue to work whether or not they report their position and destination. And vehicles are still operated for their owner's benefit, not for the benefit of some "collective" of road users. So, if we are architecting a collaborative traffic control system, we have to explicitly consider how the three conditions above needed to gain the emergent capabilities are ensured.

One way to ensure them is to not make the system collaborative. Under some social conditions, we can ensure conditions one to three by making them legally mandatory and providing enforcement. It is a matter of judgment whether or not such a mandatory regime could be imposed.

If one judges that a mandatory regime is impossible, then the system must be collaborative. Given that it is collaborative, there are many architectural choices that can enhance the cooperation of the participants. For example, we can break apart the functions of traffic prediction, routing advice, and traditional controls and allocate some to private markets. Imagine an urban area with several "Traffic Information Provider" services. These services are private and subscription based, receive the position and destination messages, and disseminate the routing advice. Each driver voluntarily chooses a service, or none at all. If the service provides accurate predictions and efficient routes, it should thrive. If it cannot provide good service, it will lose subscribers and die.

Such a distributed, market-based system may not be able to implement all of the traffic management policies that a centralized system could. However, it can facilitate social cooperation in ways the centralized system cannot. A distributed, market-based system also introduces technical complexities into the architecture that a centralized system does not. In a private system, it must be possible for service providers to disseminate their information securely to paying subscribers. In a public, centralized system, information on conditions can be transmitted openly.

Joint Air Defense Systems

A military system may seem like an odd choice for collaborative systems. After all, military systems work by command, not voluntary collaboration. Leaving aside the social issue that militaries must always induce loyalty, which is a social process, the degree to which there is a unified command on military systems or operations is variable. A system acquired and operated as a single service can count on central direction. A system belonging to a single nation but spanning multiple services can theoretically count on central direction, but in practice it is likely to be largely collaborative. A system that comes together only in the context of multiservice, multinational, coalition military operations cannot count on central control and is always a collaborative system.

All joint military systems and operations have a collaborative element, but here we consider just air defense. An air defense system must fuse a complex array of sensors (ground radars, airborne radars, beacon systems, human observers, and other intelligence systems) into a logical picture of the air space, and then allocate weapon systems to engage selected targets. If the system includes elements from several services or nations, conflicts will arise. Nations, and services, may want to preferentially protect their own assets. Their command channels and procedures may affect greater self-protection, even when ostensibly operating solely for the goals of the collective.

Taking a group of air defense systems from different nations and different services and creating an effective integrated system from them is the challenge. The obvious path might be to try and convert the collection into something resembling a single service air defense system. This would entail unifying the command, control, and communications infrastructure. It would mean removing the element of independent management that characterizes collaborative systems. If this could be done, it is reasonable to expect that the resulting integrating system would be closer to a kind of point optimum. But, the difficulties of making the unification are likely to be insurmountable.

If, instead, we accept the independence, then we can try and forge an effective collaborative system. The technical underpinnings are clearly important. If the parts are going to collaborate to create integrated capabilities greater than the sum of the parts, they are going to have to communicate. So, even if command channels are not fully unified, communications must be highly interoperable. In this example, as in other sociotechnical examples, the social side should not be ignored. It is possible that the most important unifying elements in this example will be social. These might include shared training or educational background, shared responsibility, or shared social or cultural background.

Analogies for Architecting Collaborative Systems

One analogy that may apply is the urban planner. The urban planner, like the architect, develops overall structures. The architect structures a building for effective use by the client; the urban planner structures effective communities. The client of an urban planner is usually a community government, or one of its agencies. The urban planner's client and the architect's client differ in important respects. The architect's client is making a value judgment for himself or herself, and presumably has the resources to put into action whatever plan is agreed to with the architect. When the architect's plan is received, the client will hire a builder. In contrast, the urban planner's client does not actually build the city. The plan is to constrain and guide many other developers and architects who will come later, and hopefully guide their efforts into a whole greater than if there had been no overall plan. The urban planner and client are making value judgments for other people, the people who will one day inhabit the community being planned. The urban planner's client usually lacks the resources to build the plan, but can certainly stop something from being built if it is not in the plan. To be successful, the urban planner and client have to look outward and sell their vision. They cannot bring it about without the other's aid, and they normally lack the resources and authority to do it themselves.

Urban planning also resembles architecting in the spiral or evolutionary development process more than in the waterfall. An urban plan must be continuously adapted as actual conditions change. Road capacity that was adequate at one time may be inadequate at another. The mix of businesses that the community can support may change radically. As actual events unfold, the plan must adapt and be resold to those who participate in it, or it will be irrelevant.

Another analogy for collaborative systems is in business relationships. A corporation with semi-independent division is a collaborative system if the divisions have separate business lines, individual profit and loss responsibilities, and also collaborate to make a greater whole. Now consider the problem of a postmerger company. Before the merger, the components (the companies who are merging) were probably centrally run. After the merger, the components may retain significant independence but be part of a greater whole. Now if they are to jointly create something greater, they must go through a collaborative system instead of their traditional arrangement. If the executives do not recognize this and adapt, it is likely to fail. A franchise that grants its franchisees significant independence is also like a collaborative system. It is made up of independently owned and operated elements, which combine to be something greater than they would achieve individually.

Collaborative System Heuristics

As with builder-architecting, manufacturing, sociotechnical, and software-intensive systems, collaborative systems have their own heuristics. The heuristics discussed here have all been given previously, either in this book or its predecessor. But saying that they have been given previously does not mean that they have been explored for their unique applications in collaborative systems. For most people, heuristics do not stand alone as some sort of distilled wisdom. They function mainly as guides or "outline headings" to relevant experience. What is different here is their application — or the experience with specific respect to collaborative systems that generated the heuristic. Looking at how heuristics are applied to different domains gives a greater appreciation for their use and applicability in all domains.

Stable Intermediate Forms

The heuristic on stable intermediate forms is given originally as:

> *Complex systems will develop and evolve within an overall architecture much more rapidly if there are stable intermediate forms than if there are not.*

The original source of this heuristic is the notion of self-support during construction. It is good practice in constructing a building or bridge to have a structure that is self-supporting during construction rather than requiring extensive scaffolding or other weight-bearing elements that are later removed. The idea generalizes to other systems where it is important to design them to be self-supporting before they reach the final configuration. In the broader context, "self-supporting" can be interpreted in many ways beyond physical self-support. For example, we can think of economic and political notions of "self-support."

Stability in the more general context means that intermediate forms should be technically, economically, and politically self-supporting. Technical stability means that the system operates to fulfill useful purposes. Economic stability means that the system generates and captures revenue streams adequate to maintain its operation. Moreover, it should be in the economic interests of each participant to continue to operate rather than disengage. Political stability can be stated as the system has a politically decisive constituency supporting its continued operation, a subject we return to in Chapter 13. In collaborative systems, it cannot be assumed that all participants will continue to collaborate. The system will evolve based on continuous self-assessments of the desirability for collaboration by the participants:

- Integrated air defense systems are subject to unexpected and violent "reconfiguration" in typical use. As a result, they are designed with numerous fall-back modes, down to the anti-aircraft gunner working on his own with a pair of binoculars. Air defense systems built from weapon systems with no organic sensing and targeting capability have frequently failed in combat when the network within which they operate has come under attack.
- The Internet allows components nodes to attach and detach at will. Routing protocols adapt their paths as links appear and disappear. The protocol encapsulation mechanisms of IP allow an undetermined number of application layer protocols to simultaneously coexist.

Policy Triage

This heuristic gives guidance in selecting components and in setting priorities and allocating resources in development. It is given originally as:

> *The triage: Let the dying die. Ignore those who will recover on their own. And treat only those who would die without help.*

Triage can apply to any systems, but especially applies to collaborative systems. Part of the scope of a collaborative system is deciding what not to control. Attempting to overcontrol will fail for lack of authority. Undercontrol will eliminate the system nature of the integrated whole. A good choice enhances the desired collaboration.

- The Motion Picture Experts Group (MPEG), when forming their original standard from video compression, chose to standardize only the information needed to decompress a digital video stream.[10] The standard defines the format of the data stream and the operations required to reconstruct the stream of moving picture frames. However, the compression process is deliberately left undefined. By standardizing decompression, the usefulness of the standard for interoperability was assured. By not standardizing compression, the standard leaves open a broad area for the firms collaborating on the standard to continue to compete. Interoperability increases the size of the market, a benefit to the whole collaborative group, while retaining a space for competition eliminates a reason to not collaborate with the group. Broad collaboration was essential both to ensure a large market and to ensure that the requisite intellectual property would be offered for license by the participants.

Leverage at the Interfaces

Two heuristics, here combined, discuss the power of the interfaces:

> *The greatest leverage in system architecting is at the interfaces. The greatest dangers are also at the interfaces.*

When the components of a system are highly independent, operationally and managerially, the architecture of the system *is* the interfaces. The architect is trying to create emergent capability. The emergent capability is the whole point of the system. But, the architect may only be able to influence the interfaces among the nearly independent parts. The components are outside the scope and control of an architect of the whole.

- The Internet oversight bodies concern themselves almost exclusively with interface standards. Neither physical interconnections nor applications above the network protocol layers is standardized. Actually, both areas are the subject of standards, but not the standards process of the IETF.

One consequence is attention to different elements than in a conventional system development. For example, in a collaborative system, issues like life-cycle cost are of low importance. The components are developed collaboratively by the participants, who make choices to do so independently of any central oversight body. The design team for the whole cannot choose to minimize life-cycle cost, nor should they, because the decisions that determine costs are outside their scope. The central design team can choose interface standards, and can choose them to maximize the opportunities for participants to find individually beneficial investment strategies.

Ensuring Cooperation

> *If a system requires voluntary collaboration, the mechanism and incentives for that collaboration must be designed in.*

In a collaborative system, the components actively choose to participate or not. Like a free market, the resulting system is the web of individual decisions by the participants. Thus, an economists' argument that the costs and benefits of collaboration should be superior to the costs and benefits of independence for each participant individually should apply. As an example, the Internet maintains this condition, because the cost of collaboration is relatively low (using compliant equipment and following addressing rules) and the benefits are high (access to the backbone networks). Similarly in MPEG video standards, compliance costs can be made low if intellectual property is pooled, and the benefits are high if the targeted market is larger than the participants could achieve with proprietary products. Without the ability to retain a competitive space in the market (through differentiation on compression in the case of MPEG), the balance might have been different. Alternatively, the cost of noncompliance can be made high, though this method is less used.

An alternative means of ensuring collaboration is to produce a situation in which each participant's well-being is partially dependent on the well-being of the other participants. This joint utility approach is known, theoretically, to produce consistent behavior in groups. A number of social mechanisms can be thought of as using this principle. For example, strong social indoctrination in military training ties the individual to the group and serves as a coordinating operational mechanism in integrated air defense.

Another way of looking at this heuristic is through the metaphor of the franchise. The heuristic could be rewritten for collaborative systems as follows:

> *Consider a collaborative system a franchise. Always ask why the franchisees choose to join, and then choose to remain as members.*

Variations on the Collaborative Theme

The two criteria provide a sharp definition of a collaborative system, but they still leave open many variations. Some collaborative systems are really centrally controlled, but the central authority has decided to devolve authority in the service of system goals. In some collaborative systems a central authority exists, but power is expressed only through collective action. The participants have to mutually decide and act to take the system in a new direction. And, finally, some collaborative systems lack any central authority. They are entirely emergent phenomena.

We call a collaborative system where central authority exists and can act a *closed collaborative system*. Closed collaborative systems are those in which the integrated system is built and managed to fulfill specific purposes. It is centrally managed during long-term operation to continue to fulfill those purposes, and any new purposes the system owners may wish to address. The component systems maintain an ability to operate independently, but their normal operational mode is subordinated to the centrally managed purpose. For example, most single service air defense networks are centrally managed to defend a region against enemy systems, although the component systems retain the ability to operate independently, and do so when needed under the stress of combat.

Open collaborative systems are distinct from the closed variety in that the central management organization does not have coercive power to run the system. The component systems must, more or less, voluntarily collaborate to fulfill the agreed upon central purposes. The Internet is an open collaborative system. The IETF works out standards but has no power to enforce them. IETF standards work because the participants choose to implement them without proprietary variations, at least for the most part.

As the Internet becomes more important in daily life, in effect, as it becomes a new utility like electricity or the telephone, it is natural to wonder whether or not the current arrangement can last. Services on which public safety and welfare depends are regulated. Public safety and welfare, at least in industrial countries, are likely to depend on Internet operation in the near future, if they do not already. So, will the Internet and its open processes eventually come under regulation? To some extent, in some countries, it already has. In other ways, the movement is toward further decentralization in international bodies. Clearly, the international governing bodies have less control today over the purposes for which the Internet is used than did U.S. authorities when it was being rapidly developed in the 1990s.

Virtual collaborative systems lack both a central management authority and centrally agreed upon purposes. Large-scale behavior emerges, and may be desirable, but the overall system must rely upon relatively invisible mechanisms to maintain it.

A virtual system may be deliberate or accidental. Some examples are the current form of the World Wide Web and national economies. Both "systems" are distributed physically and managerially. The World Wide Web is even more distributed than the Internet in that no agency ever exerted direct central control, except at the earliest stages. Control has been exerted only through the publication of standards for resource naming, navigation, and document structure. Although, essentially just by social agreement, major decisions about Web architecture are filtered through very few people. Web sites choose to obey the standards or not at their own discretion. The system is controlled by the forces that make cooperation and compliance to the core standards desirable. The standards do not evolve in a controlled way, rather they emerge from the market success of various innovators. Moreover, the purposes the system fulfills are dynamic and change at the whim of the users.

National economies can be thought of as virtual systems. There are conscious attempts to architect these systems, through politics, but the long-term nature is determined by highly distributed, partially invisible mechanisms. The purposes expressed by the system emerge only through the collective actions of the system's participants.

Misclassification

Two general types of misclassification are possible. One is to incorrectly regard a collaborative system as a conventional system, or the reverse. Another is to misclassify a collaborative system as directed, voluntary, or virtual.

In the first case, system versus collaborative system, consider open-source software. Open-source software is often thought of as synonymous with Linux (or, perhaps more properly, GNU/Linux), a particular open-source operating system. Actually, there is a great deal of open-source, "free" software not related to Linux in any way. The success of the Linux model has spawned an open-source model of development now widely used for other software projects and some nonsoftware projects. Software is usually considered open source if the source code is freely available to a large audience, who can use it, modify it, and further redistribute it under the same open conditions by which they obtained it. Because Linux has been spectacularly successful, many others have tried to emulate the open-source model. The open-source model is built on a few basic principles,[11] perhaps heuristics. These include, from Eric Raymond:

1. Designs, and initial implementations, should be carried out by gifted individuals or very small teams.
2. Software products should be released to the maximum possible audience, as quickly as possible.
3. Users should be encouraged to become testers, and even codevelopers, by providing them source code.
4. Code review and debugging can be arbitrarily parallelized, at least if you distribute source code to your reviewers and testers.
5. Incremental delivery of small increments, with a very large user/tester population, leads to very rapid development of high quality software*

Of course, a side effect of the open-source model is losing the ability to make any significant amount of money distributing software you have written. The open-source movement advocates counter that effective business models may still be built on service and customization, but some participants in the process are accustomed to the profit margins normally had from manufacturing software. A number of companies and groups outside of the Linux community have tried to exploit the success of the Linux model for other classes of products, with mixed results. But as of the time of this writing, there are some success stories.

Some of this can be understood by realizing that open-source software development is a collaborative system. Companies or groups that have open-sourced their software without success typically run into one of two problems that limits collaboration. First, many of the corporate open-source efforts are not fully open. For example, both Apple and Sun Microsystems have open-sourced large pieces of strategic software. But both have released them under licenses that significantly restrict usage compared to the licenses in the Linux community. They (Apple and Sun) have argued that their license structure is necessary to their corporate survival and can lead to a more practical market for all involved. Their approach is more of a cross between traditional proprietary development and true open-source development. However, successful open-source development is a social phenomenon, and even the perception that it is

---

* The speed and quality of Linux releases can be measured, and it is clearly excellent. Groups of loosely coordinated programmers achieve quality levels equivalent to those of well-controlled development processes in corporations. This point is even admitted in the Microsoft "Halloween" memos on Linux, published at www.opensource.org/

less attractive or unfair may be sufficient to destroy the desired collaboration. In both cases, they later had to alter their strategy: in Sun's case, more toward full openness, and in Apple's case, backing away from it.

Second, the hypothesis that the quality of open-source software is due to the breadth of its review may simply be wrong. The real reason for the quality may be that Darwinian natural selection is eliminating poor-quality packages — the disappointed companies among them. In a corporation, a manager can usually justify putting maintenance money into a piece of software the company is selling even when the piece is known to be of very low quality. It will usually seem easier, and cheaper, to pay for "one more fix" than to start over and rewrite the bad software from scratch — this time correctly. But in the open-source community, there are no managers who can insist that a programmer maintain a particular piece of code. If the code is badly structured, hard to read, prone to failure, or otherwise unattractive, it will not attract the volunteer labor needed to keep in the major distributions, and it will effectively disappear. If nobody works on the code, it does not get distributed and natural selection has culled it.

For the second case, classification within the types of collaborative systems, consider a multiservice integrated battle management system. Military C4I systems are normally thought of as closed collaborative systems. As the levels of integration cross higher and higher administrative boundaries, the ability to centrally control the acquisition and operation of the system lessen. In a multiservice battle management system, there is likely to be much weaker central control across service boundaries than within those boundaries. A mechanism that ensures components will collaborate within a single service's system-of-systems, say a set of command operational procedures, may be insufficient across services.

In general, if a collaborative system is misclassified as closed, the builders and operators will have less control over purpose and operation than they may believe. They may use inappropriate mechanisms for insuring collaboration and may assume cooperative operations across administrative boundaries that will not reliably occur in practice. The designer of a closed collaborative system can require that an element behave in a fashion not to its own advantage (at least to an extent). In a closed collaborative system, the central directive mechanisms exist, but in an open collaborative system, the central mechanisms do not have directive authority. In an open collaborative system, it is unlikely that a component can be induced to behave to its own detriment. In an open collaborative system, the central authority lacks real authority and can proceed only through the assembly of voluntary coalitions.

A virtual collaborative system misclassified as open may show very unexpected emergent behaviors. In a virtual collaborative system, neither the purpose nor structure are under direct control, even of a collaborative body. Hence, new purposes and corresponding behaviors may arise at any time. The large-scale distributed applications on the Internet, for example USENET and the World Wide Web, exhibit this. Both were originally intended for exchange of research information in a collaborative environment but are now used for diverse purposes, some undesired and even illegal.

Standards and Collaborative Systems

The development of multicompany standards is a laboratory for collaborative systems. A standard is a framework for establishing some collaborative systems. The standard (for example, a communication protocol or programming language standard) creates the environment within which independent implementations can coexist and compete.

> *Example*: Telephone standards allow equipment produced by many companies in many countries to operate together in the global telephone network. A call placed in country and traverse switches from different manufacturers and media in different countries with nearly the same capabilities as if the call were within a single country on one company's equipment.

> *Example*: Application programming interface (API) standards allow different implementations of both software infrastructure and applications to coexist. So, operating systems from different vendors can support the same API and allow compliant applications to run on any systems from any of the vendors.

Historically, there has been a well-established process for setting standards. There are recognized national and international bodies with the responsibility to set standards, such as the International Standards Organization (ISO), the American National Standards Institute (ANSI), and so forth. These bodies have a detailed process that has to be followed. The process defines how standards efforts are approved, how working groups operate, how voting is carried out, and how standards are approved. Most of these processes are rigorously democratic (if significantly bureaucratic). The intention is that a standard should reflect the honest consensus of the concerned community and is thus likely to be adopted.

Since 1985, this established process has been run over, at least within the computer field, by Internet, Web, and open-source processes. The IETF, which never votes on a standard in anything like the same sense as ISO or ANSI, has completely eclipsed the laboriously constructed Open

Systems Interconnect (OSI) networking standard. Moreover, the IETF model has spread to a wide variety of networking standards. As another example, in operating systems the most important standards are either proprietary (from Microsoft, Apple, and others) or defined by open-source groups (Linux and BSD). Again, the traditional standards bodies and their approaches have played only a little role.

Because the landscape is still evolving, it may be premature to conclude what the new rules are. It may be that we are in a time of transition, and that after the computing market settles down we will return to more traditional methods. It may be that when the computer and network infrastructure is recognized as a central part of the public infrastructure (like electricity and telephones), it will be subjected to similar regulation and will respond with similar bureaucratic actions. Or, it may be that traditional standards bodies will recognize the principles that have made the Internet efforts so successful and will adapt. Some fusion may prove to be the most valuable yet. In that spirit, we consider what heuristics may be extracted from the Internet experience. These heuristics are important not only to standards efforts, but to collaborative systems as a whole because standards are a special case of collaborative system.

Economists call something a "network good" if it increases in value the more widely it is consumed. So, for example, telephones are network goods. A telephone that does not connect to anybody is not valuable. Two cellular telephone networks that cannot interoperate are much less valuable than if they can interoperate. The central observation is that:

> Standards are network goods, and must be treated
> as such.

Standards are network goods because they are useful only to the extent that other people use them. One company's networking standard is of little interest unless other companies support it (unless, perhaps, that company is a monopoly). What this tells standards groups is that achieving large market penetration is critically important. Various practices flow from this realization. The IETF, in contrast to most standards groups, gives its standards away for free. A price of zero encourages wide dissemination. Also, the IETF typically gives away reference implementations with its standards. That is, a proposal rarely becomes a standard unless it has been accompanied by the release of free source code that implements the standard. The free source code may not be the most efficient, may not be fully featured, probably does not have all the extras in interface that a commercial product should have, but it is free and it does provide a reference case against which everybody else can work. The IETF culture is that proponents of an approach are rarely given much credibility unless they are distributing implementations.

The traditional standards organizations protest that they cannot give standards away because the revenue from standard sales is necessary to support their development efforts. But, the IETF has little trouble supporting its efforts. Its working conferences are filled to overflowing and new proposals and working groups are appearing constantly. Standards bodies do not need to make a profit, indeed should not. If they can support effective standards development they are successful, though removing the income of standards sales might require substantial organizational change.

Returning to collaborative systems in general, the example of standards shows the importance of focusing on real collaboration, not the image of it. Commitment to real participation in collaboration is not indicated by voting; it is indicated by taking action that costs something. Free distribution of standards and reference implementations lowers entrance costs. The existence of reference implementations provides clear conformance criteria that can be explicitly tested.

Conclusion

Collaborative systems are those that exist only through the positive choices of component operators and managers. These systems have long existed as part of the civil infrastructure of industrial societies, but have come into greater prominence as high-technology communication systems have adopted similar models, as centralized systems have been decentralized through deregulation or divestiture, and as formerly independent systems have been loosely integrated into larger wholes. What sets these systems apart is their need for voluntary actions on the part of the participants to create and maintain the whole. This requires that the architect revisit known heuristics for greater emphasis and additional elaboration. Among the heuristics that are particularly important are:

1. *Stable Intermediate Forms*: A collaborative system designer must pay closer attention to the intermediate steps in a planned evolution. The collaborative system will take on intermediate forms dynamically and without direction, as part of its nature.
2. *Policy Triage*: The collaborative system designer will not have coercive control over the system's configuration and evolution. This makes choosing the points at which to influence the design more important.
3. *Leverage at the Interfaces*: A collaborative system is defined by its emergent capabilities, but its architects have influence on its interfaces. The interfaces, whether thought of as the actual physical interconnections or as higher-level service abstractions, are the primary points at which the architect can exert control.
4. *Ensuring Cooperation*: A collaborative system exists because the partially independent elements decide to collaborate. The designer must

consider why they will choose to collaborate and foster those reasons in the design.

5. A collaboration is a network good; the more of it there is, the better. Minimize entrance costs and provide clear conformance criteria.

Exercises

1. The Internet, multimedia video standards (MPEG), and the GSM digital cellular telephone standard are all collaborative systems. All of them also have identifiable architects, a small group of individuals who carried great responsibility for the basic technical structures. Investigate the history of one of these cases and consider how the practices of the collaborative system architect differ from architects of conventional systems.

2. In a collaborative system, the components can all operate on their own whether or not they participate in the overall system. Does this represent a cost penalty to the overall system? Does it matter? Discuss from the perspective of some of the examples.

3. Collaborative systems in computing and communication usually evolve much more rapidly than those controlled by traditional regulatory bodies, and often more rapidly than those controlled by single companies. Is this necessary? Could regulatory bodies and companies adopt different practices that would make their systems as evolvable as collaborative (for example, Internet or Linux) while retaining the advantages of the traditional patterns of control?

Exercises to Close Part II

Explore another domain much as builder-architected, sociotechnical, manufacturing, software, and collaborative systems are explored in this part. What are the domain's special characteristics? What more broadly applicable lessons can be learned from it? What general heuristics apply to it? Some suggested, heuristic-domains to explore include the following:

1. *Telecommunications* in its several forms: point-to-point telephone network systems, broadcast systems (terrestrial and space), and packet-switched data (the Internet).

2. *Electric power*, which is widely distributed with collaborative control, is subject to complex loading phenomena (with a social component), and is regulated. (Hill, David J., Special Issue on Nonlinear Phenomena in Power Systems: Theory and Practical Implications, *Proceedings of the IEEE*, Vol. 83, Number 11, November, 1995.)

3. *Transportation*, in both its current form and in the form of proposed intelligent transportation systems.

4. *Financial systems*, including global trading mechanisms and the operation of regulated economics as a system.
5. *Space systems*, with their special characteristics of remote operation, high initial capital investment, vulnerability to interference and attack, and their effects on the design and operation of existing earth-borne system performing similar functions.
6. *Existing and emerging media systems*, including the collection of competing television systems of private broadcast, public broadcast, cable, satellite, and video recording.

Notes and References

1. IVHS America, Strategic Plan for Intelligent Vehicle-Highway Systems in the United States, IVHS America, Report IVHS-AMER-92-3, Intelligent Vehicle-Highway Society of America (ITS), Washington, DC, 1992; U.S. Department of Transportation (USDOT), National Program Plan for ITS, 1995.
2. Butler, S., D. Diskin, N. Howes, and K. Jordan, The Architectural Design of the Common Operating Environment for the Global Command and Control System, *IEEE Software*, pp. 57–66, November 1996.
3. Hayes, Robert H., S. C. Wheelwright, and K. B. Clark, *Dynamic Manufacturing.* New York: The Free Press, 1988.
4. Maier, M. W., Architecting Principles for Systems-of-Systems, *Systems Engineering*, Vol. 2, Number 1, pp. 1–18, 1999.
5. DeMarco, T., On Systems Architecture, Monterey Workshop on Specification-Based Software Architectures, U.S. Naval Postgraduate School, Monterey, CA, September 1995.
6. Lambot, I., and G. Girard, *City of Darkness — Life in Kowloon City.* San Francisco, CA: Watermark Press, 1999. The book contains extensive photographs and observations on the development of Kowloon Walled City outside of Hong Kong, an area that could be said to have "organic" or uncontrolled architecture.
7. Modeled after Peterson, L., and B. Davie, *Computer Networks: A Systems Approach.* San Francisco, CA: Morgan Kaufman, 1996.
8. See Bersekas, D., and R. Gallager, *Data Networks*, 2nd ed. New York: Prentice Hall, 1992, particularly Chapter 6.
9. See the IVHS America and USDOT references above. Also, Maier, M. W., On Architecting and Intelligent Transport Systems, Joint Issue *IEEE Transactions on Aerospace and Electronic Systems/System Engineering*, AES33:2, pp. 610–625, April 1997, by one of the present authors discusses the architectural issues specifically.
10. Chiariglione, L., Impact of MPEG Standards on Multimedia Industry, *IEEE Proceedings*, Vol. 86, Number 6, pp. 1222–1227, June, 1998.
11. The Cathedral and the Bazaar by Eric Raymond was the original source for these heuristics, referenced at www.catb.org/~esr/writings/cathedral-bazaar/ The open-source initiative at www.opensource.org has other additional details.

Models and Modeling

Introduction to Part III

What is the product of an architect? Although it is tempting to regard the building or system as the architect's product, the relationship is necessarily indirect. The system is actually built by the developer. The architect acts to translate between the problem domain concepts of the client and the solution domain concepts of the builder. Great architects go beyond the role of intermediary to make a visionary combination of technology and purpose that exceeds the expectation of builder or client. But the system cannot be built as envisioned unless the architect has a mechanism to communicate the vision and track construction against it. The concrete, deliverable products of the architect, therefore, are models of the system.

Individual models alone are point-in-time representations of a system. Architects need to see and treat each as a member of one of several progressions. The architect's first models define the system concept. As the concept is found satisfactory and feasible, the models progress to the detailed, technology-specific models of design engineers. The architect's original models come into play again when the system must be certified.

A Civil Architecture Analogy

Once again, civil architecture provides a familiar example of modeling and progression. An architect is retained to ensure that the building is pleasing to the client in all senses (aesthetically, functionally, and financially). One product of the architect is intangible; it is the conceptual vision that the physical building embodies and that satisfies the client. But the intangible product is worthless without a series of increasingly detailed tangible products, all models of some aspect of the building. Table III.1 lists some of the models and their purposes.

Table III.1 Models and Purposes in Civil Architecture

Model	Purpose
Physical scale model	Convey look and site placement of building to architect, client, and builder
Floor plans	Work with client to ensure building can perform basic functions desired
External renderings	Convey look of building to architect, client, and builder
Budgets, schedules	Ensure building meets client's financial performance objectives, manage builder relationship
Construction blueprints	Communicate design requirements to builder, provide construction acceptance criteria

The progression of models during the design life cycle can be visualized as a steady reduction of abstraction. Early models may be quite abstract. They may convey only the basic floor plan, associated order-of-magnitude budgets, and renderings encompassing only major aesthetic elements. Early models may cover many disparate designs representing optional building structures and styles. As decisions are made, the range of options narrows and the models become more specific. Eventually, the models evolve into construction drawings and itemized budgets and pass into the hands of the builders. As the builders work, models are used to control the construction process and to ensure the integrity of the architectural concept. Even when the building is finished, some of the models will be retained to assist in future project developments and to act as an as-built record for building alterations.

Making the key design decisions and building the models are obviously intertwined but still distinct activities. One could build a fine set of models that embodied terrible decisions, and excellent decisions could be embodied in an incompetently built set of models. The first case will undoubtedly lead to disappointment (or disaster), and the second case very likely will. The only saving grace in the second case is that later implementers might recognize the situation and work to correct it. The focus of this book is on decisions over descriptions, but in this part we address the issues of modeling and description directly.

Guide to Part III

Although the form of the models differs greatly from civil architecture to aerospace, computer, or software architectures, their purposes and relationships remain the same. Part III discusses the concrete elements of architectural practice, the models of systems, and their development. The discussion is from two perspectives broken into three chapters. First, models are treated as the concrete representations of the various views that define a system. This perspective is treated in general in Chapter 8, and through domain-specific examples in Chapter 10. Second, the evolution and development of models are treated as the core of the architecting process. Chapter 9 develops the idea of progressive design as an organizing principle for the architecting process. A community effort at standardizing architecture representation models, called architecture description frameworks, is the subject of Chapter 11.

Chapter 8 covers the types of models used to represent systems and their roles in architecting. Because architecting is multidimensional and multidisciplinary, an architecture may require many partially independent views. The chapter proposes a set of six basic views and reviews major categories of models for each view. It also introduces viewpoint as an organizing abstraction for writing architecture description standards. Because a coherent and integrated product is the ultimate goal, the models chosen must also be designed to integrate with each other. That is, they must define and resolve their interdependencies and form a complete definition of the system to be constructed.

Chapter 9 looks for principles to organize the eclectic architecting process. A particularly useful principle is that of progression — the idea that models, heuristics, evaluation criteria, and many other aspects of the system evolve on parallel tracks from the abstract to the specific and concrete. Progression also helps tie architecting into the more traditional engineering design disciplines. This book largely treats system architecting as a general process, independent of domain, but in practice it is necessarily strongly tied to individual systems and domains. Nevertheless, each domain contains a core of problems not amenable to rational, mechanistic solution that are closely associated with reconciling customer or client need and with technical capability. This core is the province of architecting. Architects are not generalists; they are specialists in systems, and their models must refine into the technology-specific models of the domains in which their systems are to be realized.

Chapter 10 returns to models, now tying the previous two chapters together by looking at specific modeling methods. Examined in the chapter is a series of integrating methodologies that illustrate the attributes discussed in the previous chapters: multiple views, integration across views, and progression from abstract to concrete implementation. Examples of integrated models and methods are given for computer-based systems, performance-oriented systems, software-intensive systems, manufacturing systems, and sociotechnical systems. Described in the first part of Chapter 10 are two general-purpose integrated modeling methods, Hatley-Pirbhai and Quantitative Quality Function Deployment. The former specializes in combining behavioral and physical implementation models. The latter specializes in integrating quantitative performance requirements with behavioral and implementation models. Subsequent sections describe integrated models for software, manufacturing systems, and sociotechnical systems.

Chapter 11 looks outward to the community interested in architecture to review recent work in standardizing architecture descriptions. Standards for architecture description are usually referred to as architecture description frameworks. The chapter reviews three of the leading ones, with some mention of others. They are the U.S. Department of Defense Architecture Framework (DODAF), the ISO Reference Model for Open Distributed Processing, and the IEEE's 1471 Recommended Practice for Architectural Description of Software-Intensive Systems. This chapter continues by discussing some of the current controversies in frameworks and possible resolutions.

chapter 8

Representation Models and Systems Architecting

> By relieving the mind of all unnecessary work, a good
> notation sets it free to concentrate on more advanced
> problems, and in effect increases the mental power
> of the [human] race.

Alfred North Whitehead

Introduction: Roles, Views, and Models

Models are the primary means of communication with clients, builders, and users; models are the language of the architect. Models enable, guide, and help assess the construction of systems as they are progressively developed and refined. After the system is built, models, from simulators to operating manuals, help describe and diagnose its operation.

To be able to express system imperatives and objectives, and manage system design, the architect should be fluent, or at least conversant, with all the languages spoken in the long process of system development. These languages are those of system specifiers, designers, manufacturers, certifiers, distributors, and users.

The most important models are those that define the critical acceptance requirements of the client and the overall structure of the system. The former are a subset of the entirety of the requirements, and the latter are a subset of the complete, detailed system design. Because the architect is responsible for total system feasibility, the critical portions may include highly detailed models of components on which success depends and abstract, top-level models of other components.

Models can be classified by their role or by their content. Role is important in relating models to the tasks and responsibilities not only of architects, but of many others in the development process. Of special importance to architects are modeling methods that tie together otherwise separate models into a consistent whole.

Roles of Models

Models fill many roles in systems architecting, including the following:

1. Communication with client, users, and builders.
2. Maintenance of system integrity through coordination of design activities.
3. Design assistance by providing templates, and organizing and recording decisions.
4. Exploration and manipulation of solution parameters and characteristics; guiding and recording of aggregation and decomposition of system functions, components, and objects.
5. Performance prediction; identification of critical system elements.
6. Provision of acceptance criteria for certification for use.

These roles are not independent; each relates to the other. But the foremost is to communicate. The architect discusses the system with the client, the users (if different), the builders, and possibly many other interest groups. Models of the system are the medium of all such communication. After all, the system will not come into being for some time to come. The models used for communication become documentation of decisions and designs and thus vehicles for maintaining design integrity. Powerful, well-chosen models will assist in decision making by providing an evocative picture of the system in development. They will also allow relevant parameters and characteristics to be manipulated and the results seen in terms relevant to client, user, or builder.

Communication with the client has two goals. First, the architect must determine the client's objectives and constraints. Second, the architect must insure that the system to be built reflects the value judgments of the client where perfect fulfillment of all objectives is impossible. The first goal requires eliciting information on objectives and constraints and casting it into forms useful for system design. The second requires that the client perceive how the system will operate (objectives and constraints) and that the client can have confidence in the progress of design and construction. In both cases, models must be clear and understandable to the client, expressible in the client's own terminology. It is desirable that the models also be expressive in the builder's terms, but because client expressiveness must take priority, proper restatement from client to builder language usually falls to the architect.

User communication is similar to client communication. It requires the elicitation of needs and the comparison of possible systems to meet those needs. When the client is the user, this process is simplified. When the client and the users are different (as discussed in Chapter 5 on sociotechnical systems), their needs and constraints may conflict. The architect is in the position to attempt to reconcile these conflicts.

In two-way communication with the builder, the architect seeks to insure that the system will be built as conceived and that system integrity is maintained. In addition, the architect must learn from the builder those technical constraints and opportunities that are crucial in insuring a feasible and satisfactory design. Models that connect the client and the builder are particularly helpful in closing the iterations from builder technical capability to client objectives.

One influence of the choice of a model set is the nature of its associated "language" for describing systems. Given a particular model set and language, it will be easy to describe some types of systems and awkward to describe others, just as natural languages are not equally expressive of all human concepts. The most serious risk in the choice is that of being blind to important alternate perspectives due to long familiarity (and often success) with models, languages, and systems of a particular type.

Models, Viewpoints, and Views

Chapters 8 through 10 discuss this book's approach to modeling in systems architecting. Chapter 11 looks outward to the community to review other important approaches and draw contrasts. Unfortunately, there is a lot of variation in the usage of important terms. There are three terms that are important in setting up a modeling framework: *model*, *view*, and *viewpoint*. We use the definitions of model, view, and viewpoint taken from the Institute of Electrical and Electronics Engineers (IEEE) standards:

> *Model*: An approximation, representation, or idealization of selected aspects of the structure, behavior, operation, or other characteristics of a real-world process, concept, or system (IEEE 610.12-1990).

> *View*: A representation of a system from the perspective of related concerns or issues (ANSI/IEEE 1471-2000).

> *Viewpoint*: A template, pattern, or specification for constructing a view (ANSI/IEEE 1471-2000).

As discussed above, a model is just a representation of something; in our case some aspect of the architecture of a system. The modeling languages of interest have a vocabulary and a grammar. The words are the parts of a model; the grammar defines how the words can be linked.

Beyond that, a modeling language has to have a method of interpretation; the models produced have to mean something, typically within some domain. For example, in a block diagramming method, the words are the kinds of blocks and lines and the grammar are the allowed patterns by which they can be connected. The method also has to define some correspondence between the blocks, lines, and connections to things in the world. A physical method will have a correspondence to physically identifiable things. A functional diagramming technique has a correspondence to more abstract entities — the functions that the system does.

A view is just a collection of models that share the property that they are relevant to the same concerns of a system stakeholder. For example, a functional view collects the models that represent a system function. An objectives view collects the models that define the objectives to be met by building the system. The idea of view is needed because complex systems tend to have complex models and require a higher-level organizing element.

View is inspired by the familiar idea of architectural views. An architect produces elevations, floor plans, and other representations that show the system from a particular perspective. The idea of view here generalizes this when physical structure is no longer primary.

Viewpoint is an abstraction of view across many systems. It is important only in defining standards for architecture description, so we defer its use until later.

These concepts are depicted schematically in Figure 8.1.

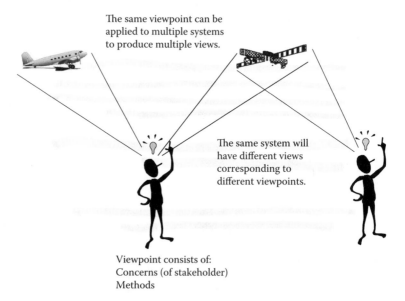

The same viewpoint can be applied to multiple systems to produce multiple views.

The same system will have different views corresponding to different viewpoints.

Viewpoint consists of:
Concerns (of stakeholder)
Methods

Figure 8.1 The concept of viewpoint and view.

Classification of Models by View

A view describes a system with respect to some set of attributes or concerns. The set of views chosen to describe a system is variable. A good set of views should be complete (cover all concerns of the architect's stakeholders) and mostly independent (capture different pieces of information). Table 8.1 lists the set of views chosen here as most important to architecting. A system can be "projected" into any view, possibly in several ways. The projection into views and the collection of models by views is shown schematically in Figure 8.2. Each system has some behavior (abstracted from implementation), has a physical form, and retains data. Views are composed of models. Not all views are equally important to system developmental success, and the set will not remain constant over

Table 8.1 Major System or Architectural Views

Perspective or View	Description
Purpose/objective	What the client wants
Form	What the system is
Behavioral or functional	What the system does
Performance objectives or requirements	How effectively the system does it
Data	The information retained in the system and its interrelationships
Managerial	The process by which the system is constructed and managed

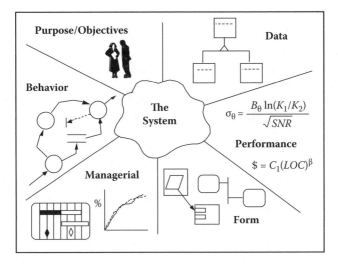

Figure 8.2 The six views. All views are representations of some aspect of the actual system. Each view may contain several models, as needed to capture the information of the view.

time. For example, a system might be behaviorally complex but have relatively simple form. Views that are critical to the architect may play only a supporting role in full development.

Although any system can be described in each view, the complexity and value of each view's description can differ considerably. Each class of systems emphasizes particular views and has favored modeling methods, or methods of representation within each view. The architect must determine which views are most important to the system and its environment and be expert in the relevant models. The views are chosen to be reasonably independent, but there is extensive linkage among views. For example, the behavioral aspects of the system are not independent of the system's form. The system can produce the desired behavior only if the system's form supports it. This linkage is conceptually similar to a front and side view being linked (both show vertical height) even though they are observations from orthogonal directions.

The following sections describe models used for representing a system in each of the views of Table 8.1. The section for each view defines what information is captured by the view, describes the modeling issues within that view, and lists some important modeling methods. Part of the architect's role is to determine which views are most critical to system success, build models for those views, and then integrate as necessary to maintain system integrity. The integration across views is a special concern of the architect.

Note to the Reader

The sections to follow, which describe models for each view, are difficult to understand without examples meaningful to each reader. Rather than trying to present detailed examples of each for each of several system domains (a task that might require its own book), we suggest the reader does so on his or her own. The examples given in the chapter are not detailed and are chosen to be understandable to the widest possible audience. Chapter 10 describes, in detail, specific modeling methods that span and integrate multiple views. The methods of Chapter 10 are what the architect should strive for, an integrated picture of all important aspects of a system.

As stated in the Introduction, Part III can be read several ways. The chapters can be read in order, which captures the intellectual thread of model concepts, modeling processes and heuristics, specific modeling methods, and organizing frameworks. In this case, it is useful to read ahead to exercises 1 and 2 at the end of this chapter and work them while reading each section to follow. The remaining exercises are intended for after the chapter is read, although some may be approached as each section is completed. An alternative is to read Chapters 8 and 10 in parallel, reading the specific examples of models in Chapter 10 as the views are covered in Chapter 8. Because the approach of Chapter 10 is to look at integrated models, models that span views, a one-for-one correspondence is impossible. The linear approach is probably best for those without extensive background in modeling methods. Those with a good background in integrated modeling methods can use either.

Objectives and Purpose Models

The first modeling view is that of objectives and purposes. Systems are built for useful purposes — that is, for what the client *wants*. Without them the system cannot survive. The architect's first and most basic role is to match the desirability of the purposes with the practical feasibility of a system to fulfill those purposes. Given a clearly identifiable client, the architect's first step is to work with that client to identify the system's objectives and priorities. Some objectives can be stated and measured precisely. Others will be quite abstract, impossible to express quantitatively. A civil architect is not surprised to hear a client's objective is for the building to "be beautiful" or to "be in harmony with the natural state of the site." The client will be very unhappy if the architect tells the client to come back with unambiguous and testable requirements. The architect must prepare models to help the client to clarify abstract objectives. Abstract objectives require provisional and exploratory models, models that may fall by the wayside later as the demands and the resulting system become well understood. Ideally, all iterations and explorations become part of the systems document set. However, to avoid drowning in a sea of paper, it may be necessary to focus on a limited set. If refinement and trade-off decisions (the creation of concrete objectives from abstract ones) are architectural drivers, they must be maintained, as it is likely the key decisions will be repeatedly revisited.

Modeling therefore begins by restating and iterating those initial unconstrained objectives from the client's language until a modeling language and methodology emerges, the first major step closer to engineering development. Behavioral objectives are restated in a behavioral modeling language. Performance requirements are formulated as measurable satisfaction models. Some objectives may translate directly into physical form, others into patterns of form that should be exhibited by the system. Complex objectives almost invariably require several steps of refinement and indeed may evolve into characteristics or behaviors quite different from their original statement.

A low-technology example (though only by modern standards) is the European cathedrals of the Middle Ages. A cathedral architect considered a broad range of objectives. First, a cathedral must fulfill well-defined community needs. It must accommodate celebration-day crowds, serve as a suitable seat for the bishop, and operate as a community centerpiece.

But, in addition, cathedral clients of that era emphasized that the cathedral "communicate the glory of God and reinforce the faithful through its very presence."

Accommodation of holiday celebrants is a matter of size and floor layout. It is an objective that can be implemented directly and requires no further significant refinement. The clients — the church and community leaders — because of their personal familiarity with the functioning of a cathedral, could determine for themselves the compliance of the cathedral by examining the floor plan. But what of defining a building that "glorifies God?" This is obviously a property only of the structure as a whole — its scale, mass, space, light, and integration of decoration and detail. Only a person with an exceptional visual imagination is able to accurately envision the aesthetic and religious impact of a large structure documented only through drawings and renderings. Especially in those times, when architectural styles were new and people traveled little, an innovative style would be an unprecedented experience for perhaps all but the architect.

In this example, we also see the interaction of heuristics and modeling. Models define the architect's approach to the cathedral, but heuristics would be needed to guide decision making. How does the architect know what building features produce the emotional effect that will be regarded as glorifying God and reinforcing the faithful? The architect can know only through induction (experience with other buildings) and the generalization of that induction through theory. Our own experiences should be enough to suggest the elements of appropriate heuristics (for example, great vertical height, visual and auditory effects, integration of iconography and visual teachings).

Refinement of objectives through models is central to architecting, but it is also a source of difficulty. A design that proceeds divorced from direct client relevance tends to introduce unnecessary requirements that complicate its implementation. Experience has shown that retaining the client's language throughout the acquisition process can lead to highly efficient, domain-specific architectures, for example, in communication systems.

> *Example*: Domain Specific Software Architectures[1] are software application generation frameworks in which domain concepts are embedded in the architectural components. The framework is used to generate a product line of related applications in which the client language can be used nearly directly in creating the product. For a set of message handler applications within command and control systems, the specification complexity was reduced 50:1.

One measure of the power of a design and implementation method is its ability to retain the original language. But this poses a dilemma. Retention implies the availability of proven, domain-specific methods and engineering tools. But unprecedented systems by definition are likely to be working in new domains, or near the technical frontiers of existing domains. By the very nature of unprecedented system development, such methods and tools are unlikely to be available. Consequently, models and methodologies must be developed and pass through many stages of abstraction, during which the original relevance can easily be lost. The architect must therefore search out domain-specific languages and methods that can somehow maintain the chain of relevance throughout.

An especially powerful, but challenging, form of modeling converts the client or user's objectives into a meta-model or metaphor that can be directly implemented. A famous example is the desktop metaphor adopted for Macintosh computers. The user's objective is to use a computer for daily, office-oriented task automation. The solution is to capture the user's objectives directly by presenting a simulation of a desktop on the computer display. Integrity with user needs is automatically maintained by maintaining the fidelity of a desktop and file system familiar to the user.

Models of Form

Models of form represent physically identifiable elements of, and interfaces to, what will be constructed and integrated to meet client objectives. Models of form are closely tied to particular construction technologies, whether the concrete and steel of civil architecture or the less tangible codes and manuals of software systems. Even less tangible physical forms are possible, such as communication protocol standards, a body of laws, or a consistent set of policies.

Models of form vary widely in their degree of abstraction and role. For example, an abstract model may convey no more than the aesthetic feel of the system to the client. A dimensionally accurate but hollow model can assure proper interfacing of mechanical parts. Other models of form may be tightly coupled to performance modeling, as in the scale model of an airplane subjected to wind tunnel testing. The two categories of models of form most useful in architecting are scale models and block diagrams.

Scale Models

The most literal models of form are scale models. Scale models are widely used for client and builder communication and may function as part of behavioral or performance modeling as well. Some major examples include the following:

1. Civil architects build literal models of buildings, often producing renderings of considerable artistic quality. These models can be abstracted to convey the feel and style of a building or can be precisely detailed to assist in construction planning.
2. Automobile makers mock up cars in body-only or full running trim. These models make the auto show circuit to gauge market interest or are used in engineering evaluations.
3. Naval architects model racing yachts to assist in performance evaluation. Scale models are drag tested in water tanks to evaluate drag and handling characteristics. Reduced or full-scale models of the deck layout are used to run simulated sail handling drills.
4. Spacecraft manufacturers use dimensionally accurate models in fit compatibility tests and in crew extravehicular activity rehearsals. Even more important are ground simulators for on-orbit diagnostics and recovery planning.
5. Software developers use prototypes that demonstrate limited characteristics of a product that are equivalent to scale models. For example, user interface prototypes that look like the planned system but do not possess full functionality, non-real-time simulations that carry extensive functionality but do not run in real-time, or just a set of screen shots with scenarios for application use.

Physical scale models are gradually being augmented or replaced by virtual reality systems. These "scale" models exist only in a computer and the viewer's mind. They may, however, carry an even stronger impression of reality than a physical scale model because of the sensory immersion achievable.

Block Diagrams

A scale model of a circuit board or a silicon chip is unlikely to be of much interest alone, except for expanded-scale plots used to check for layout errors. Nonetheless, physical block diagrams are ubiquitous in the electronics industry. To be a model of form, as distinct from a behavioral model, the elements of the block diagram must correspond to physically identifiable elements of the system. Some common types of block diagrams include the following:

1. System interconnect diagrams that show specific physical elements (modules) connected by physically identifiable channels. On a high-level diagram, a module might be an entire computer complex and a channel might be a complex internetwork. On a low level, the modules could be silicon chips with specific part numbers and the channels pin-assigned wires.

2. System flow diagrams that show modules in the same fashion as interconnect diagrams but illustrate the flow of information among modules. The abstraction level of information flow defined might be high (complex messaging protocols) or low (bits and bytes). The two types of diagrams (interconnect and flow) are contrasted in Chapter 10, Figure 10.3.

3. Structure charts,[2] task diagrams,[3] and class and object diagrams[4] that structurally define software systems and map directly to implementation. A software system may have several logically independent such diagrams, each showing a different aspect of the physical structure. Take for example, diagrams that show the invocation tree, the inheritance hierarchy, or the "withing" relationships in an Ada program. Examples of several levels of physical software diagram are given in Figure 10.5 and Figure 10.6 in Chapter 10.

4. Manufacturing process diagrams are drawn with a standardized set of symbols. These represent manufacturing systems at an intermediate level of abstraction, showing specific classes of operation but not defining the machine or the operational details.

Several authors have investigated formalizing block diagrams over a flexible range of architectural levels. The most complete, with widely published examples, is that of Hatley and Pirbhai.[5] Their method is discussed in more depth in Chapter 10 as an example of a method for integrating a multiplicity of architectural views across models. A number of other methods and tools that add formalized physical modeling to behavioral modeling are appearing. Many of these are commercial tools so the situation is fluid and their methodologies are often not fully defined outside of the tools documentation. Some other examples are the system engineering extensions to ADARTS (described later in the context of software), RDD-100,[6] and StateMate.[7]

An attribute missing in most block diagram methods is the logic of data flow. The diagram may show that a data item flows from module A to module B, but it does not show who controls the flow. Control can be of many types. A partial enumeration includes the following:

Soft push: The sender sends and the item is lost if the receiver is not waiting to receive it.

Hard push: The sender sends and the act of sending interrupts the receiver who must take the data.

Blocking pull: The receiver requests the data and waits until the sender responds.

Nonblocking pull: The receiver requests the data continues on without it if the sender does not send.

Hard pull: When the receiver requests the data, the sender is interrupted and must send.

Queuing channel: The sender can push data onto the channel without interrupting the receiver and with data being stored in the channel. The receiver can pull data from the channel's store.

Of course, there are many other combinations as well. The significance of the control attribute is primarily in interfacing to disciplinary engineers, especially software engineers. In systems whose development cost is dominated by software, which is now virtually all complex systems, it is essential that systems activities provide the information needed to enable software architecting as quickly as possible. One of the elements of a software architecture is the concurrency and synchronization model. The constraints on software concurrency and synchronization are determined by the data flow control logic around the software–hardware boundary. So, it is just the kind of information on data flow control that is needed to better match systems activities to software architecture.

Behavioral (Functional) Models

Functional or behavioral models describe specific patterns of behavior by the system. These are models of what the system *does* (how it behaves) as opposed to what the system *is* (which are models of form). Architects increasingly need behavioral models as systems become more intelligent and their behavior becomes less obvious from the systems form. Unlike a building, a client cannot look at a scale model of a software system and infer how the system behaves. Only by explicitly modeling the behavior can it be understood by the client and builder.

Determining the level of detail or rigor in behavioral specification needed during architecting is an important choice. Too little detail or rigor will mean the client may not understand the behavior being provided (and possibly be unsatisfied) or the builder may misunderstand the behavior actually required. Too much detail or rigor may render the specification incomprehensible — leading to similar problems — or unnecessarily delay development. Eventually, when the system is built, its behavior is precisely specified (if only by the actual behavior of the built system).

From the perspective of architecting, what level of behavioral refinement is needed? The best guidance is to focus on the system acceptance requirements and to ensure the acceptance requirements are passable but complete. Ask what behavioral attributes of the system the client will demand be certified before acceptance, and determine through what tests those behavioral attributes can be certified. The certifiable behavior is the behavior the client will get, no more and no less.

Example: In software systems with extensive user interface components, it has been found by experience that only a prototype of the interface adequately conveys to users how the system will work. Hence, to ensure not just client acceptance but also user satisfaction, an interface prototype should be developed very early in the process. Major office application developers have videotaped office workers as they use prototype applications. The tapes are then examined and scored to determine how effective various layouts were at encouraging users to make use of new features, how rapidly they were able to work, and so on.

Example: Hardware and software upgrades to military avionics almost always must remain backward compatible with other existing avionics systems and maintain support for existing weapon systems. The architecture of the upgrade must reflect the behavioral requirements of existing system interface. Some may imply very simple behavioral requirements, like providing particular types of information on a communication bus. Others may demand complex behaviors, such as target handover to a weapon requiring target acquisition, queuing of the weapon sensor, real-time synchronization of the local and weapon sensor feeds, and complex launch codes. The required behavior needs to be captured at the level required for client acceptance, and at the level needed to extract architectural constraints.

Behavioral tools of particular importance are threads or scenarios, data and event flow networks, mathematical systems theory, autonomous system theory, and public choice and behavior models.

Threads and Scenarios

A thread or scenario is a sequence of system operations. It is an ordered list of events and actions that represents an important behavior. It normally does not contain branches; that is, it is a single serial scenario of operation, a stimulus and response thread. Branches are represented by additional threads. Behavioral requirements can be of two types. The first type is to require that the system *must* produce a given thread — that is, to require a particular system behavior. The alternative is to require that a particular thread not occur — for example, that a hazardous command never be issued without a positive confirmation having occurred first. The former is more common, but the latter is just as important.

Threads are useful for client communication. Building the threads can be a framework for an interactive dialogue with the client. For each input, pose the question "When this input arrives what should happen?" Trace the response until an output is produced. In a similar fashion, trace outputs backward until inputs are reached. The list of threads generated in this way becomes part of the behavioral requirements.

Threads are also useful for builder communication. Even if not complete, they directly convey desired system behavior. They also provide useful tools during design reviews and for test planning. Reviewers can ask that designers walk through their design as it would operate in each of a set of selected threads. This provides a way for reviewers to survey a design using criteria very close to the client's own language. Threads can be used similarly as templates for system tests, ensuring that the tests are directly tied to the client's original dialog.

Another name for behavioral specification by threads and scenarios is use-cases. Use-case has become the popular term for behavioral specification by example. The term originally comes from the object-oriented software community, but it has been applied much more widely. The normal form of a use-case is the listing of an example dialogue between the system and an actor. An actor is a human user of the system. The use-case consists of the sequence of messages passed between the system and actor, augmented by additional explanation in ways specific to each method. Use-cases are intended to be narrative. That is, they are specifically intended to be written in the language of users and to be understandable by them. When a system is specified by many use-cases, and the use-cases interact, there are a number of diagrams that can be used to specify the connections. Chapter 10 briefly discusses these within Unified Modeling Language (UML).

Data and Event Flow Networks

A complex system can possess an enormous (perhaps infinite) set of threads. A comprehensive list may be impossible, yet without it, the behavioral specification is incomplete. Data and event flow networks allow threads to be collapsed into more compact but complete models. Data flow models define the behavior of a system by a network of functions or processes that exchange data objects. The process network is usually defined in a graphical hierarchy, and most modern versions add some component of finite state machine description. Current data flow notations are descendants either of DeMarco's data flow diagram (DFD) notation[8] or Functional Flow Block Diagrams (FFBD).[9] Chapter 10 gives several examples of data flow models and their relationships with other model types. Figure 10.1 and

Figure 10.2 show examples of data flow diagrams for an imaging system. Both the DFD and FFBD methods are based on a set of root principles:

1. The system functions are decomposed hierarchically. Each function is composed of a network of subfunctions until a "simple" description can be written in text.
2. The decomposition hierarchy is defined graphically.
3. Data elements are decomposed hierarchically and are separately defined in an associated "data dictionary."
4. Functions are assumed to be data triggered. A process is assumed to execute anytime its input data elements are available. Finer control is defined by a finite state control model (DFD formalism) or in the graphical structure of the decomposition (FFBD formalism).
5. The model structure avoids redundant definition. Coupled with graphical structuring, this makes the model much easier to modify.

Mathematical Systems Theory

The traditional meaning of system theory is the behavioral theory of multidimensional feedback systems. Linear control theory is an example of system theory on a limited, well-defined scale. Models of macroeconomic systems and operations research are also system theoretic models, but on a much larger scale.

System theoretic formalisms are built from two components:

1. A definition of the system boundary in terms of observable quantities, some of which may be subject to user or designer control.
2. Mathematical machinery that describes the time evolution (the behavior) of the boundary quantities given some initial or boundary conditions and control strategies.

There are three main mathematical system formalisms distinguished by how they treat time and data values:

1. Continuous systems: These systems are classically modeled by differential equations, linear and nonlinear. Values are continuous quantities and are computable for all times.
2. Temporally discrete (sampled data) systems: These systems have continuously valued elements measured at discrete time points. Their behavior is described by difference equations. Sampled data systems are increasingly important because they are the basis of most computer simulations and nearly all real-time digital signal processing.
3. Discrete event systems: A discrete event system is one in which some or all of the quantities take on discrete values at arbitrary points in

time. Queuing networks are the classical example. Asynchronous digital logic is a pure example of a discrete event system. The quantities of interest (say data packets in a communication network) move around the network in discrete units, but they may arrive or leave a node at an arbitrary, continuous time.

Continuous systems have a large and powerful body of theory. Linear systems have comprehensive analytical and numerical solution methods and an extensive theory of estimation and control. Nonlinear systems are still incompletely understood, but many numerical techniques are available, some analytical stability methods are known, and practical control approaches are available. The very active field of dynamical systems addresses nonlinear as well as control aspects of systems. Similar results are available for sampled data systems. Computational frameworks exist for discrete event systems (based on state machines and Petri Nets), but are less complete than those for differential or difference equation systems in their ability to determine stability and synthesize control laws. A variety of simulation tools are available for all three types of systems. Some tools attempt to integrate all three types into a single framework, though this is difficult.

Many modern systems are a mixture of all three types. For example, consider a computer-based temperature controller for a chemical process. The complete system may include continuous plant dynamics, a sampled data system for control under normal conditions, and discrete event controller behavior associated with threshold crossings and mode changes. A comprehensive and practical modern system theory should answer the classic questions about such a mixed system — stability, closed-loop dynamics, and control law synthesis. No such comprehensive theory exists, but constructing one is an objective of current research. Manufacturing systems are a special example of large-scale mixed systems for which qualitative system understanding can yield architectural guidance.

Autonomous Agent, Chaotic Systems

System-level behavior, as defined in Chapter 1, is behavior not contained in any system component but which emerges only from the interaction of all the components. A class of system of recent interest is that in which a few types of multiply replicated, individually relatively simple, components interact to create essentially new (emergent) behaviors. Ant colonies, for example, exhibit complex and highly organized behaviors that emerge from the interaction of behaviorally simple, nearly identical, sets of components (the ants). The behavioral programming of each individual ant, and its chaotic local interactions with other ants and the environment, is sufficient for complex high-level behaviors to emerge from the colony as a whole. There is considerable interest in using this truly distributed architecture, but traditional top-down, decomposition-oriented models and their bottom-up integration-oriented complements do not describe it. Some attempts have been made to build theories of such systems from chaos methods. Attempts have also been made to find rules or heuristics for the local intelligence and interfaces necessary for high-level behaviors to emerge.

> *Example*: In some prototype flexible manufacturing plants, instead of trying to solve the very complex work scheduling problem, autonomous controllers schedule through distributed interaction. Each work cell independently "bids" for jobs on its input. Each job moving down the line tries to "buy" the production and transfer services it needs to be completed.[10] Instead of central scheduling, the equilibrium of the pseudo-economic bid system distributes jobs and fills work cells. Experiments have shown that rules can be designed that result in stable operation, near optimality of assignment, and very strong robustness to job changes and work cell failure. But the lack of central direction makes it difficult to assure particular operational aspects (for example, to assure that "oddball" jobs will not be ignored for the apparent good of the mean).

Public Choice and Behavior Models

Some systems depend on the behavior of human society as part of the system. In such cases, the methods of public choice and consumer analysis may need to be invoked to understand the human system. These methods are often ad hoc, but many have been widely used in marketing analysis by consumer product companies.

> *Example*: One concept in intelligent transportation systems proposals (recall the discussion in "Case Study 3" on ITS before Chapter 5) is the use of centralized routing. In a central routing system, each driver would inform the center (via some data network) of his or her beginning location and his or her planned destination for each trip. The center would use that information to compute a route for each vehicle and communicate the selected route back to the driver. The route might be dynamically updated in response to accidents or other incidents. In principle, the routing center could adjust routes to optimize the performance of the network as a whole. But would drivers accept centrally selected

routes, especially if they thought the route benefited
the network but not them? Would they even bother
to send in route information?

A variety of methods could be used to address such questions. At the
simplest level are consumer surveys and focus groups. A more involved
approach is to organize multiperson driving simulations with the performance of the network determined from individual driver decisions. Over
the course of many simulations, as drivers evaluate their own strategies,
stable configurations may emerge.

Performance Models

A performance model describes or predicts how effectively an architecture satisfies some objective, either functional or not. Performance models
are usually quantitative, and the most interesting performance models
are those of system-level functions — that is, properties possessed by the
system as a whole but by no subsystem. Performance models describe
properties like overall sensitivity, accuracy, latency, adaptation time,
weight, cost, reliability, and many others. Performance requirements are
often called "nonfunctional" requirements because they do not define a
functional thread of operation, at least not explicitly. Cost, for example,
is not a system *behavior*, but it is an important property of the system.
Detection sensitivity to a particular signal, however, does carry with it
implied functionality. Obviously, a signal cannot be detected unless the
processing is in place to produce a detection. It will also usually be impossible to formulate a quantitative performance model without constraining
the system's behavior and form.

Performance models come from the full range of engineering and
management disciplines. But the internal structure of performance models
generally falls into one of three categories:

1. *Analytical*: Analytical models are the products of the engineering
   sciences. A performance model in this category is a set of lower-level
   system parameters and a mathematical rule of combination that predicts the performance parameter of interest from lower-level values.
   The model is normally accompanied by a "performance budget"
   or a set of nominal values for the lower-level parameters to meet a
   required performance target.
2. *Simulation*: When the lower-level parameters can be identified, but an
   easily computable performance prediction cannot, a simulation can
   take the place of the mathematical rule of combination. In essence,
   a simulation of a system is an analytical model of the system's behavior
   and performance in terms of the simulation parameters. The connection

is just more complex and difficult to explicitly identify. A wide variety of continuous, discrete time, and discrete event simulators are available, many with rich sets of constructs for particular domains.

3. *Judgmental:* Where analysis and simulation are inadequate or infeasible, human judgment may still yield reliable performance indicators. In particular, human judgment, using explicit or implicit design heuristics, can often rate one architecture as better than another even where a detailed analytical justification is impossible.

Formal Methods

The software engineering community has taken a specialized approach to performance modeling known as formal methods. Formal methods seek to develop systems that provably produce formally defined functional and nonfunctional properties. In formal development, the team defines system behavior as sets of allowed and disallowed sequences of operation, and may add further constraints, such as timing, to those sequences. They then develop the system in a manner that guarantees compliance to the behavioral and performance definition. Roughly speaking, the formal methods approach is as follows:

1. Identify the inputs and outputs of the system. Identify a set of mathematical and logical relations that must exist between the input and output sequences when the system is operating as desired.
2. Decompose the system into components, identifying the inputs and outputs of each component. Determine mathematical relations on each component such that their composition is equivalent to the original set of relations one level up.
3. Continue the process iteratively to the level of primitive implementation elements. In software, this would be programming language statements. In digital logic, this might be low-level combinational or sequential logic elements.
4. Compose the implementation backward up the chain of inference from primitive elements in a way that conserves the decomposed correctness relations. The resulting implementation is then equivalent to the original specification.

From the point of view of the architect, the most important applications of formal methods are in the conceptual phases and in the certification of high-assurance and ultraquality systems. Formal methods require explicit determination of allowed and disallowed input and output sequences. Trying to make that determination can be valuable in eliciting client information, even if the resulting information is not captured in precise mathematical terms. Formal methods also hold out the promise of being able to certify system characteristics that can never be tested. No set of tests can certify that certain event chains cannot occur, but theorems to that effect are provable within a formal model.

Various formal and semiformal versions of the process are in limited use in software and digital system engineering.[11] Although a fully formal version of this process is apparently impractical for large systems at the present time (and is definitely controversial), semiformal versions of the process have been successfully applied to commercial products.

A fundamental problem with the formal methods approach is that the system can never be more "correct" than the original specification. Because the specification must be written in highly mathematical terms, it is particularly difficult to use in communication with the typical client.

Data Models

The next dimension of system complexity is retained data. What data does the system retain and what relationships among the data does it develop and maintain? Many large corporate and governmental information systems have most of their complexity in their data and the data's internal relationships. The most common data models have their origins in software development, especially large database developments. Methods for modeling complex data relationships were developed in response to the need to automate data-intensive, paper-based systems. Although data-intensive systems are most often thought of as large, automated database systems, many working examples are actually paper based. Automating legacy paper-based systems requires capturing the complex interrelationships among large amounts of retained data.

Data models are of increasing importance because of the greater intelligence being embedded in virtually all systems and the continuing automation of legacy system. In data-intensive systems, generating intelligent behavior is primarily a matter of finding relationships and imposing persistent structure on the records. This implies that the need to find structure and relationships in large collections of data will be determinants of systems architecture.

> *Example*: Manufacturing software systems are no longer responsible just for control of work cells. They are part of integrated enterprise information networks in which real-time data from the manufacturing floor, sales, customer operations, and other parts of the enterprise are stored and studied. Substantial competitive advantages accrue to those who can make intelligent judgments from these enormous data sets.

> *Example:* Intelligent transport systems are a complex combination of distributed control systems, sensor networks, and data fusion. Early deployment stages will emphasize only simple behavioral adaptation, as in local intelligent light and on-ramp controllers. Full deployment will fuse data sources across metropolitan areas to generate intelligent prediction and control strategies. These later stages will be driven by problems of extracting and using complex relationships in very large databases.

The basis for modern data models are the Entity-Relationship diagrams developed for relational databases. These diagrams have been generalized into a family of object-oriented modeling techniques. An object is a set of "attributes" or data elements and a set of "methods" or functions that act upon the attributes (and possibly other data or objects as well). Objects are instances of classes that can be thought of as templates for specific objects. Objects and classes can have relationships of several types. Major relationship types include aggregation (or composition); generalization, specialization, or inheritance; and association (which may be two-way or M-way). Object-oriented modeling methods combine data and behavioral modeling into a single hierarchy organized along and driven by data concerns. Behavioral methods like those described earlier also include data definitions, but the hierarchy is driven by functional decomposition.

One might think of object-oriented models as turning functional decomposition models inside out. Functional decomposition models like data flow diagramming describe the system as a hierarchy of functions, and hang a data model onto the functional skeleton. The only data relationship supported is aggregation. An object-oriented model starts with a decomposition of the data and hangs a functional model on it. It allows all types of data relationships. Some problems decompose cleanly with functional methods and only with difficulty in object-oriented methods, and some other problems are the opposite.

An example of a well-developed, object-oriented data modeling technique (OMT) is given in Chapter 10. Figure 10.7 shows a typical example of the type of diagram used in that method, which combines conventional entity relationship diagram and object-oriented abstraction. OMT has further evolved into UML, which is discussed in Chapter 10.

Data-oriented decompositions share the general heuristics of systems architecture. The behavioral and physical structuring characteristics have direct analogs — composing or aggregation, decomposition, and minimal communications. There are also similar problems of scale. Very large data models must be highly structured with limited patterns of relationship (analogous to limited interfaces) to be implementable.

Managerial Models

To both the client and architect, a project may be as much a matter of planning milestones, budgets, and schedules as it is a technical exercise. In sociotechnical systems, planning the system deployment may be more difficult than assembling its hardware. The managerial or implementation view describes the process of building the physical system. It also tracks construction events as they occur.

Most of the models of this view are the familiar tools of project management. In addition, management-related metrics that can be calculated from other models are invaluable in efforts to create an integrated set of models. Some examples include the following:

1. The waterfall and spiral system development meta-models — the templates on which project-specific plans are built
2. Program Evaluation and Review Technique/Critical Path Method (PERT/CPM) and related task and scheduling dependency charts
3. Cost and progress accounting methods
4. Predictive cost and schedule metrics calculable from physical and behavioral models
5. Design or specification time quality metrics — defect counts, post-simulation design changes, rate of design changes after each review

The architect has two primary interests in managerial models. First, the client usually cannot decide to go ahead with system construction without solid cost and schedule estimates. Usually producing such estimates requires a significant effort in management models. Second, the architect may be called upon to monitor the system as it is developed to ensure its conceptual integrity. In this monitoring process, managerial models will be very important.

Examples of Integrated Models

As noted earlier, models that integrate multiple views are the special concern of the architect. These integrating models provide the synthesized view central to the architect's concerns. An integrated modeling method is a system of representation that links multiple views. The method consists of a set of models for a subset of views and a set of rules or additional models to link the core views. Most integrated modeling methods apply to a particular domain. Listed in Table 8.2 are some representative methods. These models are described in greater detail, with examples, in Chapter 10. The references are given there, as well.

These methods use different models and cover different views. Their components and dimensions are summarized in Table 8.3.

Table 8.2 Integrated Modeling Methods and Their Domains

Method	Reference	Domain
Hatley-Pirbhai (H/P)	Hatley[12,13]	Computer-based reactive or event-driven systems
Quantitative Quality Function Deployment (Q²FD)	Maier[14]	Systems with extensive quantitative performance objectives and understood performance models
Object Modeling Technique (OMT)	Rumbaugh[15]	Large-scale, data-intensive software systems, especially those implemented in modern object languages
ADARTS	SPC[16]	Large-scale, real-time software systems
Manufacturing System Analysis (MSA)	Baudin[17]	Intelligent manufacturing systems

Conclusion

An architect's work revolves around models. Because the architect does not build the system directly, its integrity during construction must be maintained through models acting as surrogates. Models will represent and control the specification of the system, its design, and its production plan. Even after the system is delivered, modeling will be the mechanism for assessing system behavior and planning its evolution. Because the architect's concerns are broad, architecting models must encompass all views of the system. The architect's knowledge of models, like an individual's knowledge of language, will tend to channel the directions in which the system develops and evolves.

Modeling for architects is driven by three key characteristics:

1. Models are the principal language of the architect. Their foremost role is to facilitate communication with client and builder. By facilitating communication, they carry out their other roles of maintaining design integrity and assisting synthesis.
2. Architects require a multiplicity of views and models. The basic ones are objective, form, behavior, performance, data, and management. Architects need to be aware of the range of models that are used to describe each of these views within their domain of expertise, and the content of other views that may become important in the future.
3. Multidisciplinary, integrated modeling methods tie together the various views. They allow the design of a system to be refined in steps from conceptually abstract to the precisely detailed necessary for construction.

The next chapter reconsiders the use of modeling in architecture by placing modeling in a larger set of parallel progressions from abstract to

Table 8.3 Comparison of Representative Modeling Methods

View	H/P	OMT	ADARTS	Q²FD	MSA
Objectives	Text	Text	Text	Numbers	Text
Behavior	Data/control flow	Class diagrams, data flow, StateCharts	Data/event flow	Links only	Data flow
Performance	Text (timing only)	Text	Text	Satisfaction models, QFD matrices	Text, links to standard scheduling models
Data	Dictionary	Class/object diagrams	Dictionary	N/A	Entity–relationship diagrams
Form	Formalized block diagrams	Object diagrams	Task-object-structure charts (multilevel)	Links by allocation	SME process flow diagrams
Managerial	N/A (link via metrics)	N/A	N/A (link via metrics)	N/A	Funds flow model, scheduling behavior

Notes: H/P, Hatley-Pirbhai; OMT, Object Modeling Technique; ADARTS, Ada-Based Design Approach for Real-Time Systems; Q²FD, Quantitative Quality Function Deployment; MSA, Manufacturing System Analysis.

concrete. There the field of view will expand to the whole architectural design process and its parallel progressions in heuristics, modeling, evaluation, and management.

Exercises

1. Choose a system familiar to you. Formulate a model of your system in each of the views discussed in the chapter. How effectively does each model capture the system in that view? How effectively do the models define the system for the needs of initial concept definition and communication with clients and builders? Are the models integrated? That is, can you trace information across the models and views?

2. Repeat exercise 1, but with a system unfamiliar to you, and preferably embodying different driving issues. Investigate models used for the views most unfamiliar to you. In retrospect, does your system in exercise 1 contain substantial complexity in the views you are unfamiliar with?

3. Investigate one or more popular computer-aided systems or software engineering (CASE) tools. To what extent do they support each of the views? To what extent do they allow integration across views?

4. A major distinction in behavioral modeling methods and tools is the extent to which they support or demand executability in their models. Executability demands a restricted syntax and up-front decision about data and execution semantics. Do these restrictions and demands help or hinder initial concept formulation and communication with builders and clients? If the answer is variable with the system, is there a way to combine the best aspects of both approaches?

5. Models of form must be technology specific because they represent actual systems. Investigate modeling formalisms for domains not covered in the chapter, for example, telecommunication systems, business information networks, space systems, integrated weapon systems, chemical processing systems, or financial systems.

Notes and References

1. Balzer, B., and D. Wile, Domain Specific Software Architectures, Technical Reports, Information Sciences Institute, University of Southern California, Los Angeles, California, 1996.

2. Yourdon, Edward, and Larry L. Constantine, *Structured Design: Fundamentals of a Discipline of Computer Program and Systems Design.* New York: Yourdon Press, 1979.

3. ADARTS Guidebook, SPC-94040-CMC, Version 2.00.13, Vols. 1–2, September 1991. Available through the Software Productivity Consortium, Herndon, Virginia.

4. Rumbaugh, J. et. al., *Object-Oriented Modeling and Design.* Upper Saddle River, NJ: Prentice Hall, 1991.
5. Hatley, D. J., and I. Pirbhai, *Strategies for Real-Time System Specification.* New York: Dorset House, 1988.
6. A comprehensive system modeling tool marketed by Ascent Logic Corporation, Princeton, New Jersey.
7. A tool with both discrete event behavioral modeling and physical block diagrams marketed by i-Logix.
8. DeMarco, T., *Structured Analysis and System Specification.* New York: Yourdon Press, 1979.
9. Functional Flow Diagrams, AFSCP 375-5 MIL-STD-499, USAF, DI-S-3604/ S-126-1, Form DD 1664, June, 1968. Much more modern implementations exist, for example, the RDD-100 modeling and simulation tool developed and marketed by Ascent Logic Corporation, Princeton, New Jersey.
10. Morley, R. E., The Chicken Brain Approach to Agile Manufacturing, *Proceedings Manufacturing, Engineering, Design, Automation Workshop,* Stanford, Palo Alto, California, pp. 19–24, 1992.
11. Two references can be noted: for theory, Hoare, C. A. R., *Communicating Sequential Processes.* Upper Saddle River, NJ: Prentice Hall, 1985; for application in software, Mills, H. D., Stepwise Refinement and Verification in Box-Structured Systems, *IEEE Computer,* pp. 23–36, June 1988.
12. Hatley, D., and I. Pirbhai, *Strategies for Real Time System Specification.* Hoboken, NJ: John Wiley and Sons, 1988.
13. Hatley, D., P. Hruschka, and I. Pirbhai, *Process for System Architecture and Requirements Engineering,* New York: Dorset House, 2000.
14. Maier, M. W., Quantitative Engineering Analysis with QFD, *Quality Engineering,* 7:4, pp. 733–746, 1995.
15. Rumbaugh, J. et al, *Object-Oriented Modeling and Design.* Upper Saddle River, NJ: Prentice Hall, 1991.
16. ADARTS Guidebook, SPC-94040-CMC, Version 2.00.13, Vols. 1–2, September 1991.
17. Baudin, M., *Manufacturing Systems Analysis.* New York: Yourdon Press Computing Series, 1990.

chapter 9

Design Progression in Systems Architecting

Introduction: Architecting Process Components

Having outlined the products of architecting (models) in Chapter 8, this chapter turns to its process. The goal is not a formal process definition. Systems are too diverse to allow a fixed or dogmatic approach to architecting. Instead of trying for a formal process definition, developed in this chapter is a set of meta-process concepts for architecting activities and their relationships. Architectural design processes are inherently eclectic and wide ranging, going abruptly from the intensely creative and individualistic to the more prescribed and routine. Even though the processes may be eclectic, they can be organized. Of the various organizing concepts, one of the most useful is stepwise progression or "refinement."

First, a brief review of the architecting process: The architect develops system models that span the range of system concerns, from objectives to implementation. The architectural approach is from beginning to end concerned with the feasibility as well as the desirability of the system implementation. An essential characteristic that distinguishes architecting from other engineering is the parallel development of problem and solution. Architecting does not assume the problem is fixed. It strives for fit, balance, and compromise between client preferences and builder capabilities. Compromise can only be assured by an interplay of activities, including both high-level structuring and such detailed design as is critical to overall success.

This chapter presents a three-part approach to the process of systems architecting:

1. A conceptual model that connects the unstructured processes of architecture to the rigorous engineering processes of the specialty domains or disciplines. This model is based on stepwise reduction of abstraction (or progression) in models, evaluation criteria, heuristics, and purposes from initial architecting to formal systems engineering.
2. An introduction to and review of the general concepts of design, including theories of design, the elements of design, and the processes

of creating a design. These frame the activities that make up the progressions and organize much of the conceptual framework on which this book is built.

3. A guide to the organization of architecting and its methods, including the placement of specialized design domains and the evolutionary development of domain-specific methods. Architecting is recursive within a system as it is defined in terms of its implementation domains. A split between architecting and engineering is an irreducible characteristic of every domain, though the boundaries of that split cannot be clear until the scientific basis for the methods in a domain are known.

The progressions of architecting are inextricably bound up with the progressions of all system development. Architecting is not only iterative, it can be recursive. As a system progresses, architecting may reoccur on subsystems. The goal here is to understand the intellectual nature of its conduct, whether it happens at a very high level or within a subsystem.

Design Progression

Progressive refinement of design is one of the most basic patterns of engineering practice. It permeates the process of architecting from models to heuristics, information acquisition, and management. Its real power, especially in systems architecting, is that it provides a way to organize the progressive transition from the ill-structured, chaotic, and heuristic processes needed at the beginning to the rigorous engineering and certification processes needed later. All can be envisioned as a stepwise reduction of abstraction, from mental concept to delivered physical system.

In software, the process is known as stepwise refinement. Stepwise refinement is a specific strategy for top-down program development. The same notion applies to architecting but is applied in-the-large to complex, multidisciplinary system development. Stepwise refinement is the progressive removal of abstraction in models, evaluation criteria, and goals. It is accompanied by an increase in the specificity and volume of information recorded about the system and a flow of work from general to specialized design disciplines. Within the design disciplines, the pattern repeats as disciplinary objectives and requirements are converted into the models of form of that discipline. In practice, the process is neither so smooth nor continuous. It is better characterized as episodic, with episodes of abstraction reduction alternating with episodes of reflection and purpose expansion.

Stepwise refinement can be thought of as a meta-process model, much as the waterfall and spiral. It is not an enactable process for a specific project, but it is a model for building a project-specific process.

Systems are too diverse to follow a fixed process or dogmatic formula for architecting.

Introduction by Examples

Before treating the conceptually difficult process of general systems architecting, look to the roots. When a civil architect develops a building, does he or she go directly from client words to construction drawings? Obviously not; there are many intermediate steps. The first drawings are rough floor plans showing the spatial relationships of rooms and sizes and external renderings showing the style and feel of the building. Following these are intermediate drawings giving specific dimensions and layouts. The construction drawings with full details for the builder follow on after. The architect's role does not have a universally applicable stopping point, but the normal case is based on the needs of the client. The client hired the architect to accomplish a specific portion of the overall development and construction process. When the designs are sufficiently refined (in enough views) for the client to make the decision to proceed with construction, the architect's conceptual development job is complete. The architect may be busy with the project for some time to come in shepherding the conceptual design through detailed design, overseeing construction, and advising the client on certification, but the initial concept role is complete when the client can make the construction decision.

In a different domain, the beginning computer programmer is taught a similar practice. Stepwise refinement in programming means to write the central controlling routine first. Anywhere high complexity occurs, ignore it by giving it a descriptive name and making it a subroutine or function. Each subroutine or function is "stubbed" — that is, given a dummy body as a placeholder. When the controlling routine is complete, it is compiled and executed as a test. Of course, it does not do anything useful because its subroutines are stubbed. The process is repeated recursively on each subroutine until routines can be easily coded in primitive statements in the programming language. At each intermediate step, an abstracted version of the whole program exists that has the final program's structure but lacks internal details.

Both examples show progression of system representation or modeling. Both examples embed strategy, in terms of ordering of decisions in ways that meet client or sponsor needs, into the progressive development process. In building, the sponsor needs to set up a distinct decision point where financing is resolved and a building contractor is hired. In software development, top-down stepwise refinement assembles a program in a fashion that facilitates continuous testing and incremental delivery. The building sponsor needs a development process (and a relationship with the architect) that supports the customary financial arrangements and the limitations of the contracting industry. Software developers, especially in commercial markets, prefer mechanisms that facilitate incremental delivery and provide full program level "test harnesses."

Progression also occurs along other dimensions. For example, both the civil architect and the programmer may (should) create distinct alternative designs in their early stages. How are these partial designs evaluated to choose the superior approach? In the earliest stages, both the programmer and the civil architect use heuristic reasoning. The civil architect can measure rough size (to estimate cost), judge the client's reaction, and ask the aesthetic opinion of others. The programmer can judge code size, heuristically evaluate the coupling and cohesion of the resulting subroutines and modules, review applicable patterns from catalogs, and review functionality with the client. As their work progresses, both will be able to make increasing use of rational and quantitative evaluation criteria. The civil architect will have enough details for proven cost models; the programmer can measure execution speed, compiled size, and behavioral compliance, and invoke quantitative software quality metrics. Programmers will also have improved cost models as progression continues. Software cost models are predominantly based on code size, and the progressive development of the top-down structure supports improved estimates of code size.

Design as the Evolution of Models

All architects, indeed all designers, manipulate models of the system. These models become successively less abstract as design progresses. The integrated models discussed in Chapter 10 exhibit stepwise reduction of abstraction in representation and in their design heuristics.

In Hatley-Pirbhai, the reduction of abstraction is from behavioral model, to technology-specific behavioral model, to architecture model. There is also hierarchical decomposition within each component. The technology of modules is indeterminate at the top level and becomes technology specific as the hierarchy develops. The Quantitative Quality Function Deployment (Q^2FD) performance modeling technique shows stepwise refinement of customer objectives into engineering parameters. As the Q^2FD chain continues, the engineering parameters get closer to implementation until, ultimately, they may represent machine settings on the factory floor. Likewise, the structure of integrated models in software and manufacturing systems follow the same logic or progression.

Evaluation Criteria and Heuristic Refinement

The criteria for evaluating a design progress or evolve in the same manner as design models. In evaluation, the desirable progression is from general to system specific to quantitative. For heuristics, the desirable progression is from descriptive and prescriptive qualitatives to domain-specific quantitatives and rational metrics. This progression is best illustrated by following the progression of a widely recognized heuristic into quantitative metrics within a particular discipline. Start with the partitioning heuristic:

> *In partitioning, choose the elements so that they are as independent as possible — that is, elements with low external complexity and high internal complexity.*

This heuristic is largely independent of domain. It serves as an evaluation criteria and partitioning guide whether the system is digital hardware, software, human driven, or otherwise. But, the guidance is nonspecific; neither independence nor complexity is defined. By moving to a more restricted domain, computer-based systems in this example, this heuristic refines into more prescriptive and specific guidelines. The literature on structured design for software (or, more generally, computer-based systems) includes several heuristics directly related to the partitioning heuristic.[1] The structure of the progression is illustrated in Figure 9.1.

1. Module fan-in should be maximized. Module fan-out should generally not exceed 7 ± 2.
2. The coupling between modules should be, in order of preference, data, data structure, control, common, and content.
3. The cohesion of the functions allocated to a particular module should be, in order of preference, functional/control, sequential, communicational, temporal, periodic, procedural, logical, and coincidental.

These heuristics give complexity and independence more specific form. As the domain restricts even farther, the next step is to refine into quantitative design quality metrics. This level of refinement requires a specific domain and detailed research and is the concern of specialists in each domain. But, to finish the example, the heuristic can be formulated into a quantitative software complexity metric. A very simple example is as follows:

> *Compute a complexity score by summing: One point for each line of code, 2 points for each decision point, 5 points for each external routine call, 2 points for each write to a module variable, 10 points for each write to a global variable.**

---

* Much more sophisticated complexity metrics have been published in the software engineering literature. One of the most popular is the McCabe metric, for which there is a large automated tool set.

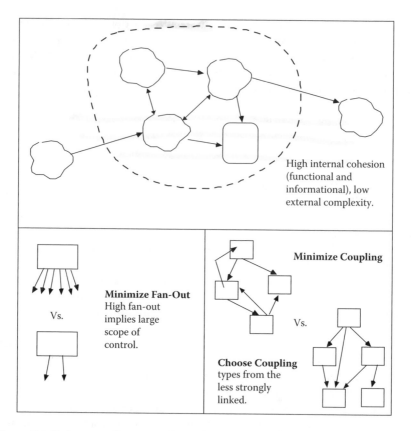

Figure 9.1 Software refinement of coupling and cohesion heuristic. The general heuristic is refined into a domain-specific set of heuristics.

Early evaluation criteria or heuristics must be as unbounded as the system choices. As the system becomes constrained, so do the evaluation criteria. What was a general heuristic judgment becomes a domain-specific guideline and, finally, a quantitative design metric.

Progression of Emphasis

On a more abstract level, the social or political meaning of a system to its developers also progresses. A system goes from being a product (something new), to a source of profit or something of value, to a policy (something of permanence). In the earliest stages of a system's life, it is most likely viewed as a product. It is something new, an engineering challenge. As it becomes established and its development program progresses, it becomes an object of value to the organization. Once the system exists, it acquires an assumption of permanence. The system, its capabilities, and its actions become part of the organization's nature. To have and operate the system becomes a policy that defines the organization.

With commercial systems, the progression is from product innovation to business profit to corporate process.[2] Groups innovate something new, successful systems become businesses, and established corporations perpetuate a supersystem that encompasses the system, its ongoing development, and its support. Public systems follow a similar progression. At their inception they are new, at their development they acquire a constituency, and they eventually become a bureaucratic producer of a commodity.

Concurrent Progressions

Other concurrent progressions include risk management, cost estimating, and perceptions of success. Risk management progresses in specificity and goals. Early risk management is primarily heuristic with a mix of rational methods. As prototypes are developed and experiments conducted, risk management mixes with interpretation. Solid information begins to replace engineering estimates. After system construction, risk management shifts to postincident diagnostics. System failures must be diagnosed, which is a process that should end in rational analysis but may have to be guided by heuristic reasoning.

Cost estimating goes through an evolution similar to other evaluation criteria. Unlike other evaluation criteria, cost is a continually evolving characteristic from the systems inception. At the very beginning, the need for an estimate is highest and the information available is lowest. Little information is available because the design is incomplete and no uncertainties have been resolved. As development proceeds, more information is available, both because the design and plans become more complete and because actual costs are incurred. Incurred costs are no longer estimates. When all costs are in (if such an event can actually be identified), there is no longer a need for an estimate. Cost estimating goes through a progression of declining need but of continuously increasing information.

All of the "ilities" are part of their own parallel progressions. These system characteristics are known precisely only when the system is deployed. Reliability, for example, is known exactly when measured in the field. During development, reliability must be estimated from models of the system. Early in the process, the customer's desires for reliability may be well known, but the reliability performance is quite uncertain. As the design progresses to lower levels, the information needed to refine reliability estimates becomes known, including information like parts counts, temperatures, and redundancy.

Perceptions of success evolve from architect to client and back to architect. The architect's initial perception is based on system objectives determined through client interaction. The basic measure of success for the architect becomes successful certification. But once the system is delivered, the client will perceive success on his or her own terms. The project may produce a system that is successfully certified but that nonetheless becomes a disappointment. Success is affected by all other conditions affecting the client at delivery and operation, whether or not anticipated during design.

Episodic Nature

The emphasis on progression appears to define a monotonic process. Architecting begins in judgment, rough models, and heuristics. The heuristics are refined along with the models as the system becomes bounded until rational, disciplinary engineering is reached. In practice, the process is more cyclic or episodic with alternating periods of synthesis, rational analysis, and heuristic problem solving. These episodes occur during system architecting and may appear again in later, domain-specific stages.

The occurrence of the episodes is integral to the architect's process. An architect's design role is not restricted solely to "high-level" considerations. Architects dig down into specific subsystem and domain details where necessary to establish feasibility and determine client-significant performance (see Chapter 1, Figure 1.1 and the associated discussion). The overall process is one of high-level structuring and synthesis (based on heuristic insight) followed by rational analysis of selected details. Facts learned from those analyses may cause reconsideration of high-level synthesis decisions and spark another episode of synthesis and analysis. Eventually, there should be convergence to an architectural configuration and the driving role passes to subsystem engineers.

Design Concepts for Systems Architecture

Although systems design is an inherently complicated and irregular practice, it has well-established and identifiable characteristics and can be organized into a logical process. As was discussed in the Preface, the activities of architecting can be distinguished from other engineering activities, even if not crisply. Architecting is characterized by the following:

1. Architecting is, predominantly, an eclectic mix of rational and heuristic engineering. Other elements, such as normative rules and group processes, enter in lesser roles (recall the discussion of the four theories in Chapter 1).
2. Architecting revolves around models but is composed of the basic processes of scoping, synthesis, and certification. Few complete rational methods exist for these processes, and the principal guidelines are heuristic.

3. Synthesis can be considered as creative invention and can be usefully broken down into iterative design activities.
4. Uncertainty is inherent in complex system design. Heuristics are specialized tools to reduce or control but not eliminate uncertainty.
5. Continuous progression on many fronts is an organizing principle of architecting, architecture models, and supporting activities.

Civil engineering and architecture are perhaps the most mature of all engineering disciplines. Mankind has more experience with engineering civil structures than any other field. If any area could have the knowledge necessary to make it a fully rational and scientific endeavor, it should be civil engineering. But it is in civil practice that the distinction between architecture and engineering is best established. Both architects and engineers have their roles, often codified in law, and their professional training programs emphasize different skills. Architects deal particularly with those problems that cannot be entirely rationalized by scientific inquiry. The architect's approach does not ignore science; it combines it with art. Civil engineers must likewise deal with unrationalizable problems, but the focus of their concerns is with well-understood rational design and specification problems. By analogy, this suggests that all design domains contain an irreducible kernel of problems that are best addressed through creative and heuristic approaches that combine art and science. This kernel of problems, it might be called the architectonic kernel, is resistant to being subsumed into engineering science because it inherently binds together social processes (client interaction) with engineering and science. The social side is how we determine and understand people's needs. The engineering and science side is determining the feasibility of a system concept. The bridge is the creative process of imagining system concepts in response to expressions of client need.

Note that the kernel is independent of modeling or description processes. Using a framework-centric process does not relieve us of the kernel. The kernel is decision centric, not model centric. An approach through modeling can, at best, clarify the decisions.

Historical Approaches to Architecting

As indicated in the introduction to Part I, civil architects recognize four basic theories of design: the normative or pronouncement, the rational, the argumentative or participative, and the heuristic. Although all have their roots in the civil architecture practice, they are recognizable in modern complex systems as well. They have been discussed before, in particular in Rechtin 1991,[3] and in the Introduction to Part I. The purpose in returning to them here is to indicate their relationship to progressive modeling and to bring in their relevance to software-oriented development.*

To review, normative theory is built from pronouncements (statements of what should be — a set of hard rules), most often given as restrictions on the content of particular views (usually form). A pronouncement demands that a particular type of form be used, essentially unchanged, throughout. Alternatively, one may pronounce the reverse and demand that certain types of form *not* be used. In either case, success is *defined* by accurate implementation of the normative pronouncements, not by measures of fitness. In the normative theory, success is defined as following the rules. Building codes are a prominent example of the normative theory at work, in a positive sense.

Consensual or participative system design uses models primarily as a means of communicating alternative designs for discussion and negotiation among participants. From the standpoint of modeling, consensuality is one of several techniques for obtaining understanding and approval of stakeholders, rather than of itself a structured process of design.

Rational system design is tightly integrated with modeling because it seeks to derive optimal solutions, and optimality can be defined only within a structured and mathematical framework. To be effective, rational methods require modeling methods that are broad enough to capture all evaluation aspects of a problem, deep enough to capture the characteristics of possible solutions, and mathematically tractable enough to be solved for problems of useful size. Given these, rational methods "mechanically" synthesize a design from a series of modeled problem statements in progressively more detailed subsystems.

General heuristics are guides to — and sometimes obtained from — models, but they are not models themselves. Heuristics are employed at all levels of design, from the most general to domain specific. Heuristics are needed whenever the complexity of the problem, solutions, and issues overwhelms attempts at complete rational modeling. This occurs as often in software or digital logic design as in general system design. Within a specific domain, the heuristic approach can be increasingly formalized, generating increasingly prescriptive guidance. This formalization is a reflection of the progression of all aspects of design — form, evaluation, and emphasis — from abstract to concrete.

The power of heuristics in architecting, as discussed in Chapter 2, comes by cataloging those that apply in many domains, giving them generality in those domains, likely extensibility in others, and a system-level credibility. Applied consistently through the several levels of system

---

* Stepwise refinement is a term borrowed from software that describes program development by sequential construction of programs, each complete into itself but containing increasing fractions of the total desired system functionality.

architecture, they help insure system integrity. For example, "Quality cannot be tested in; it must be designed in" is equally applicable from the top-level architectural sketch to the smallest detail. However, the general heuristic relies on an experienced system-level architect to select the ones appropriate for the system at hand, interpret their application-specific meaning, and promulgate them throughout its implementation. A catalog of general heuristics is of much less use to the novice; indeed, an uninformed selection among them could be dangerous. For example, "If it ain't broke, don't fix it," which is questionable at best, can mislead one from making the small incremental changes that often characterize successful continuous improvement programs and can block one from recognizing the qualitative factors, like ultraquality, that redefine the product line completely.

Specialized and Formalized Heuristics

Although there are many very useful general heuristics, there really is not a general heuristic method as such.* Heuristics most often are formalized as part of more formalized methods within specific domains. A formalized-heuristic method gives up generality for more direct guidance in its domain. Popular design methods often contain formalized heuristics as guidelines for design synthesis. A good example is the ADARTS† software engineering methodology. ADARTS provides an extensive set of heuristics to transform a data-flow-oriented behavioral model into a multitasking, modular software implementation. Some examples of formalized ADARTS prescriptive heuristics include the following:

> *Map a process to an active I/O process if that transformation interfaces to an active I/O device.*[4]

> *Group processes that read or update the same data store or data from the same I/O device into a single process.*[5]

As the ADARTS method makes clear, these are recommended guidelines and not the success-*defining* pronouncements of the normative approach. These heuristics do not produce an optimal, certifiable, or even unique result, much less success-by-definition. There is ambiguity in their application. Different heuristics may produce conflicting software

---

* On the other hand, knowledge of a codified set of heuristics can lead to new ways of thinking about problems. This could be described as heuristic thinking or qualitative reasoning.

† This method is described in the ADARTS Guidebook, SPC-94040-CMC, Version 2.00.13, Volume 1, September 1991, available from the Software Productivity Consortium, now the Systems and Software Consortium, Herndon, Virginia.

structuring. The software engineer must select from the heuristic list and interpret the design consequences of a given heuristic with awareness of the specific demands of the problem and its implementation environment.

Conceptually, general and domain-specific formalized heuristics might be arranged in a hierarchy. In this hierarchy, domain-specific heuristics are specializations of general heuristics, and the general are abstractions of the specific. Architecting in general and architecting in specific domains may be linked through the progressive refinement and specialization of heuristics. To date, this hierarchy can be clearly identified only for a limited set of heuristics. In any case, the pattern of refining from abstract to specific is a broadly persistent pattern, and it is essential for understanding life-cycle design progression.

Scoping, Synthesis, and Certification

A development can be envisioned as the creation and transformation of a series of models. For example, to develop the systems requirements is to develop a model of what the system should do and how effectively it should do it. To develop a system design is to develop a model of what the system is. In a pure waterfall development, there is rough alignment between waterfall steps and the views defined in Chapter 8. Requirements development develops models for objectives and performance. Functional analysis develops models of behavior, and so on down the waterfall chain. Architects develop models for all views, though not at equal levels of detail. In uncritical views or uncritical portions of the system, the models will be rough. In some areas, the models may need to be quite detailed from the beginning.

Models are best understood by view because the views reflect their content. Although architecting is consistent with waterfall or spiral development, it does not traverse the steps in the conventional manner. Architects make use of all views and traverse all development steps, but at varying levels of detail and completeness. Because architects tend to follow complex paths through design activities, some alternative characterization of design activities independent of view is useful. The principal activities of the architect are scoping, synthesis, integration, and certification. Figure 9.2 lists typical activities in each category and suggests some relationships. In a subsequent section, we will reconsider these activities in a process model.

Scoping

Scoping procedures are methods for selecting and rejecting problem statements, of defining constraints, and of deciding on what is "inside" or "outside" the system. Scoping implies the ability to rank alternative statements and priorities on the basis of overall desirability or feasibility.

Scoping	
Purpose Expansion/Contraction	
Behavioral Definition/Analysis	
Large Scale Alternative Consideration	
Client Satisfaction–Builder Feasibility	

Synthesis	
Problem Reformulation/Replacement	
Creative Invention	
Iteration	

Aggregation	**Partitioning**
Functional Aggregation (abstract)	Behavioral–Functional Decomposition
Functional Aggregation (to physical units)	Physical Decomposition (to lower level design)
Physical Components to Subsystems	Performance Model Construction
Interface Definition/Analysis	Interface Definition/Analysis
Assembly on Timelines or Behavioral Chains	Decomposition to Cyclic Processes
Collection into Decoupled Threads	Decomposition into Threads

Certification	
Operational Walkthroughs	
Test and Evaluation	
Verification	
Formal Methods Verification	
Failure Assessment	

Figure 9.2 Typical activities within scoping, synthesis, aggregation, partitioning, and certification.

Scoping should not design system internals, though some choices may implicitly do so for lack of design alternatives. Desirably, scoping limits what needs to be considered and why. Scoping is dominantly a problem domain activity.

Scoping is central in orientation and purpose analysis, the activities illustrated in our process model. Purpose analysis is an inquiry into why someone wants the system. Purpose precedes requirements, at least it precedes requirements in the sense of specific acquisition requirements. Requirements are determined by understanding how having a system is valuable to the client, and what combination of fulfilled purposes and systems costs represents a satisfactory and feasible solution.*

Scoping is the heart of front-end architecting. A well-scoped system is one that is both desirable and feasible, the essential definition of success in system architecting. As the project begins, the scope forms, at least implicitly. All participants will form mental models of the system and its characteristics; in doing so, the system's scope is being defined.

---

* Kevin Kreitman has pointed out the extensive literature in soft systems theory that applies to purpose analysis.

If incompatible models appear, scoping has failed through inconsistency. Heuristics suggest that scoping is among the most important of all system design activities. One of the most popular heuristics in Rechtin (1991) was: *All the really important mistakes are made the first day*. Its popularity certainly suggests that badly chosen system scope is a common source for system disasters.

Of course, it is as impossible to prevent mistakes on the first day as it is on any other day. What the heuristic indicates is that mistakes of initial conception will have the worst long-term impact on the project. Therefore, one must be particularly careful that a mistake of scope is discovered and corrected as soon as possible.* One way of doing this is to defer absolute decisions on scope by retaining expansive and restrictive options as long as possible — a course of action recommended by other heuristics (the options heuristics of Appendix A).

In principle, scope can be determined rationally through decision theory. Decision theory applies to any selection problem. In this case, the things being selected are problem statements, constraints, and system contexts. In practice, the limits of decision theory apply especially strongly to scoping decisions. These limits, discussed in greater detail in a subsequent section, include the problems of utility for multiple system stakeholders, problem scale, and uncertainty. The judgments of experienced architects, at least as expressed through heuristics (see Appendix A for a detailed list), is that the most useful techniques to establish system scope are qualitative.

Scoping heuristics and decision theory share an emphasis on careful consideration of who will use the system and will judge success. Decision theory requires a utility function, a mathematical representation of system value as a function of its attributes. A utility function can be determined only by knowing whose judgments of system value will have priority and what the evaluation criteria are. Compare the precision of the utility method to related heuristics:

> *Success is defined by the beholder, not by the architect.*

> *The most important single element of success is to listen closely to what the customer perceives as his requirements and to have the will and ability to be responsive.*
> (J. E. Steiner, 1978)

---

* A formalized heuristic with a similar idea comes from software engineering. It says: *The cost of removing a defect rises exponentially with the time (in project phases) between its insertion and discovery.* Hence, mistakes of scope (the very earliest) are potentially dominant in defect costs.

> *Ask early about how you will evaluate the success of your*
> *efforts.* (F. Hayes-Roth et al., 1983)

Scoping heuristics suggest approaches to setting scope that are outside the usual compromise procedures of focused engineering. One way to resolve intractable problems of scope is to expand. The heuristic is as follows:

> *Moving to a larger purpose widens the range of solutions.*
> (Gerald Nadler, 1990)

The goal of scoping is to form a concept of what the system will do, how effectively it will do it, and how it will interact with the outside world. The level of detail required is the level required to gain customer acceptance, first of continued development and ultimately of the built system. Thus, the scope of the architect's activities is governed not by the ultimate needs of system development, but by the requirements of the architect's further role. The natural conclusion to scoping is certification, where the architect determines that the system is fit for use. Put another way, the certification is that the system is appropriate for its scope.

Scoping is not solely a requirements-related activity. For scope to be successfully determined, the resulting system must be both satisfactory and feasible. The feasibility part requires some development of the system design. The primary activities in design by architects are aggregation and partitioning, the basic structuring of the system into components.

Synthesis

Synthesis is creation. Specifically, synthesis is constructing new solution concepts, and sometimes new problem concepts, in response to the understanding of client purpose. Because synthesis is fundamentally a creative act, we can go to the literature on inventive creativity for heuristics and processes. That literature is very large, and so we will not attempt to review it here. We will highlight some key heuristics, first addressing the more pure synthesis or creativity oriented side, than the more building block side (aggregation and partitioning).

> *Often the most striking and innovative solutions come from*
> *realizing that your concept of the problem was wrong.*[6]

One of the authors was once involved in an assessment of the risks of implementing some very advanced database technology. The technology was being considered in order to synchronize databases distributed over several globally distributed sites. It was important to the researcher-users that the databases each saw in his own location be accurately synchronized with the databases seen by other researchers elsewhere. Because the databases were quite large and in nearly continuous use, the problem of synchronizing them was considerable. After extended discussion, one of the outside participants asked "How did we get into this mess? Is it impossible to just use one database and have everybody access it?" The reason that was "not possible" was because there was insufficient international communications capacity. So, the natural question was why not buy more? Granted, international capacity is expensive, but the database solution being considered was likewise expensive and technologically risky as well. The answer was "International communication capacity comes out of a different budget, and we can't trade that budget for this." In this particular case, the ultimate owner of both was a single, commercial company, and so those budgets could be traded, if one went high enough in the corporate hierarchy. Study of the solution revealed that the problem was not database synchronization; it was the operating pattern of the researchers and their inability to purchase certain types of assets because of internal rules.

Plan to throw one away, you will anyway.[7]

In coming up with great solutions, we rarely, if ever, come up with one right away. We usually come up with some bad ideas, and do not address the problems with our ideas until we have explored them quite a ways. The more innovative the system concept is, the more likely it will have to be exposed to the market and users for an extended period before the desirable approach is revealed. The more innovative the solution is, the more likely it is that extensive, linked operational changes will have to be made before anything like full value can be realized. Those changes will often involve "throwing away" the early attempts. Consider some of the following examples:

1. Personal digital assistant (PDA) devices were on the market for many years before becoming popular. A wide variety of form factors and user interface styles were tried. It was not until the Palm Pilot® (Palm, Inc., Sunnyvale, California) hit the market with its key combination of features (form factor that fit into a shirt pocket, usable stylus-based input, and one-touch computer synchronization) that the market began growing very rapidly. Ironically, but not surprisingly, the pre-Palm leaders were generally unable to follow Palm's lead even after the Pilot had shown the market-winning combination of features.
2. The Global Positioning System (GPS) became a huge success only after two events not linked to the GPS program office occurred. First, commercial companies invested in GPS chipsets that drove the cost down, and volume up, for commercial receivers. Second, the

U.S. military pioneered an entirely new bombing CONOPS based on locating targets to GPS coordinates with high precision and receivers so cheap they could be put on the weapons.

3. In the DC-3 story (Part II) we note that it was the DC-3 (and not the DC-1 or DC-2) that revolutionized the airline business. Intermediate systems had to be thrown away.

4. Although the original Macintosh computer could be said to have revolutionized personal computing, that revolution was dependent on extensive evolution. First came the Macintosh with the product of failed systems by Apple (the Apple III and Lisa), not to mention its Xerox precursors. Second, the original Macintosh had to be rather extensively reengineered to accommodate the desktop publishing market (that it had almost single-handedly created). Third, the revolution truly began to be global only when the key interface ideas were ported to Microsoft operating system based personal computers.

These large-scale examples of the heuristic, where the program that surrounds the system "throws one away," either intentionally or not, are mirrored in small-scale design activity. It is rare that we can derive a best solution or representation, much like the first drafts of written works are rarely very good. We improve our designs, like we improve our writing, and like we improve our systems, by iterative development.

Much of synthesis is in the detail rather than grand visions and strategies. We can usefully classify the details as aggregation and partitioning.

Aggregation and Partitioning

Aggregation and partitioning are the grouping and separation of related solutions and problems. They are two sides of the same coin. Both are the processes by which the system is defined as components. One can argue about which precedes the other, but in fact they are used so iteratively and repeatedly that neither can be usefully said to precede the other. Conventionally, the components are arranged into hierarchies with a modest number of components at each level of the hierarchy (the famous 7 ± 2 structuring heuristic). The most important aggregation and partitioning heuristics are to minimize external coupling and maximize internal cohesion, usually worded as follows[8]:

> *In partitioning, choose the elements so that they are as independent as possible — that is, elements with low external complexity and high internal cohesion.*

> *Group elements that are strongly related to each other; separate elements that are unrelated.*

These two heuristics are especially interesting because they are part of the clearest hierarchy in heuristic refinement. Design is usefully viewed as progressive or stepwise refinement. Models, evaluation criteria, heuristics, and other factors are all refined as design progresses from abstract to concrete and specific. Ideally, heuristics exist in hierarchies that connect general design guidance, such as the two preceding heuristics, to domain-specific design guidelines. The downward direction of refinement is the deduction of domain-specific guidelines from general heuristics. The upward abstraction is the induction of general heuristics from similar guidelines across domains.

The deductive direction asks, taking the coupling heuristic as an example, how can coupling be measured? Or, for the cohesion heuristic, given alternative designs, which is the most cohesive? Within a specific domain, the questions should have more specific answers. For example, within the software domain, these questions are answered with greater refinement, though still heuristically. Studies have demonstrated quantitative impact on system properties as the two heuristics are more and less embodied in a systems design. A generally accepted software measure of partitioning is based on interface characterization and has five ranked levels. A related metric for aggregation quality (or cohesion) has seven ranked levels of cohesion.* Studies of large software systems show a strong correlation between coupling and cohesion levels, defect rates, and maintenance costs. A major study[9] found that routines with the worst coupling-to-cohesion ratios (interface complexity to internal coherence) had seven times more errors and 20 times higher maintenance costs than the routines with the best ratios.

Aggregation and partitioning with controlled fan-out and limited communication is a tested approach to building systems in comprehensible hierarchies. Numerous studies in specific domains have shown that choosing loosely coupled and highly cohesive elements leads to systems with low maintenance cost and low defect rates. However, nature suggests that much flatter hierarchies can yield systems of equal or greater robustness, in certain circumstances.

Chapter 8 introduced an ant colony as an example of flat hierarchy system that exhibits complex and adaptive behavior. The components of an ant colony, a few classes of ants, interact in a very flat system hierarchy. Communication is loose and hierarchically unstructured. There is no intelligent central direction. Nevertheless, the colony as a whole produces complex system-level behavior. The patterns of local, nonhierarchical

---

* The cohesion and coupling levels are carefully discussed in Yourdon, E., and L. L. Constantine, *Structured Design: Fundamentals of a Discipline of Computer Program and Systems Design.* New York: Yourdon Press, 1979. They were introduced earlier by the same authors and others in several papers.

interaction produce very complex operations. The colony is also very robust in that large numbers of its components (except the queen) can be removed catastrophically and the system will smoothly adapt. A perhaps related technological example is the Internet. Again, the system as a whole has a relatively flat hierarchy and no strong central direction. However, the patterns of local communication and resulting collaboration are able to produce complex, stable, and robust system-level behavior.

The observations that controlled and limited fan-out and interaction (the 7 ± 2 heuristic and coupling and cohesion studies) and that extreme fan-out and high distributed communication and control (ant colonies and the Internet) can both lead to high-quality systems is not contradictory. Rather they are complementary observations of the refinement of aggregation and partitioning into specific domains. In both cases, a happy choice of aggregations and partitions yields good systems. But the specific indicators of what constitutes good aggregation and partitioning vary with the domain. The general heuristic stands for all, but prescriptive or formalized guidance must be adapted for the domain.

Certification

To certify a system is to give an assurance to the paying client that the system is fit for use. Certifications can be elaborate, formal, and very complex or the opposite. The complexity and thoroughness are dependent on the system. A house can be certified by visual inspection. A computer flight control system might require highly detailed testing, extensive product and process inspections, and even formal mathematical proofs of design elements. Certification presents two distinct problems. The first is determining that the functionality desired by the client and created by the builder is acceptable. The second is the assessment of defects revealed during testing and inspection and the evaluation of those failures with respect to client demands.

Whether or not a system possesses a desired property can be stated as a mathematically precise proposition. Formal methods develop and track and verify such propositions throughout development, ideally leading to a formal proof that the system as designed possesses the desired properties. However, architecting practice has been to treat such questions heuristically, relying on judgment and experience to formulate tests and acceptance procedures. The heuristics on certification criteria do not address what such criteria should be, but they address the process for developing the criteria. Essentially, certification should not be treated separately from scoping or design. Certifiability must be inherent in the design. Two summarizing heuristics — actually on scoping and planning — are as follows:

For a system to meet its acceptance criteria to the satis-
faction of all parties, it must be architected, designed and
built to do so — no more and no less.

Define how an acceptance criterion is to be certified at the
same time the criterion is established.

The first part of certification is intimately connected to the conceptual phase. The system can be certified as possessing desired criteria
only to the extent it is designed to support such certification. The second
element of certification, dealing with failure, carries its own heuristics.
These heuristics emphasize a highly organized and rigorous approach
to defect analysis and removal. Once a defect is discovered, it should
not be considered resolved until it has been traced to its original source,
corrected, the correction tested at least as thoroughly as was needed to
find the defect originally, and the process recorded. Deming's famous
heuristic summarizes:

Tally the defects, analyze them, trace them to the source,
make corrections, keep a record of what happens after-
wards and keep repeating it.

A complex problem in ultraquality systems is the need to certify levels
of performance that cannot be directly observed. Suppose a missile system
is required to have a 99% success rate with 95% confidence. Suppose further that only 50 missiles are fired in acceptance tests (perhaps because
of cost constraints). Even if no failures are experienced during testing,
the requirement cannot be quantitatively certified. Even worse, suppose
a few failures occurred early in the 50 tests but were followed by flawless
performance after repair of some design defects. How can the architect
certify the system? It is quite possible that the system meets the requirement, but it cannot be proven within statistical criteria.

Certification of ultraquality might be deemed a problem of requirements. Many would argue that no requirement should be levied that cannot
be quantitatively shown. But the problem will not go away. The only acceptable failure levels in one-of-a-kind systems and those with large public
safety impacts will be immeasurable. No such systems can be certified if
certification in the absence of quantitatively provable data is not possible.

Some heuristics address this problem. The Deming approach, given
as a heuristic above, seeks to achieve any quality level by continuous
incremental improvement. Interestingly, there is a somewhat contradictory heuristic in the software domain. When a software system is tested,
the number of defects discovered should level off as testing continues.
The amount of additional test time to find each additional defect should

increase, and the total number of discovered defects will level out. The leveling out of the number of defects discovered gives an illusion that the system is now defect free. In practice, testing or reviews at any level rarely consistently find more than 60% of the defects present in a system. But if testing at a given level finds only a fixed percentage of defects, it likewise leaves a fixed percentage undiscovered. And the size of that undiscovered set will be roughly proportional to the number found in that same level of test or review. The heuristic can be given in two forms:

> *The number of defects remaining in a system after a given level of test or review (design review, unit test, system test, etc.) is proportional to the number found during that test or review.*

> *Testing can indicate the absence of defects in a system only when: (1) The test intensity is known from other systems to find a high percentage of defects, and (2) Few or no defects are discovered in the system under test.*

So the discovery and removal of defects is not necessarily an indication of a high-quality system. A variation of the "zero-defects" philosophy is that ultraquality requires ultraquality throughout all development processes. That is, certify a lack of defects in the final product by insisting on a lack of defects anywhere in the development process. The ultraquality problem is a particular example of the interplay of uncertainty, heuristics, and rational methods in making architectural choices. That interplay needs to be examined directly to understand how heuristic and rational methods interact in the progression of system design.

Organization into a Process Model

Even though effective architecting rarely proceeds on a linear, predetermined course, many people have found it convenient to have some reference model for how to organize the activities. A model found particularly useful is illustrated in Figure 9.3, originally introduced in Chapter 2. The figure illustrates a core set of activities associated with architecting and their relationships. The presence of multiple feedback channels shows that one does not proceed linearly through the activities. In practice, even within a cycle the path is often not linear, a subject we take up subsequently. The activities in Figure 9.3 are defined as orientation, purpose analysis, problem structuring, solution structuring, harmonization, selection or abstraction, architecture description, and supporting study.

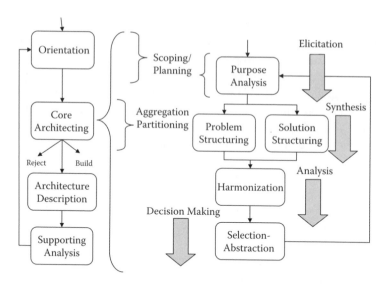

Figure 9.3 Expanded activities in an Architecting Process Model.

Orientation

Orientation is the process of understanding the context of an architecting project. When one finds oneself beginning what appears to be an architecting project, orientation is the set of activities necessary to make a preliminary definition of the project (although the system of interest presumably emerges). A simple heuristic to guide orientation is to ask the following questions:

1. What is the system-of-interest to the architecting effort (at least, the assumed system-of-interest)? What sort of system does the sponsor believe will eventually emerge?
2. What is the scope of the system-of-interest (and of the overall effort)? Is the system-of-interest a narrowly defined system with a single mission, a complex multimission system, or some assemblage of multiple systems (for example, a family of systems or a collaborative system)?
3. What is the apparent technology level? That is, is everything needed to accomplish the basic purpose well within current state-of-practice, pushing state-of-practice, or well beyond it?
4. What hard constraints are believed to exist (like a fixed delivery date)? Are they really hard constraints or just assumptions?
5. What resources are available for the effort, can more be acquired, if so how, and what are the expectations for interacting with the sponsor? Is the sponsor prepared to engage in value discussions with the team, and is sponsor time available to have the discussions?

6. When the architecture effort is complete, what will be done with its products? Will they be used to start a system acquisition, to guide other acquisitions, to guide research and development (R&D) activities, to mark off completion of a bureaucratic requirement, or for some other purpose?

7. Are the purposes of the system-of-interest, the architecting effort, and the architecture documentation to be developed all consistent with each other?

8. What is motivating the investigation into constructing the system-of-interest? Is it the sponsor needs, a new technology believed to be able to create value, or some other reason?

Purpose Analysis

Purpose analysis is the process of determining what the system of interest is for, and why the sponsor wants it (or at least believes he or she wants it). Purpose analysis is a broad-scoped investigation in the system-of-interest. It does not consist of just trying to discover and record assumed requirements or objectives. The intent is to delve more deeply into why the sponsor believes that having the system-of-interest will create value. One of the most useful heuristics in Purpose analysis is the *Four Who's*: *Always ask who benefits, who supplies, who pays, and who loses?*

Problem Structuring

Where purpose analysis is broad based and inclusive, problem structuring seeks to narrow. Purpose analysis accepts the full range of stakeholder inputs, whether precisely stated or not, whether unambiguous or not, and whether feasible or not. Problem structuring seeks to convert the rich picture of stakeholder concerns from purpose analysis into more rigorously structured models. To accomplish that without losing the richness of the original presentations, it may be necessary to spawn multiple problem descriptions. There may not be a single problem to solve; perhaps the best representation of the problem space is as multiple problems that will eventually be separately (if at all) addressed. The most useful heuristics and techniques in problem structuring include problem framing, expansion and contraction heuristics, use-case analysis, and functional decomposition.

Solution Structuring

In parallel with problem structuring, we can synthesize solutions. We can do this in parallel (that is, without full knowledge of objectives or requirements) because we assume that exposure to solution alternatives will affect sponsor beliefs about the nature of his or her problem. This is one of the basic tenets of ill-structured problem solving, that exposure to solutions changes perceptions of problems, and because we wish to embrace that change, we must let the two sides of the process influence each other. Solution structuring makes use of the synthesis heuristics, including those for aggregation and partitioning. The products of solution structuring are models of the system of interest, and so usually include block diagrams in all their forms and other models of form (discussed elsewhere in this book).

Harmonization

Harmonization is where we match up problem and solution descriptions to determine what can go together and what cannot. Harmonization is analytical, if not always rigorously so. The most useful techniques in harmonization are thus analytical and include functional walkthroughs, performance analysis, and executable simulations.

Selection or Abstraction

At some point, we have to make choices. One choice might be to drop the whole pursuit (perhaps a very wise choice in some circumstances, and one best made early). If the fundamental purpose is to emerge from architecting and arrange for construction of the system-of-interest, at some point we must select the configuration desired. In family-of-system and collaborative system cases, "abstraction" may be a better concept. By abstraction we mean selecting from the family or collaboration the things that are common, and likewise leaving out the things that are not common and leaving those to performing individuals or programs.

Architecture Description

Architecture description moves from collections of working models to more formalized groupings structuring as reference documents. Architecture models are organized into a formal architecture description, often using an "Architecture Framework," a concept defined subsequently. The key here is to avoid confusing the architecture description from the work process that precedes it. An architecture description is (or should be) a consequence of good architecting work.

Supporting Study

In practice, effective architecting often depends on relatively specialized and in-depth data. Recall the heuristic of variable technical depth. Good architecting is typified by deep investigation of particular, narrow areas in subsystems or subdisciplines (the heuristic of *Variable Technical Depth*). These deeper investigations are typically done separately from the core process of architecting. A cycle of architecting reveals the areas needing in-depth investigation. The architect sets up the in-depth study, and those results are fed back into further architecting.

Certainty, Rationality, and Choice

All of the design processes — scoping, partitioning, aggregation, and certification — require decisions. They require decisions on which problem statement to accept, what components to organize the system into, or when the system has reached an acceptable level of development. A by-product of a heuristic-based approach is continuous uncertainty. Looking back on their projects, most architects interviewed for the University of Southern California program concluded that the key choices they made were rarely obvious decisions at the time. Although, in retrospect, it may be obvious that a decision was either a very good or a very bad one, at the time the decision was actually made it was not clear at all. The heuristic summarizing is: *Before the flight it was opinion; after the flight it was obvious.* The members of the teams were constantly arguing and a decision was reached only through the authority of the leading architect for the project.*

A considerable effort has been made to develop rational decision-making methods. The goal of a fully rational or scientific approach is to make decisions optimally with respect to rigorously determined criteria. Again, the model of architecting practice presented here is a pragmatic mixture of heuristic and rigor. Decision theory works well when the problem can be parameterized with a modest number of values, uncertainty is limited, and estimates are reliable, and the client or users possess consistent utility functions with tractable mathematical expression. The absence of any of these conditions weakens or precludes the approach. Unfortunately, some or all of the conditions are usually absent in architecting problems (and even in more restricted disciplinary design problems). To understand why, one must understand the elements of the decision theoretic approach:

1. Identify the attributes contributing to client satisfaction and an algorithm for estimating the value of sets of attributes. More formally, this is determining the set over which the client will express preference.
2. Determine a utility function, a function that combines all the attributes and represents overall client satisfaction. Weighted, additive utility functions are commonly used, but not required. The utility function converts preferences into mathematically useful objective function.
3. Include uncertainty by determining probabilities, calculating the utility probability distribution, and determining the client's risk aversion curve. The risk aversion curve is a utility theory quantity that measures the client's willingness to trade risk for return.
4. Select the decision with the highest weighted expected utility.

* Comments by Harry Hillaker at USC on his experience as YF-16 architect.

The first problem in applying this framework to architecting problems is scale. To choose an optimum, the decision theory user must be able to maximize the utility functions over the decision set. If the set is very large, the problem is computationally infeasible. If the relationship between the parameters and utility is nonlinear, only relatively small problems are solvable. Unfortunately, both conditions commonly apply to the architecting and creation of complex systems.

The second problem is to workably and rationally include the effects of uncertainty or risk. In principle, uncertainty and unreliability in estimates can be folded into the decision theoretic framework through probability and assessment of the client's risk aversion curve. The risk aversion curve measures the client's willingness to trade risk and return. A risk-neutral client wants the strategy that maximizes expected return. A risk-averse client prefers a strategy with certainty over opportunity for greater return. A risk-disposed client prefers the opposite — wanting the opportunity for greater return even if the expectation of the return is less.

In practice, however, the process of including uncertainty is heavily subjective. For example, how can one estimate probabilities for unprecedented events? If the probabilities are inaccurate, the whole framework loses its claim to optimality. Estimation of risk aversion curves is likewise subjective, at least in practice. When so much subjective judgment has been introduced, it is unclear if maintaining the analytical framework leads to much benefit or if it is simply a gloss.

One clear benefit of the decision theory framework is that it makes the decision criteria explicit and, thus, subject to direct criticism and discussion. This beneficial explicitness can be obtained without the full framework. This approach is to drop the analytic gloss, make decisions based on heuristics and architectural judgment, but (and this is more honored in the breach) require the basis be explicitly given and recorded.

A third problem with attempting to fully rationalize architectural decisions is that for many of them there will be multiple clients who have some claim to express a preference. Single clients can be assumed to have consistent preferences and, hence, consistent utility functions. However, consistent utility functions do not generally exist when the client or user is a group, as in sociotechnical systems.* Even with single clients, value judgments may change, especially after the system is delivered and the client acquires direct experience.†

---

* This problem with multiple clients and decision theory has been extensively studied in literature on public choice and political philosophy. A tutorial reference is Mueller, D. C., *Public Choice*. London; New York: Cambridge University Press, 1979.

† Nonutility theory based decision methods, such as the Analytic Hierarchy Process, have many of the same problems. Most writers have discussed that the primary role of decision theoretic methods should be to elucidate the underlying preferences. See Saaty, T., *The Analytic Network Process*. Pittsburgh, PA: RWS Publications, 1996, Preface and Chapter 1.

An observation about decision theory, paraphrased from Keeney, is that decision analysis is most applicable when it is least important. When doing decision analysis, one often finds one of the following two situations:

1. Analysis of the objective function shows that one alternative is much better than the rest. So, choosing the optimum is of high value. But, it is rarely hard to find such a clear winner; it usually stands out obviously from analysis of the objectives. Clear analysis of the objectives leads to the inevitable conclusion even without the full formal machinery.
2. The optimal choice cannot be found without the full machinery, but the true optimum is not much better than nearby choices. Any of the nearby choices would be almost as good, and given uncertainties, not distinguishable.

Thus, pursuit of the true optimum is of much less importance than the supporting reasoning, in objectives, and in fully exploring the alternative space. We get more benefit from the systematic thinking associated with decision analysis than we get from the machinery itself.[10]

Rational and analytical methods produce a gloss of certainty, but often hide highly subjective choices. No hard and fast guideline exists for choosing between analytical choice and heuristic choice when unquantified uncertainties exist. Certainly, when the situation is well understood and uncertainties can be statistically measured, the decision theoretic framework is appropriate. When even the right questions are in doubt, it adds little to the process to quantify them. Intermediate conditions call for intermediate criteria and methods. For example, a system might have as client objectives "be flexible" and "leave in options." Obviously, these criteria are open to interpretation. The refinement approach is to derive or specialize increasingly specific criteria from very general criteria. This process creates a continuous progression of evaluation criteria from general to specific and eventually measurable.

> *Example*: The U.S. Department of Transportation has financed an Intelligent Transport System (ITS) architecture development effort. Among their evaluation criteria was "system flexibility," obviously a very loose criteria.[11] An initial refinement of the loose criteria could be as follows:
>
> 1. Architecture components should fit in many alternative architectures.
> 2. Architecture components should support multiple services.

3. Architecture components should expand with linear or sublinear cost to address greater load.
4. Components should support non-ITS services.

These refined heuristic evaluation criteria can be applied directly to candidate architectures. Or they can be further refined into quantitative and measurable criteria. The intermediate refinement on the way to quantitative and measurable criteria creates a progression that threads through the whole development process. Instead of thinking of design as beginning and stopping, it continuously progresses. Sophisticated mixtures of the heuristics and rational methods are part of architecting practice in some domains. This progression is the topic of the next section.

Although architecting problems rarely can be effectively modeled and resolved as simple decision theoretic problems, the decision theoretic process holds much value, if used appropriately. The decision theoretic process of building a decision model is a valuable guide to good architecting:

1. Make objectives or attributes explicit and visible to all stakeholders (build a value model). Encourage debate, and hopefully agreement, on objectives.
2. Use the objectives to search for solutions better than any currently known. Instead of using weights and trades to "pick the best of a bad lot," use the objectives to focus the search for higher-valued possibilities (the "Value Focused Thinking" notion of Keeney[12]).
3. Build explicit models of uncertainty. Use those models to search for ways that uncertainty can be exploited instead of merely adapted to.

Stopping or Progressing?

When does architecting and modeling stop? The short answer is that given earlier: they never stop; they progress. The architecting process (along with many other parallel tracks) continuously progresses from the abstract to the concrete in a steady reduction of abstraction. In a narrow sense, there are recognizable points at which some aspects of architecting and modeling must stop. To physically fabricate a component of a system, its design must be frozen. It may not stop until the lathe stops turning or the final line of code is typed in, but the physical object is the realization of *some* design. In the broader sense, even physical fabrication does not stop architecting. Operations can be interpreted only through recourse to models, though those models may be quite precise when driven by real data. In some systems, such as distant space probes, even operational modeling is still somewhat remote.

The significant progressions in architecting are promoted by the role of the architect. The architect's role makes two decisions foremost: the selection of a system concept and the certification of the built system. The former decision is almost certain to be driven by heuristic criteria; the latter is more open, depending on how precisely the criteria of fitness for use can be defined. A system concept is suitable when it is both satisfactory and feasible. Only the client can judge the system "satisfactory," though the client will have to rely on information provided by the architect. In builder-architected systems, the architect must often make the judgment for the client (who will hopefully appear after the system reaches the market). Feasible means the system can be developed and deployed with acceptable risk. Certification requires that the system as built adequately fulfills the client's purposes, including cost, as well as the contract with the builder.

Risk, a principal element in judging feasibility, is almost certain to be judged heuristically. The rational means of handling risk is through probability, but a probabilistic risk assessment requires some set of precedents to estimate over — that is, a series of developments of similar nature for which the performance and cost history is known. By definition, such a history cannot be available for unprecedented systems. So the architect is left to estimate risk by other means. In well-defined domains, past history should be able to provide a useful guide; it certainly does for civil architects. Civil architects are expected to control cost and schedule risk for new structures and can do so because construction cost estimation methods are reasonably well developed. The desired approach is to use judgment, and perhaps a catalog of domain-specific heuristics, to size the development effort against past systems, and use documented development data from those past systems to estimate risk. For example, in software systems, cost models based on code size estimates are known and are often calibrated against past development projects in builder organizations. If the architect can deduce code size, and possible variation in code size, reliably, a traceable estimate of cost risk is possible.

The judgment of how satisfactory a concept is and the certification process both depend on how well customer purposes can be specified. Here there is great latitude for both heuristic and rational means. If customer purposes can be precisely specified, it may be possible to precisely judge how well a system fulfills them, either in prospect or retrospect. In prospect, it depends on having behavior and performance models that are firmly attached to customer purposes. With good models with certain connection between the models and implementation technology, the architect can confidently predict how well the planned system will fulfill the desired purposes. The retrospective problem is that of certification, of determining how well the built system fulfills customer purposes. Again, well-founded, scientific models and mature implementation technologies make system assessment relatively certain.

More heuristic problems arise when the same factors do not apply. Mainly this occurs when it is hard to formulate precise customer purpose models, when it is hard to determine whether or not a built system fulfills a precisely stated purpose, or when there is uncertainty about the connection between model and implemented system. The second two are related because they both concern retrospective assessment of architecture models against a built system in the presence of uncertainty.

The first case applies when customer purposes are vague or likely to change in response to actual experience with the system. When the customer is relatively inexperienced with systems of the type, his or her perception of the system's value and requirements is likely to change, perhaps radically, with experience. Vague customer purposes can be addressed through the architecture. Take, for example, an emphasis on options in the architecture and a development plan that includes early user prototypes with the ability to feed prototype experience back into the architecture. This is nicely captured in two heuristics:

> *Firm commitments are best made after the prototype works.*[13]

> *Hang on to the agony of decision as long as possible.*[14]

The second case, problems in determining whether or not a built system fulfills a given purpose, is mainly a problem when requirements are fundamentally immeasurable or when performance is demanded in an environment that cannot be provided for test. For example, a space system may require a failure rate so low it will never occur during any practical test (the ultraquality problem). Or, a weapon system may be required to operate in the presence of hostile countermeasures that will not exist outside a real combat environment. Neither of these requirements can be certified by test or analysis. To certify a system with requirements like these, it is necessary to either substitute surrogate requirements agreed to by the client or to find alternative certification criteria.

To architect-in certifiable criteria essentially means to substitute a refined set of measurable criteria for the client's immeasurable criteria. This requires that the architect be able to convince the client of the validity of a model for connecting the refined criteria to the original criteria. One advantage of a third-party architect is the independent architect's greater credibility in making just such arguments, which may be critical to developing a certifiable system. A builder-architect, with an apparent conflict of interest, may not have the same credibility. The model that connects the surrogate criteria to the real, immeasurable criteria may be a detailed mathematical model or may be quite heuristic. An example of the former category is failure tree analysis that tries to certify untestable reliability levels from testable subsystem reliability levels. A more heuristic model may be more appropriate for certifying performance in uncertain combat environments. Although the performance required is uncertain, criteria like flexibility, reprogrammability, performance reserve, fallback modes, and ability to withstand damage can be specified and measured.

Rational and heuristic methods can be combined to develop ultraquality systems. A good example is a paper by Jaynarayan.[15] This paper discusses the architectural principles for developing flight control computers with failure rates as low as 10^{-10} per hour. Certification of such systems is a major problem. The authors discuss a two-pronged approach. First, instead of using a brute-force failure modes and effects analysis with its enormous fault trees, they design for "Byzantine failure." Byzantine failure means failure in which the failed element actively, intelligently, and malevolently attempts to cause system failure. They go on to describe formal methods for designing systems resistant to a given number of Byzantine faults, thus replacing the need to trace fault trees for each type of failure. The analysis of failure trees is then brought down to tractable size. The approach is based on designs that do not allow information or energy from a possibly failed element to propagate outside an error confinement region. The second prong is a collection of guidelines for minimizing common mode failures. In a common mode failure, several nominally independent redundant units fail simultaneously for the same reason. These are the system failures due to design errors rather than component failures. Because one cannot design-in resistance to design failure, other means are necessary. The guidelines, partially a set of heuristics, provide guidance in this otherwise nonmeasurable area.

The third and last case is uncertainty about the connection between the model and the actual system. This is an additional case where informed judgment and heuristics are needed. To reduce the uncertainty in modeling requires tests and prototypes. The best guidance on architecting prototypes is to realize that all prototypes should be purpose driven. Even when the purposes of the system are less than clear, the purposes of the prototype should be quite clear. Thus, the architecting of the prototype can be approached as architecting a system, with the architect as the client.

Architecture and Design Disciplines

Not very many years ago, the design of a system of the complexity of several tens-of-thousands of logic gates was a major undertaking. It was an architectural task in the sense it was probably motivated by an explicit purpose and required the coordination of a multidisciplinary design effort. Today, components of much higher complexity are the everyday building blocks of the specialized digital designer. No architectural effort is required to use such a component, or even to design a new one. In principle, art has been largely removed from the design process because the discipline has a firm scientific basis. In other words, the design discipline or domain is well worked out, and the practitioners are recognized specialists. Today it is common to discuss digital logic synthesis directly from fairly level specifications, even if automated synthesis is not yet common practice. So, there should be no surprise if systems that today tax our abilities and require architectural efforts one day become routine with recognized design methodologies taught in undergraduate courses.

The discussion of progression leads to further understanding of the distinctions between architecture and engineering. The basic distinction was reviewed in the Preface, along with the types of problems addressed and the tools used to address them. A refinement of the distinction was discussed in Chapter 1, the extent to which the practitioner is primarily concerned with scoping, conceptualizing, and certification. By looking at the spectrum of architecture and engineering across systems disciplines, these distinctions become clearer and can be further refined. First, the methods most associated with architecting (heuristics) work best one step beyond where rational design disciplines have been worked out. This may or may not be at the forefront of component technology. Large-scale systems, by their nature, push the limits of scientific engineering at whatever level of technology development is current. But, as design and manufacturing technology change the level of integration that is considered a component, the relative working position of the architect inevitably changes. Where the science does not exist, the designer must be guided by art. With familiarity and repetition, much that was done heuristically can now be done scientifically or procedurally.

However, this does not imply that where technology is mature, architecting does not exist. If it did, there would be no need for civil architects. Only systems that are relatively unique need to be architected. Development issues for unique systems contain a kernel of architectural concerns that transcend whatever technology or scientific level is current. This kernel concerns the bridge between human needs (which must be determined through social interaction and are not the domain of science) and technological systems. In low-technology systems, like buildings, only the nonroutine building needs to be architected. But dealing with the nonroutine, the unique, the client/user customized, is different from other engineering practices. It contains an irreducible component of art. A series of related unprecedented systems establishes a precedent. The precedent establishes recognized patterns, sometimes called architectures, of recognized worth. Further systems in the field will commonly use those established architectures, with variations more on style than in core structure.

Current development in software engineering provides an example of evolution to a design discipline. Until relatively recently, the notion of software engineering hardly existed; there was only programming. Programming is the process of assembling software from programming language statements. Programming language statements do not provide a very rich language for expressing system behaviors. They are constrained to basic arithmetic, logical, and assignment operations. To build complex system behaviors, programs are structured into higher-level components that begin to express system domain concepts. But in traditional programming, each of the components must be handcrafted from the raw material of programming languages.

The progression in software is through the construction and standardization of components embodying behaviors closer and closer to problem domains. Instead of programming in what was considered a "high-level language," the engineer can now build a system from components close to the problem domain. Where the programming language is still, but it may be used primarily to knit together prebuilt components. Programming libraries have been in common use for many years. The libraries shipped with commercial software development environments are often very large and contain extensive class or object libraries. In certain domains, the gap has grown very small.

> *Example*: The popular mathematics package MATLAB®
> (Natick, Massachusetts) allows direct manipulation
> of matrices and vectors. It also provides a rich library
> of functions targeted at control engineers, image
> processing specialists, and other fields. One can
> dispense with the matrices and vectors all together
> by "programming" with a graphical block diagram
> interface that hides the computational details and
> provides hundreds of prebuilt blocks. Further exten-
> sions allow the block diagram to be compiled into
> executable programs that run on remote machines.
> Increasingly it is possible to compile for direct con-
> nection to real-time embedded systems.

Wherever a family of related systems is built, a set of accepted models and abstractions appears and forms the basis for a specialized design discipline. If the family becomes important enough, the design discipline will attract enough research attention to build scientific foundations. It will truly become a design discipline when universities form departments devoted to it. At the same time, a set of common design abstractions will be recognized as "architectures" for the family. Mary Shaw, observing the software field, finds components and patterns constrained by component and connector vocabulary, topology, and semantic constraints. These patterns can be termed "styles" of architecture in the field, as was discussed in Chapter 6 for software.

Architecture and Patterns

The progression from "inspired" architecture to formal design method is through long experience. Long experience in the discipline by its practitioners eventually yields tested patterns of function and form. Patterns, pattern languages, and styles are a formalization of this progression. Architecting in a domain matures as architects identify reusable components and repeating styles of connection. Put another way, they recognize recurring patterns of form and their relationships to patterns in problems. In a mature domain, patterns in both the problem and solution domains develop rigorous expression. In digital logic (a relatively mature design domain), problems are stated in formal logic and solutions in equally mathematically well-founded components. In a less mature domain, the patterns are more abstract or heuristic.

A formalization of patterns in architecture is due to Christopher Alexander.[16] Working within civil architecture and urban design, Alexander developed an approach to synthesis based on the composition of formalized patterns. A pattern is a reoccurring structure within a design domain. A pattern consists of both a problem or functional objective for a system and a solution. Patterns may be quite concrete (such as "A sunny corner") or relatively abstract (such as "Masters and apprentices"). A template for defining a pattern is as follows:

1. A brief name that describes what the pattern accomplishes.
2. A concise problem statement.
3. A description of the problem including the motivation for the pattern and the issues in resolving the problem.
4. A solution, preferably stated in the form of an instruction.
5. A discussion of how the pattern relates to other patterns in the language.

A pattern language is a set of patterns complete enough for design within a domain. It is a method for composing patterns to synthesis solutions to diverse objectives. In the Alexandrian method, the architect consults sets of patterns and chooses from them those patterns that evoke the elements desired in a project. The patterns become the building blocks for synthesis, or suggest important elements that should be present in the building. The patterns each suggest instructions for solution structure, or contain a solution fragment. The fragments and instructions are merged to yield a system design.

Because the definition of a pattern and a pattern language are quite general, they can be applied to other forms of architecture. The ideas of patterns and pattern languages are now a subject of active interest in software engineering.* Software architects often use the term *style* to refer to reoccurring patterns in high-level software design. Various authors have suggested patterns in software using a pattern template similar to that of Alexander. An example of a software pattern is "Callbacks and Handlers," a commonly used style of organizing system-dependent bindings of code to fixed behavioral requirements.

The concept of a style is related to Alexandrian patterns because each style can be described using the pattern template. Patterns are also a special class of heuristic. A pattern is a prescriptive heuristic describing particular choices of form and their relationship to particular problems. Unlike patterns, heuristics are not tied to a particular domain.

Although the boundaries are not sharp, heuristics, patterns, styles, and integrated design methods can be thought to form a progression. Heuristics are the most general, spanning domains and categories of guidance. However, they are also the least precise and give the least guidance to the novice. Patterns are specially documented, prescriptive heuristics of form. They prescribe (perhaps suggest) particular solutions to particular problems within a domain. A style is still more precisely defined guidance, this time in the form of domain-specific structure. Still farther along the maturity curve are fully integrated design methods. These have domain-specific models and specific procedures for building the models, transforming models of one type into another type, and implementing a system from the models.

Thus, the largest-scale progression is from architecting to a rigorous and disciplined design method; one that is essential to the normative theory of design. Along the way, the domain acquires heuristics, patterns, and styles of proven worth. As the heuristics, patterns, and styles become more specific, precise, and prescriptive, they give the most guidance to the novice and come closest to the normative (what *should* be) theory of design. As design methods become more precise and rigorous, they also become more amenable to scientific study and improvement. Thus, the progression carries from a period requiring (and encouraging) highly creative and innovative architecting to one defined by quantifiable and provable science.

Civil architecture experience suggests that at the end of the road there will still be a segment of practice that is best addressed through a fusion of art and science. This segment will be primarily concerned with the clients of a system and will seek to reconcile client satisfaction and

---

* A brief summary with some further references is Bercuzk, C., Hot Topics, Finding Solutions through Pattern Languages, *IEEE Computer*, Vol. 27, Number 12, pp. 75–76, 1995.

technical feasibility. The choice of method will depend on the question. If you want to know how a building will fare in a hurricane, you know to ask a structural engineer. If you want the building to express your desires, and do so in a way beyond a rote calculation of floor space and room types, you know to ask an architect.

Conclusion

A fundamental challenge in defining a systems architecting method or a systems architecting tool kit is its unstructured and eclectic nature. Architecting is synthesis oriented and operates in domains and with concerns that preclude rote synthesis. Successful architects proceed through a mixture of heuristic and rational or scientific methods. One meta-method that helps organize the architecting process is that of progression.

Architecting proceeds from the abstract and general to the domain specific. The transition from the unstructured and broad concerns of architecting to the structured and narrow concerns of developed design domains is not sharp. It is progressive as abstract models are gradually given form through transformation to increasingly domain-specific models. At the same time, all other aspects of the system undergo concurrent progressions from general to specific.

The emphasis has been on the heuristic and unstructured components of the process, but that is not to undervalue the quantitative and scientific elements required. The rational and scientific elements are tied to the specific domains where systems are sufficiently constrained to allow scientific study. The broad outlines of architecting are best seen apart from any specific domain. A few examples of the intermediate steps in progression were given in this chapter. The next chapter brings these threads together by showing specific examples of models and their association with heuristic progression. In part this is done for the domains of Part II, and in part for other recognized large not domains not specifically discussed in Part II.

Exercises

1. Find an additional heuristic progression by working from the specific to the general. Find one or more related design heuristics in a technology-specific domain. Generalize those heuristics to one or more heuristics that apply across several domains.
2. Find an additional heuristic progression by working from the general to the specific. Choose one or more heuristics from Appendix A. Find or deduce domain specific heuristic design guidelines in a technology domain familiar to you.

3. Examine the hypothesis that there is an identifiable set of "architectural" concerns in a domain familiar to you. What issues in the domain are unlikely to be reducible to normative rules or rational synthesis?
4. Trace the progression of behavioral modeling throughout the development cycle of a system familiar to you.
5. Trace the progression of physical modeling throughout the development cycle of a system familiar to you.
6. Trace the progression of performance modeling throughout the development cycle of a system familiar to you.
7. Trace the progression of cost estimation throughout the development cycle of a system familiar to you.

Notes and References

1. Yourdon, E., and L. L. Constantine, *Structured Design: Fundamentals of a Discipline of Computer Program and Systems Design.* New York: Yourdon Press, 1979.
2. Ben Baumeister, personal communication.
3. Rechtin, E., *Systems Architecting, Creating and Building Complex Systems.* Englewood Cliffs, NJ: Prentice Hall, 1991, pp. 14–24. (Please note that further reference to this citation will be referred to as Rechtin 1991.)
4. ADARTS Guidebook, pp. 8-4.
5. ADARTS Guidebook, pp. 8–15.
6. Raymond, Eric S., *The Cathedral and the Bazaar: Musings on Linux and Open Source by an Accidental Revolutionary.* Sebastopol, CA: O'Reilly Media, 2001.
7. Brooks, F., *The Mythical Man-Month: Essays on Software Engineering*, 2nd ed. Reading, MA: Addison-Wesley, 1995.
8. Rechtin 1991, p. 312.
9. Selby, R. W., and V. R. Basili, Analyzing Error-Prone System Structure, *IEEE Transactions on Software Engineering*, SE-17, Number 2, pp. 141–152, February 1991.
10. See Keeney, R. L., Making Better Decision Makers, *Decision Analysis*, Vol. 1, Number 4, pp. 193–204, December 2004, for the original observations.
11. Discussed further by one author in Maier, M. W., On Architecting and Intelligent Transport Systems, *Joint Issue IEEE Transactions on Aerospace and Electronic Systems/System Engineering*, AES33:2, pp. 610–625, April 1997.
12. Keeney, R. L., *Value Focused Thinking.* Cambridge, MA: Harvard University Press, 1996.
13. Comments to the authors on Rechtin (1991) by L. Bernstein in the context of large network software systems.
14. Rechtin 1991.
15. Jaynarayan, H., and R. Harper, Architectural Principles for Safety-Critical Real-Time Applications, *Proceedings of the IEEE*, Vol. 82, Number 1, pp. 25-40, January 1994.
16. Alexander, C., *The Timeless Way of Building.* Oxford: Oxford University Press, 1979; Alexander, C., *A Pattern Language: Towns, Buildings, Construction.* Oxford: Oxford University Press, 1977.

chapter 10

Integrated Modeling Methodologies

Introduction

The previous two chapters explored the concepts of model views, model integration, and progression along parallel paths. This chapter brings together these threads by presenting examples of integrated modeling methodologies. Part III concludes in the next chapter where we review the architecture community's standards for architecture description. The distinction between this chapter and the next is twofold. First, in this chapter we study modeling methods without concern for how their elements combine in formal documents. The focus is on integrated, multiview modeling methods as tools for architecting, not as descriptors within a document. Second, this chapter is concerned with methods from the literature whether or not they are formally standardized. De facto standardization or standardization in the scientific literature is sufficient in this chapter. The next chapter is concerned with how models are brought together in formalized architecture description documents and with community efforts at formal standardization. The methodologies in this chapter are further divided by domain specificity, with the first models more nearly domain independent and later models more domain specific.

Architecting clearly is domain dependent. A good architect of avionics systems, for example, may not be able to effectively architect social systems. Hence, there is no attempt to introduce a single set of models suitable for architecting everything. The models of greatest interest are those tied to the domain of interest, although they must support the level of abstraction needed in architecting. The integrated models chosen for this chapter include two principally intended for real-time, computer-based, mixed hardware/software systems (H/P and Q^{2}FD), three methods for software-based systems, one method for manufacturing systems, and, conceptually, at least, some methods for including human behavior in sociotechnical system descriptions.

The examples for each method were chosen from relatively simple systems. They are intended as illustrations of the methods and their relevance to architectural modeling and to fit within the scope of the book. They are not intended as case studies in system architecting. Brief case studies at the decision level precede each chapter in Part II.

In choosing integrated modeling methods to present, we look for the following factors:

1. The collection of models spans three or more of the baseline views illustrated in Figure 8.2.
2. The syntax and semantics of the modeling language is defined with enough formality that it supports both intraview and interview consistency checking rules.
3. The models can be used at levels of abstraction from concept presentation to transition to disciplinary engineering. That is, the models support progressive refinement from architecting through systems engineering.

General Integrated Models

Two very general integrated modeling methods are Hatley-Pirbhai (H/P) and Q²FD. The Unified Modeling Language (UML) is also quite general, although in practice it is used mostly in software systems. The more recent extensions to the UML, known as Systems Modeling Language (SysML), are of greater applicability for integrated modeling, although we shall see limitations there as well.

Hatley-Pirbhai — Computer-Based, Reactive Systems

A computer-based, reactive system senses and reacts to events in the physical world, with much of the implementation complexity in programmable computers. Multifunction automobile engine controllers, programmable manufacturing robots, and military avionics systems (among many others) all fall into this category. They are distinguished by mixing continuous, discrete, and discrete event logics and being implemented largely through modern computer technology. The integrated models used to describe these systems emphasize detailed behavior descriptions, form descriptions matched to software and computer hardware technologies, and some performance modeling. Efforts in the recent past at defining an Engineering of Computer Based Systems discipline[1] are directed at systems of this type.

Several different investigators have worked to build integrated models for computer-based reactive systems. The most complete example of such integration is the Hatley-Pirbhai (H/P) methodology.* Other methods

---

* Wood, D. P., and W. G. Wood, Comparative Evaluation of Four Specification Methods for Real-Time Systems, Software Engineering Institute Technical Report, CMU/SEI-89-TR-36, 1989. This study compared four popular system modeling methods. Their conclusion was that the Hatley-Pirbhai method was the most complete of the four, though similarities were more important than the differences. In the intervening time, many of the popular methods have been extended and additional tools reflecting multiview integration have begun to appear, although actual use seems to have faded. The Hatley-Pirbhai method has been likewise further extended.

developed contemporaneously with Hatley-Pirbhai are of similar levels of integration. The UML is also close to this level of completeness. The practical reality is that many of these methods, whether well-founded or not, have fallen out of fashion. Today, discussions of integrated modeling methods for architecting are likely to focus on UML, SysML, or one of the architecture description frameworks. However, the issues with these current approaches are better understood with some historical context on integrated modeling, as the historical approaches make both the strengths and weaknesses or the current approaches more clear. This section concentrates on the structure of H/P. With the concepts of H/P in mind, it is straightforward to make a comparative assessment of other tools and methods.

H/P defines a system through three primary models: two behavioral models (the "Requirements Model" [RM] and the "Enhanced Requirements Model" [ERM]) and a model of form called the "Architecture Model" (AM). The two behavioral models are linked through an embedding process. Static allocation tables link the behavioral and form models. The performance view is linked statically through timing allocation tables. More complex performance models have been integrated with H/P, but descriptions have only recently been published. A dictionary defines the data view. This dictionary provides a hierarchical data element decomposition but does not provide a syntax for defining dynamic data relationships. No managerial view is provided, although managerial metrics have been defined for models of the H/P type.

Both behavioral models are based on DeMarco-style data flow diagrams. The data flow diagrams are extended to include finite state and event processing through what is called the "Control Model." The control model uses data flow diagram syntax with discrete events and finite state machine processing specifications. The behavioral modeling syntax is deliberately nonrigorous and is not designed for automated execution. This "lack" of rigor is deliberate; it is intended to encourage flexibility in client and user communication. The method believes the flexibility rather than rigor at this stage enhances communication with clients and users. The method also believes, through its choice of data flow diagrams, that functional decomposition better communicates to stakeholders than does specification by example methods, such as use-cases. The ERM is a superset of the requirements model. It surrounds the core behavioral model and provides a behavioral specification of the processing necessary to resolve the physical interfaces into problem domain logical interfaces. The ERM defines implementation-dependent behaviors, such as user interface and physical I/O.

> *Example*: *Microsatellite Imaging System* — Some portions of the H/P model formulated for the imaging (camera) subsystem of a microsatellite provide an

illustration of the H/P concepts. This example is to present the flavor of the H/P idiom for architects, not to fully define the imaging system. The level chosen is representative of that of a subsystem architecture (not all architecting has to be done on systems of enormous scale). Figure 10.1 shows the top-level behavioral model of the imaging system, defined as a data flow diagram (DFD). Each circle on the diagram represents a data-triggered function or process. So, for example, process number 2, "Evaluate Image," is triggered by the presence of a "Raw Image" data element. Also from the diagram, process number 2 produces a data element of the same type (the outgoing arrow labeled "Raw Image") and another data element called "Image Evals."

Each process in the behavior model is defined either by its own data flow diagram or by a textual specification. During early development, processes may be defined with brief and nonrigorous textual specifications. Later, as processes are allocated to physical modules, the specifications are expanded in greater detail until implementation-appropriate rigor is reached. Complex processes may have more detailed specifications even early in the process. For example, in Figure 10.2, process number 1 "Form Image" is expanded into its own diagram.

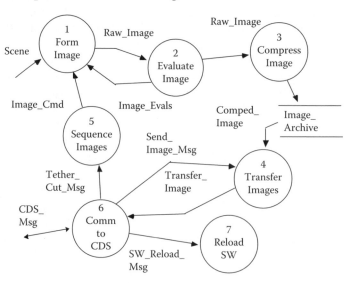

Figure 10.1 Top-level data flow diagram for a microsatellite imaging system.

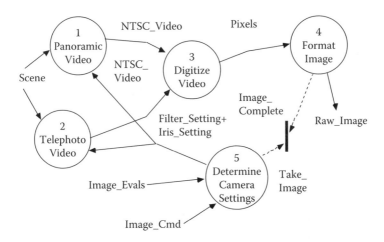

Figure 10.2 Expanded data and control flow diagram for the top-level process "Form Image."

Figure 10.2 also introduces control flow. The dotted arrows indicate flow of control elements, and the solid line into which they flow is a control specification. The control specification is shown as part of the same figure. Control flows may be interpreted either as continuous time, discrete valued data items, or discrete events. The latter interpretation is more widely used, although it is not preferred in the published H/P examples. The control specification is a finite state machine, here shown as a state transition diagram, although other forms are also possible. The actions produced by the state machine are to activate or deactivate processes on the associated data flow diagram.

All data elements appearing on a diagram are defined in the data dictionary. Each may be defined in terms of lower-level data elements. For example, the flow "Raw Image" appearing in Figure 10.2 appears in the data dictionary as follows:

```
Raw Image = 768{484{Pixel}}
```

It indicates, in this case, that Raw Image is composed of 768 × 484 repetitions of the element "Pixel." At early stages, Pixel is defined qualitatively as a range of luminance values. In later design stages, the definition will be augmented, though not replaced, by a definition in terms of implementation-specific data elements.

In addition to the two behavior models, the H/P method contains an "Architecture Model." The architecture model is the model of form that defines the physical implementation. The architecture model is hierarchical. It allows sequential definition in greater detail by expansion of

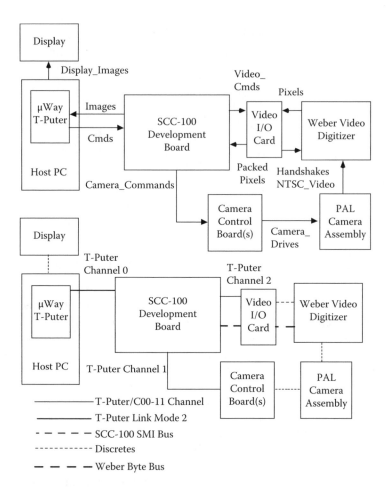

Figure 10.3 Architecture flow and interconnect diagrams for the example of microsatellite imaging system laboratory prototype.

modules. Figure 10.3 shows the paired architecture flow and interconnect models for the microsatellite imaging system.

The H/P block diagram syntax partitions a system into modules, that are defined as physically identifiable implementation elements. The flow diagram shows the exchange of data elements among the modules. Which data elements are exchanged among the modules is defined by the allocation of behavioral model processes to the modules.

The interconnection model defines the physical channels through which the data elements flow. Each interconnect is further defined in a separate specification. For example, the interconnect "Tputer channel 1" connects the processor module and the camera control module. Allocation requires camera commands to flow over the channel. Augmentations to the data dictionary define a mapping between the logical camera commands and the line codes of the channel. If the channel requires message framing, protocol processing, or the like, it is defined in the interconnection specification. Again, the level of detail provided can vary during design based on the interface's impact on risk and feasibility.

Quantitative QFD (Q²FD) — Performance-Driven Systems

Many systems are driven by quantitatively stated performance objectives. These systems may also contain complex behavior or other attributes, but its performance objectives are of utmost importance to the client. For these systems, it is common practice to take a performance-centered approach to system specification, decomposition, and synthesis. A particularly attractive way of organizing decomposition is through extended Quality Function Deployment (QFD) matrices.[2]

QFD is a Japanese-originated method for visually organizing the decomposition of customer objectives.[3] It builds a graphical hierarchy of how customer objectives are addressed throughout a system design, and carries the relevance of customer objectives throughout design. A Q²FD-based approach requires that the architect do the following:

1. Identify a set of performance objectives of interest to the customer. Determine appropriate values or ranges for meeting these objectives through competitive analysis.
2. Identify the set of system-level design parameters that determine the performance for each objective. Determine suitable satisfaction models that relate the parameters and objectives.
3. Determine the relationships of the parameters and objectives and the interrelationships among the parameters. Which affect which?
4. Set one or more values for each parameter. Multiple values may be set — for example, minimum, nominal, and target. Additional slots provide tracking from detailed design activities.
5. Repeat the process iteratively using the system design parameters as objectives. At each stage, the parameters at the next level up become the objectives at the next level down.
6. Continue the process of decomposition as many levels as desired. As detailed designs are developed, their parameter values can flow up the hierarchy to track estimated performance for customer objectives. The structure is illustrated in Figure 10.4.

Unfortunately, QFD models for real problems tend to produce quite large matrices. Because they map directly to computer spreadsheets, this causes no difficulty in modern work environments, but it does cause a problem in presenting an example. Also, the graphic of the matrix shows

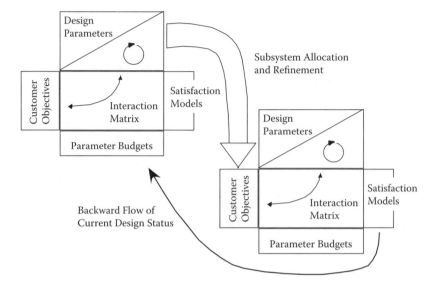

Figure 10.4 Quantitative Quality Function Deployment (QFD) Hierarchy Tree. The basic matrix shows the interrelationships of customer objectives and engineering design parameters. QFD matrices are arranged in a hierarchy that can mirror the decomposition of a system into subsystems and modules.

the result but hides the satisfaction models. The satisfaction models are equations, simulations, or assessment processes necessary to determine the performance measure value. The original reference on QFD by Hauser contains a qualitative example of using QFD for objective decomposition, as do other books on QFD. Two papers by one of the present authors[4] contain detailed, quantitative examples of QFD performance decomposition using analytical engineering models.

Integrated Modeling and Software

Chapters 8 and 9 introduced the ideas of model views and stepwise refinement-in-the-large. Both of these ideas have featured prominently in the software engineering literature. Software methods have been the principal sources for detailed methods for expressing multiple views and development through refinement. Software engineers have developed several integrated modeling and development methodologies that integrate across views and employ explicit heuristics. Three of those methods are described in detail: structured analysis and design, Ada-Based Design Approach for Real-Time Systems (ADARTS), and object modeling technique (OMT). We also take up the current direction in an integrated language for software-centric systems, the Unified Modeling Language (UML).

The three methods are targeted at different kinds of software systems. Structured analysis and design was developed in the late 1970s and early 1980s and is intended for single-threaded software systems written in structured procedural languages. ADARTS was intended for large, real-time, multithreaded systems written in Ada. OMT was intended for database-intensive systems, especially those written in object-oriented programming languages. The UML is a merger of object-oriented concepts from OMT and other sources.

Structured Analysis and Design

The first of the integrated models for software was the combination of structured analysis with structured design.[5] The software modeling and design paradigms established in that book have continued to the present as one of the fundamental approaches to software development. Structured analysis and design models two system views, uses a variety of heuristics to form each view, and connects to the management view through measurable characteristics of the analysis and design models (metrics).

The method prescribes development in three basic steps. Each step is quite complex and is composed of many internal steps of refinement. The first step is to prepare a data flow decomposition of the system to be built. The second step is to transform that data flow decomposition into a function and module hierarchy that fully defines the structure of the software in subroutines and their interaction. The design hierarchy is then coded in the programming language of choice. The design hierarchy can be mechanically converted to software code (several tools do automatic forward and backward conversion of structured design diagrams and code). The internals of each routine are coded from the included process specifications, though this requires human effort.

The first step, known as structured analysis, is to prepare a data flow decomposition of the system to be built. A data flow decomposition is a tree hierarchy of data flow diagrams, textual specifications for the leaf nodes of the hierarchy, and an associated data dictionary. This method was first popularized by DeMarco,[6] though the ideas had appeared previously and it has since been extensively modified and re-presented. Figure 10.1 and Figure 10.2, discussed in a previous section, are examples of data flow diagrams. Behavioral analysis by data flow diagram originated in software and has since been applied to more general systems as well. The basic tenets of structured analysis are as follows:

1. Show the structure of the problem graphically, engaging the mind's ability to perceive structure and relationships in graphics.
2. Limit the scope of information presented in any diagram to five to nine processes and their associated data flows.

3. Use short (<1 page) free form and textual specifications at the leaf nodes to express detailed processing requirements.
4. Structure the models so each piece of information is defined in one and only one place. This eases maintenance.
5. Build models in which the processes are loosely coupled, strongly cohesive, and which obey a defined syntax for balance and correctness.

Structured design follows structured analysis and transforms a structured analysis model into the framework for a software implementation. The basic structured design model is the structure chart. A structure chart, as illustrated in Figure 10.5, shows a tree hierarchy of software routines. The arrows connecting boxes indicate the invocation of one routine or subroutine by another. The circles, arrows, and names show the exchange of variables and are known as data couples. Additional symbols are available for pathological connection among routines, such as unconditional jumps. Each box on the structure chart is linked to a textual specification of the requirements for that routine. The data couples are linked to a data dictionary.

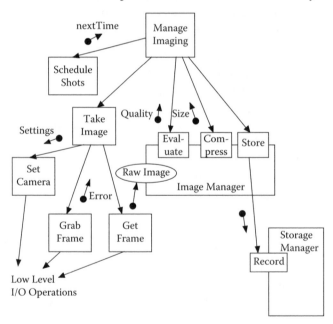

Figure 10.5 Example of a structure chart. A structure chart shows the components and interfaces for a hierarchically structured software system. This simple example illustrates routines as components (boxes), collection of routines into modules (boxes encapsulating other boxes), invocation of one routine by another (arrows connecting boxes), data elements (the labeled oval), and data passing among routines (solid circles with arrows).

Structure charts are closely aligned with the ideas and methods of structured programming, which was a major innovation at the time structured design was introduced. Structure charts can be mechanically converted to nested subroutines in languages that support the structured programming concepts. In combination, the chart structure, the interfaces shown on the chart, and the linked module specifications define a compilable shell for the program and an extended set of code comments. If the module specifications are written formally, they can be the module's program design language or can be compiled as module precondition and postcondition assertions.

The structured analysis and design method goes farther in providing detailed heuristics for transformation of an analysis model into a structure chart and for evaluation of alternative designs. The heuristics are strongly prescriptive in the sense that they are stated procedurally. However, they are still heuristics because their guidance is provisional and subject to interpretation in the overall context of the problem. The transformation is a type of refinement or reduction of abstraction. The data flow network of the analysis phase defines data exchange, but it does not define execution order beyond that implied by the data flow. Hence, the structure chart removes the abstraction of flow of control by fixing the invocation hierarchy. The heuristics provided are of two types. One type gives guidelines for transforming a data flow model fragment into a module hierarchy. The other type measures comparative design quality to assist in selection among alternative designs. The following are examples of the first type:

> *Step one: Classify each data flow diagram as "transform oriented" or "transaction oriented" (these terms are further defined in the method).*

> *Step two: In each case, find either the "transform center" or the "transaction center" of the diagram and begin factoring the modules from there.*

Further heuristics follow for structuring transform-centered and transaction-centered processes. In the second category are several quite famous design heuristics:

> *Choose designs that are loosely coupled. Coupling, from loosest to tightest, is measured as: Data, data structure, control, global, and content.*

> *Choose designs in which the modules are strongly cohesive. Cohesion is rated as (from strongest to weakest):*

Functional, sequential, communicational, procedural, temporal, logical, and coincidental.

Choose modules with high fan-in and low fan-out.

As discussed in Chapter 9, very general and domain-specific heuristics may be related by chains of refinement. In structured analysis and design, the software designer transforms rough ideas into data flow diagrams, data flow diagrams into structure charts, and structure charts into code. At the same time, heuristic guidelines like "strive for loose coupling" are given measurable form as the design is refined into specific programming constructs.

Various efforts have also been made to tie structured analysis and design to managerial models by predicting cost, effort, and quality from measurable attributes of data flow diagrams or structure charts. This is done both directly and indirectly. A direct approach computes a system complexity metric from the data flow diagrams or the structure charts. That complexity metric then must be correlated to effort, cost, schedule, or other quantities of management interest. A later work by DeMarco[7] describes a detailed approach on these lines, but the suggested metrics have not become popular nor have they been widely validated on significant projects. Other metrics, such as function or feature points, that are more loosely related to structured analysis decompositions have found some popularity. Software metrics is an ongoing research area and there is a growing body of literature on measurements that appear to correlate well with project performance.

An alternative linkage is indirect by using the analysis and design models to guide estimates of the most widely accepted metrics, the constructive cost model (COCOMO) and effective lines of code (ELOC). COCOMO is Barry Boehm's famous effort estimation formula. The model predicts development effort from a formula involving the total lines of code, an exponent dependent on the project type, and various weighting factors. One problem with the original COCOMO model is that it does not differentiate between newly written lines of code and reused code. One method (there are others) of extending the COCOMO model is to use ELOC in place of total lines of code. ELOC measures the size of a software project, giving allowance for modified and reused code. A new line of code counts for one ELOC, modified and unmodified, reused code packages count for somewhat less. The weight factors given to each are typically determined organization by organization based on past measurements. The counts by subtype are summed with their weights and the total treated as new lines in the COCOMO model.

The alternative approach is to use the models to guide ELOC estimation. Early in the process, when no code has been written, the main source of error in COCOMO is likely to be errors in the ELOC estimate. With a data flow model in hand, engineers and managers can go through it process by process and compare the requirements to past efforts by the organization. This, at least, structures the estimation problem to identifiable pieces. Similarly, the structured design model can be used in the same way, with estimates of the ELOC for each module flowing upward into a system-level estimate. As code is written, the estimates become facts, and, hopefully, the estimated and actual efforts will converge. Of course, if the organization is incapable of producing a given ELOC level predictably, any linkage of analysis and design models to managerial models is moot.

The architect needs to be cognizant of these issues insofar as they affect judgments of feasibility. As the architect develops models of the system, they should be used jointly by client and builder. The primary importance of cost models is in the effect they have on the client's willingness to go forward with a project. A client's resources are always limited, and an intelligent decision on system construction can be made only with knowledge of the resources it will consume. Of course, there will be risk, and in immature fields like software, the use of risk mitigation techniques (such as spiral development) may partially replace accurate early estimates. As the client's value judgments should be made in the context of the models, the builder's estimates should be as well. If builder organizations have a lot of variance in what effort is required to deliver a fixed complexity system, then that variance is a risk to the client.

ADARTS

Ada-Based Design Approach for Real-Time Systems (ADARTS) is an extensively documented example of a more advanced integrated modeling method for software. Even though neither classic structured analysis and design nor Ada-based development is currently cutting edge in large software systems, we examine them for the principles they elucidate. ADARTS may be obsolete, but the principles it embodies are not. The original work on data flow techniques was directly tied to the advanced implementation paradigms of the day. In a similar way, the discrete event system-oriented specification methods like H/P can be closely tied to implementation models. In the case of real-time, event-driven software, one of the most extensive methods was the ADARTS[8] methodology of the Software Productivity Consortium (SPC) (since renamed). Of course, the method still exists; it is not in the past tense, except in that Ada-based development is no longer the leading edge of software development. The ADARTS method combines a discrete-event-based behavioral model with a detailed, stepwise refined, physical design model. The behavioral model is based on data flow diagrams extended with the discrete event formalisms of Ward and Mellor[9] (which are similar to those of H/P).

The physical model includes evolving abstractions for software tasks or threads, objects, routines, and interfaces. It also includes provisions for software distributed across separate machines and their communication. ADARTS includes a catalog of heuristics for choosing and refining the physical structure through several levels of abstraction.

ADARTS links the behavioral and physical models through allocation tables. Performance decomposition and modeling is considered specifically, but only in the context of timing. There are links to sophisticated scheduling formalisms and SPC developed simulation methodologies as part of this performance link. Again, managerial views are supported through metrics, where they can be calculated from the models. Software domain-specific methods can more easily perform the management metric integration because a variety of cost and quality metrics that can be (at least roughly) calculated from software design models are known.

The example shown is a simplified version of the first two design refinements required by ADARTS applied to the microsatellite imaging system originally discussed in the Hatley-Pirbhai example. The resulting diagrams are shown in Figure 10.6. The ADARTS process takes the functional hierarchy of the behavioral model and breaks it into undifferentiated components. Each component is shown on the diagram by a cloud-shaped symbol, indicating its specific implementation structure has not yet been decided. The clouds exchange data elements dependent on the behavior allocated to each cloud. Various heuristics and engineering judgment guide the choice of clouds.

The next refinement specializes the clouds to tasks, modules or objects, and routines. ADARTS actually uses several discrete steps for this, but they are combined into one for the simple example given here. Again, the designer uses ADARTS-provided heuristics and individual judgment in making the refinements. In the example, the two tasks result from the need to provide asynchronous external communications and overall system control. The clouds that hide the physical and logical interfaces to hardware are multientry modules. The entries are chosen from the principal user functions addressed by the interface. For example, the Camera I/O module has entries that correspond to its controls (camera shutter speed, camera gain, filter wheel position, and so forth). The single thread sequence of taking an image is implemented as a simple routine calling tree.

To avoid diagram clutter, the diagram is not fully annotated with the data elements and their flow directions. In complex systems, diagram clutter is a serious problem, and one not well addressed by existing tools. The architect needs to suppress some detail to process the larger picture. But correct software ultimately depends on getting each detail right. In the second part of the figure, the arrowed lines indicate direction of control, not direction of data flow. Additional enhancements specify flow. The

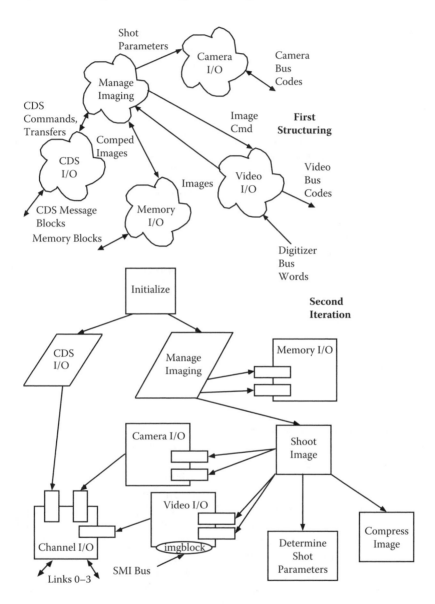

Figure 10.6 Example of a two-step design refinement in Ada-Based Design Approach for Real-Time Systems (ADARTS).

next step in the ADARTS process, not shown here, is to refine the task and module definitions once again into language- and system-specific software units. ADARTS as published assumes the use of the Ada language for implementation. When implementing in the Ada language, tasks become Ada tasks and multientry modules become packages. The public/private interface structure of the modules is implemented directly using constructs of the Ada language. Other languages can be accommodated in the same framework by working out language- and operation-specific constructs equivalent to tasks, modules, and routines. For example, in the C language there is no language construct for tasks or multientry modules. But multientry modules can be implemented in a nearly standard way using separately compilable files on the development system, the static declaration, and suitable header files. Similarly, many implementation environments support multitasking and some development environments supply task abstractions for the programmer's use. In C++ there is no direct language implementation of tasks, but multientry modules are easily implemented through classes and objects. This is similar in Java, except in Java a direct implementation of the general concept of a task does exist (called a thread), although the communication semantics for Java threads are quite different than for Ada tasks.

Once again, the pattern of stepwise reduction of abstraction is evident. Design is conducted through steps, and at each step a model of the client needs is refined in an implementation environment dependent way. In environments well matched to the problem modeling method, the number of steps is small; client relevant models can be nearly directly implemented. In less well-suited environments, layers of implementation abstraction become necessary.

OMT

The Hatley-Pirbhai method and its cousins are derived from structured functional decomposition, structured software design, and hardware system engineering practice. The object-oriented methods, of which OMT[10] is a leading member, derive from data-oriented and relational database software design practice. Relational modeling methods focus solely on data structure and content and are largely restricted to database design (where they are very powerful). Object-oriented methods package data and functional decomposition together. Where structured methods build a functional decomposition backbone on which they attempt to integrate data decomposition, the object-oriented methods emphasize a data decomposition on which the functional decomposition is arranged. Some problems naturally decompose nicely in one method and not in the other. Complex systems can be decomposed with either, but either approach will yield subsections where the dominant decomposition paradigm is awkward.

OMT combines the data (relational), behavioral, and physical views. The physical view is well captured for software-only systems, but specific abstractions are not given for hardware components. Even though, in principle, OMT and other object-oriented methods can be extended to mixed hardware/software systems and even more general systems, there is a lack of real examples to demonstrate feasibility. Broad, real experience has been obtained only for predominantly software-based systems.

Neither the OMT nor other object-oriented methods substantially integrate the performance view. Again, managerial views can be integrated to the extent that useful management metrics can be derived from the object models. Because of the software orientation of object-oriented methods, there have been some efforts to integrate formal methods into object models.

As an example of the key ideas of object-oriented methods, we present part of an object model. Object modeling starts by identifying classes. Classes can be thought of (for those unfamiliar with object concepts) as templates for objects or types for abstract data types. They define the object in terms of associated simple data items and functions associated with the object. Classes can specialize into subclasses that share the behavior and data of their parent while adding new attributes and behavior. Objects may be composed of complex elements or relate to other objects. Both composition or aggregation and association are part of a class system definition. The microsatellite imager described in the preceding section will produce images of various types. Consider an image database for storing the data produced by the imager. A basic class diagram is shown in Figure 10.7 to illustrate specific instances of some of the concepts.

A core assumption, which the model must capture, is that images are of several distinct but related types. The actual images captured by the

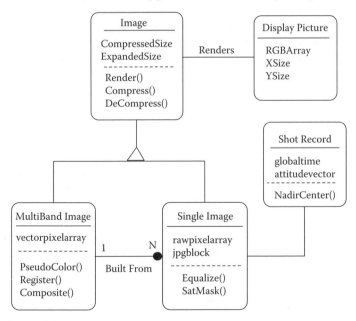

Figure 10.7 Class structure diagram for an image database.

cameras are single, gray-scale images. Varying sets of gray-scale images captured through different filters are combined into composite multi-band images, with a particular gray-scale image possibly part of several composite images. In addition, images will be displayed on multiple platforms so we demand a common "rendered" image form. Each of these considerations is illustrated in Figure 10.7.

The top box labeled "Image" indicates there is a data class "Image." That class contains two data attributes — CompressedSize and ExpandedSize — and three operations or "methods" (the functions Render(), Compress(), and Expand()). The triangle boxed lines down to the class boxes "Multi-Band Image" and "Single Image" define those two classes as subclasses of Image. As subclasses, they are different than their parent class but inherit the parent class's data attributes and associated methods.

The class Single Image is the basic image data object descriptor. It contains two data arrays: one to hold the raw image and the other to hold the compressed form. It also has basic image processing methods associated. A multiband image is actually made up of many single images suitably processed and composited. This is defined on the diagram by the round headed line connecting the two class boxes. The labeling defines a 1 to N way association named "Built From." The additional methods associated with Multi-Band Image build the image from its associated simple images.

The two additional associations define other record keeping and display. The associated line between Single Image and Shot Record associates an image with a potentially complicated data record of when it was taken and the conditions at that moment. The association line to Display Picture shows the association of an image with a common display data structure. Both associations, in these cases, are one to one.

Figure 10.7 is considerably simplified on several points. A complete definition in OMT would require various enhancements to show actual types associated with data attributes and operations. In addition, several enhancements are required to distinguish abstract methods and derived attributes. A brief explanation of the former is in order. Consider the method Compress in the class Image. The implementation of image compression may be quite different for a single gray-scale image and for a composited multiband image. A method that is reimplemented in subclasses is called either virtual or abstracted and may be noted by a diagrammatic enhancement.

The logic of object-oriented methods is to decompose the system in a data-first fashion, with functions and data tightly bound together in classes. Instead of a functional decomposition hierarchy, we have a class hierarchy. Functional definition is deferred to the detailed definition of the classes. The object-oriented logic works well where data and especially data relation complexity dominates the system.

Object-oriented methods also follow a stepwise reduction of abstraction approach to design. From the basic class model, we next add implementation specific considerations. These will determine whether or not additional model refinements or enhancements are required. If the implementation environment is strongly object oriented, there will be direct implementations for all of the model constructs. For example, in an object-oriented database system, one can declare a class with attributes and methods directly and have long-term storage (or "persistence") automatically managed. In nonobject environments, it may be necessary to manually flatten class hierarchies and add manual implementations of the model features. Manual adjustments can be captured in an intermediate model of similar type. The steps of abstraction reduction depend on the environment. In a favorable implementation environment, the model nearest to the client's domain can be implemented almost directly. In unfavorable environments, we have no choice but to add additional layers of refinement.

UML

As object-oriented methods became popular in the 1990s, there emerged several distinctive styles of notation. These notations differed enough to make tools incompatible and automated translation difficult. But the notations did not capture fundamentally different concepts. The basic concepts of class, object, and relationship were present in all of them, with only slight notational differences. The differences were more in the additional views and how the parts were integrated. They also differed somewhat more fundamentally in their approach to the design process and which portions they chose to emphasize. For example, some of the object-oriented methods emphasized front-end problem analysis through use-cases. Others were more design oriented and focused on building information models after there was a well-understood problem statement.

Because the profusion of notations was not helpful to the community, there was some pressure to settle on a collective standard. This was done, partially through several of the leading "gurus" of the different methods all moving to work for one company (the Rational Corporation). The product of their collaboration, and a large standards effort, is the Unified Modeling Language[11] (UML). Because the UML has successfully incorporated most of the best features of its roots and has gained a fairly broad industry consensus, it is increasingly popular. Probably the most significant complaint about the UML is its complexity. It is certainly true that if you tried to model a system using all the parts of the UML, the resulting model would be quite complex. But the content of the UML should not be confused with a process. A designer is no more compelled to use all the parts of the UML than a writer is compelled to use all the words in the English language. Of course, it is not simple to figure out which parts should be used in any given situation, and it can take fairly deep knowledge of the UML to know how to ignore features.

The primary importance of UML is that it may lead to more broadly accepted standardization of software and systems engineering notations. The notations are fundamentally software-centric, but as the software fraction (measured as percentage of development effort) makes up the majority of a development effort, this will seem appropriate. The two viewpoints within UML, use-cases and class-object models, most commonly discussed are the two that are the most software-centric. There are several other views that are more clearly systems oriented.

The use-case view within UML has two parts: the textual use-cases and diagrams that show the relationships among use-cases and actors. The textual form of a use-case is not strictly defined. In general, it is a narrative listing of messages that pass between an "actor," a system stakeholder, and the system. Thus, a use-case, in its pure form, follows the definition of the systems boundary. The use-case diagram shows the relationships between actors and use-cases, including linkages among use-cases.

A simple form for a textual use-case has four required parts and a group of optional parts*:

1. *Title* (preferably evocative).
2. *Actors*, a list.
3. *Purpose*, what the actors accomplish through this use-case, why the actors use the system.
4. *Dialogue*, a step-by-step sequence of messages exchanged across the actor–system boundary. The use-case gives the normal sequence. Alternative sequences (from errors or other choices) can be integrated into the use-case, given as different use-cases, or organized into the optional section.
5. *Optional material.* Some useful adjuncts include type (such as essential, optional, phase X, and so forth), an overview for a very complex use-case, and alternative paths.

UML uses class-object models similar to those described in the OMT section. The differences are primarily details of notation, such as the graphic element used to indicate a particular type of relationship. There is also a fairly complex set of textual notations for showing the components of the classes (data and methods). For example, there are textual indications for public, private, and virtual elements. The discussion of class-object

---

* There are many different formats for use-cases in use. The forms described here are inspired by various UML documents and Kevin Kreitman in private communication.

notations in the OMT section gives the flavor of how a model of the same sort would work if written in UML.

UML does introduce some modeling elements not discussed to this point and of high interest to system architects. On the behavioral side, the UML defines sequence diagrams. A sequence diagram depicts both the pattern of message passing among the system's objects and the timing relationships. The sequence diagram is useful both for specification and for diagnosis. When the client has a complex legacy system with which the new system must interface, or when the client's problems are primarily expressed in term of deficiencies in a current system, the sequence diagram is a method for visually presenting time relationships. This is often quite important in real-time software-intensive systems.

SysML

Since 2000, there has been activity on extending (or in some cases contracting) the UML for systems engineering purposes. The resulting language is known as SysML or Systems Modeling Language. SysML is, at the time of this writing, the subject of active standardization efforts with a version 1.0 specification completed. The most up-to-date information can be found at www.sysmlforum.com. The motivation of the SysML developers was primarily that the UML is weak in constructs that support traditional aspects of systems engineering (for example, requirements decomposition and allocation) but overprovides diagram types that are not relevant to systems engineering (for example, constructs to model software implementations). Nonetheless, the SysML developers adopted the UML approach to language specification and built SysML with UML extensions. SysML differs from UML primarily in the following:

1. A number of UML diagram types applicable almost exclusively to software implementations (for example, component, communication, deployment, and object diagrams) are dropped from the specification.
2. The class and structure diagram aspects of UML are heavily modified into Block Definition and Internal Block diagrams. These block diagram types more closely resemble the usual block diagram notions of systems engineering.
3. Requirement and Parametric diagram types are added to support the requirement trees and quantitative performance view breakdowns commonly used in systems engineering.
4. The Package Diagram constructs are altered to match the view and viewpoint constructs of ANSI/IEEE Std-1471-2000 (ANSI/IEEE Recommended Practice for Architecture Description of Software Intensive Systems).

For the architectural purposes of this book, SysML may or may not be an improvement over the UML for any particular problem. Software architects may apply the overall heuristics and methods described in this book to a software-only system, in which case the adaptation of SysML may seem counterproductive to just using UML methods. On the other hand, for largely hardware systems, the integrated modeling methods of SysML are likely to be much better suited.

SysML addresses the issues of multiple view and integration across views directly. Multiple views are explicitly provided, and a model type for checking across views (the allocation table) is provided. SysML intends to be applicable to general systems, but it is necessarily better suited to some cases than others. A notable issue for mixed hardware–software systems is the role of layering. A central abstraction for complex systems today is their arrangement into layers, in which the higher layers do not contain the elements of the lower layers but rather just use them. This issue was discussed in some depth in Chapter 6 and "Case Study 4." Direct support for layering abstractions is absent in SysML (as it is absent in most other methods). This lack is significant for many architecting projects because of the need for layered abstractions to control the complexity of large systems.

Performance Integration: Scheduling

One area of nonfunctional performance that is very important to software, and for which there is large body of science, is timing and scheduling. Real-time systems must perform their behaviors within a specified time-line. Absolute deadlines produce "hard real-time systems." More flexible deadlines produce "soft real-time systems." The question of whether or not a given software design will meet a set of deadlines has been extensively studied.* To integrate these timing considerations with the design requires integration of scheduling and scheduling analysis.

In spite of the extensive study, scheduling design is still at least partly art. Theoretical results yield scheduling and performance bounds, and associated scheduling rules, but can do so only for relatively simple systems. When system functions execute interchangeably on parallel processors, run times are random, and when events requiring reaction occur randomly, there are no deducible, provably optimal solutions. Some measure of insight and heuristic guidance is needed to make the system both efficient and robust.

---

* Stankovic, J. A., M. Spuri, M. Di Natale, and G. C. Buttazzo, Implications of Classical Scheduling Results for Real-Time Systems, *IEEE Computer,* pp. 16–25, June 1995, provides a good tutorial introduction to the basic results and a guide to the literature.

Integrated Models for Manufacturing Systems

The domain of manufacturing systems contains nice examples of integrated models. The modeling method of Baudin[12] integrates four modeling components (data flow, data structure, physical manufacturing flow, and cash flow) into an interconnected model of the manufacturing process. Baudin further shows how this model can then be used to analyze production scheduling under different algorithms. The four parts of the core model are as follows:

1. A data flow model using the notations of DeMarco and state transition models.
2. A data model based on entity-relationship diagrams.
3. A material flow model of the actual production process — the model of physical form — using American Society of Mechanical Engineers (ASME) and Japanese notations.
4. A funds flow model.

These parts, which mostly use the same component models familiar from previous discussion, form an integrated architect's tool kit for the manufacturing domain. They are shown in Figure 10.8. The data flow models are in the same fashion as the requirements model of Hatley-Pirbhai. The

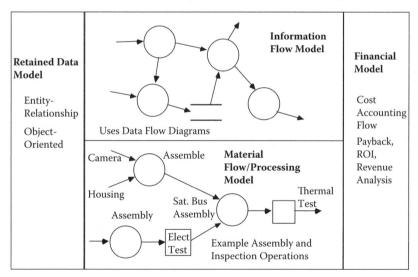

Figure 10.8 The manufacturing model elements of the method of Baudin. The information flow and retained data models are the same as before. The Material Flow and Processing model uses manufacturing-specific symbology for assembly, inspection, and other operations.

data model is more complex and uses basic object-oriented concepts. In the material flow model, the progression of removal of abstraction is taken to a logical conclusion. Because the physical architecture of manufacturing systems is restricted, the architecture model components are similarly restricted. Baudin incorporates, in fact exploits, the restricted physical structure of manufacturing systems by using a standardized notation for the physical or form model.

Baudin further integrates domain-specific performance and system models by considering the relationship to production planning in its several forms (MRP-II, OPT, JIT). As he shows, these formalisms can be usefully placed into context on the integrated models. In the terms used in Chapter 8, this is a form of performance model integration.

Integrated Models for Sociotechnical Systems

On the surface, the modeling of sociotechnical systems is not greatly different from other systems, but the deeper reality is quite different. The physical structure of sociotechnical systems is the same as of other systems, though it spans a considerable range of abstraction, from the concrete and steel of transportation networks to the pure laws and policy of communication standards. But people and their behavior are inextricably part of sociotechnical systems. Sociotechnical system models must deal with the wide diversity of views and the tension between facts and perceptions as surely as they must deal with the physics of the physical systems.

Physical system representation is the same as in other domains. A civil transport system is modeled with transportation tools. A communications network is modeled with communications tools. If part of the system is an abstract set of laws or policies, it can be modeled as proposed laws and policies. The fact that part of the system is abstract does not prevent its representation, but it does make understanding the interaction between the representation and the surrounding environment difficult. In general, modeling the physical component of sociotechnical systems does not present any insurmountable intellectual challenges. The unique complexity is in the interface to the humans who are components of the system and in their joint behavior.

In purely technical systems, the environment and the system interact. But it is uncommon to ascribe intelligent, much less purposively hostile behavior to their environments. But human systems constantly adapt. If an intelligent transport system unclogs highways, people may move farther away from work and reclog the highways until a new equilibrium is reached. A complete model of sociotechnical system behavior must include the joint modeling of system and user behavior, including adaptive behavior on the part of the users.

This joint behavioral modeling is one area where modeling tools are lacking. The tools that are available fall into a few categories: econometrics, experimental microeconomics and equilibrium theory, law and economics, and general system dynamics. Other social science fields also provide guidance, but not generally descriptive and predictive behavior.

Econometrics provides models of large-scale economic activity as derived from past behavior. It is statistically based and usually operates by trying to discover models in the data rather than imposing models on data. In contrast, general system dynamics* builds dynamic models of social behavior by analysis of what linkages should be present and then tests their aggregated models against history. System dynamics attempts to find large-scale behavioral patterns that are robust to the quantitative details of the model internals. Econometrics tries to make better quantitative predictions without having an avenue to abstract larger-scale structural behavior.

Experimental economics and equilibrium theory try to discover and manipulate a population's behavior in markets through use of microeconomic theory. As a real example, groups have applied these methods to pricing strategies for pollution licenses. Instead of setting pollution regulations, economists have argued that licenses to pollute should be auctioned. This would provide control over the allowed pollution level (by the number of licenses issued) and be economically efficient. This strategy has been implemented in some markets and the strategies for conducting the auctions were tested by experimental groups before hand. The object is to produce an auction system that results in stable equilibrium price for the licenses.

Law and economics is a branch of legal studies that applies micro- and macroeconomic principles to the analysis of legal and policy issues. It endeavors to assure economic efficiency in policies and to find least-cost strategies to fulfill political goals. Although the concepts have gained fairly wide acceptance, they are inherently limited to those policy areas, for market distribution is considered politically acceptable.

Conclusion

A variety of powerful integrated modeling methods already exist in large domains. These methods exhibit, more or less explicitly, the progressions of refinement and evaluation noted as the organizing principle of architecting. In some domains, such as software, the models are very well organized, cover a wide range of development projects, and include

---

* An introductory reference on system dynamics is Wolstenholme, E. F., *System Enquiry: A System Dynamics Approach*. Chichester: Wiley, 1990, which explains the rationale, gives examples of application, and references the more detailed writings.

a full set of views. However, even in these domains, the models are not in very wide use and have less than complete support from computer tools. In some domains, such as sociotechnical systems, the models are much more abstract and uncertain. But in these domains, the abstraction of the models matches the relative abstraction of the problems (purposes) and the systems built to fulfill the purposes.

Exercises

1. For a system familiar to you, investigate the models commonly used to architecturally define such systems. Do these models cover all important views? How are the models integrated? Is it possible to trace the interaction of issues from one model to another?
2. Build an integrated model of a system familiar to you covering at least three views. If the models in any view seem unsatisfactory, or integration is lacking, investigate other models for those views to see if they could be usefully applied.
3. Choose an implementation technology extensively used in a system familiar to you (software, board-level digital electronics, micro-waves, or any other). What models are used to specify a system to be built? That is, what are the equivalents of buildable blueprints in this technology? What issues would be involved in scaling those models up one level of abstraction so they could be used to specify the system before implementation design?
4. What models are used to specify systems (again, familiar to you) to implementation designers? What transformations must be made on those models to specify an implementation? How can the two levels be made better integrated?

Notes and References

1. White, S. et al., Systems Engineering of Computer-Based Systems, *IEEE Computer*, Vol. 26, Number 11, pp. 54–65, November 1993.
2. Maier, M. W., Quantitative Engineering Analysis with QFD, *Quality Engineering*, Vol. 7, Number 4, pp. 733–746, 1995.
3. Hauser, J. R., and D. Clausing, The House of Quality, *Harvard Business Review*, Vol. 66, Number 3, pp. 63–73, May–June 1988.
4. The above-mentioned 1995 as well as Maier, M. W., Integrated Modeling: A Unified Approach to System Engineering, *Journal of Systems and Software*, Vol. 32, Number 2, February 1996.
5. Yourdon, E., and L. L. Constantine, *Structured Design: Fundamentals of a Discipline of Computer Program and Systems Design*. New York: Yourdon Press, 1979.
6. DeMarco, T., *Structured Analysis and System Specification*. New York: Yourdon Press, 1979.

7. DeMarco, T., *Controlling Software Projects.* New York: Yourdon Press, 1982.

8. ADARTS Guidebook, SPC-94040-CMC, Version 2.00.13, Vols. 1–2, September 1991. Formerly available through the Software Productivity Consortium, Herndon, Virginia; Lykins, H., R. Kirk, and D. Smith, Using the CoRE Requirements Method with ADARTS, Report Number A151972, Software Productivity Consortium, March 1994; Cochran, M., and H. Gomaa, Validating the ADARTS Software Design Method for Real-Time Systems, *Proceedings of the Conference on TRI-Ada 1991*, San Jose, California, pp. 33–44, 1991.

9. Ward, P. T., and S. J. Mellor, *Structured Development for Real-Time Systems,* Vol. 1–3. New York: Yourdon Press, 1985.

10. Rumbaugh, J. et. al., *Object-Oriented Modeling and Design.* Upper Saddle River, NJ: Prentice Hall, 1991.

11. There is a great deal of published material on UML. The Rational Corporation Web site (www.rational.com) has online copies of the basic language guides, including the language reference and user guides.

12. Baudin, M., *Manufacturing Systems Analysis.* New York: Yourdon Press Computing Series, 1990.

chapter 11

Architecture Frameworks

Introduction

A great deal of architecture discussion in the engineering community revolves around the use of architecture frameworks. Even though discussion of architecture frameworks is widespread, and numerous architecture frameworks exist, there is actually relatively little agreement on what an architecture framework is. However, architecture frameworks are a primary vehicle for standardization.

In the terms used in this book, standards could cover architecture content, architecture description, or architecture processes. By analogy, a standard on *architecture content* is like a building code in that it would place standardized constraints on how actual systems are built. An *architecture description* standard is analogous to a blueprint standard in that it defines how a description document or model is written. An *architecture process* standard would be analogous to a development standard that defines how the design process is conducted.

The standards that have advanced to official status are primarily architecture description standards. These are often referred to as "Architecture Frameworks." Architecture frameworks are standards for the description of architectures. A framework typically defines what products the architect must deliver (to the client or to some other agency with authority) and how those products must be constructed. The framework generally does not constrain the contents of any of those products, although such constraints could be incorporated.

Architecture frameworks serve much the same purposes as blueprint standards, although their developers have had additional purposes in mind as well. It is hoped these standards will improve the quality of architecting efforts by institutionalizing best practices and fostering communication about architectures through standardizing languages. Standardized architectural description languages may also facilitate architecture evaluation by standardizing the elements that must be considered in the evaluation.

In the sections to follow, we first define the framework concept, then explore the most popular current frameworks, review some common problems in application, and then discuss research and practice development activities.

Architecture Framework

ۀ architecture description framework, we need to under-
٠s, its definition of "architectural level" (as opposed to other
ۀ), and its organizing concepts. We treat each of these in turn,
altho٠_ ۀe discussion of description concepts (viewpoints and views)
has largely been given previously in Chapter 8 and is discussed in more
detail in Appendix C.

Goals of the Frameworks

Each group developing and promulgating a standard has asserted different goals, but they generally fall into a few common categories:

1. Codify best practices for architectural description and by so doing
   improve the state of the practice.
2. Ensure that the sponsors of the framework receive architectural
   information in the format they desire.
3. Facilitate comparative evaluation of architectures through standardization of their means of description.
4. Improve the productivity of development teams by presenting basic
   designs in a standard way.
5. Improve interoperability of information systems by requiring that
   interoperation critical elements be described, and be described in a
   common way.

The fairest way of evaluating different frameworks is against their
own goals. If an architecture description is developed under the constraints (or with the guidelines) of a framework, and such a description
reliably fulfills the purposes of the framework, than we can say the framework is successful. Likewise, if following the framework does not reliably
produce description documents that fulfill the identified purpose, the
framework is poorly constructed. Even a well-conceived and constructed
framework may be inappropriate for a given project.

Understanding "Architectural Level"

An architecture framework specifies information about architectures, as
opposed to about detailed design, program management, or some other
set of concerns. So, a framework needs to distinguish what information is
"architectural" as opposed to something else. In this book, the separation
is connected to purpose. Information is architectural if it is needed to
resolve the purposes of the client, particularly with respect to fitness or
use or feasibility to build. The distinction is pragmatic not theoretical.

Moreover, we recognize that architecting and engineering are on a continuum of practice and sharp distinctions cannot be drawn. Other frameworks take different positions.

Organization of an Architecture Description Framework

The architecture frameworks described here use a few basic concepts, though they use them differently and sometimes with different terminology. All of them organize architecture descriptions into collections of related models. The obvious question is, by what relationship should models be gathered into collections? Following the terminology use of several standards, and our previous definitions in this book (though not necessarily the exact meaning), we will call a collection of models that represent the *whole system* from the *perspective* of a set of related *stakeholder concerns* a "view." We refer to the stakeholder concerns that define the perspective and the model language rules used within the view as a "viewpoint." Thus, a viewpoint is the template or specification for a view, and a view is a particular instance of description for a given system. Thus, an architecture framework consists of (or should consist of) the following:

1. A purpose and audience for which the compliant architecture description is to be written.
2. A set of viewpoints that when used should satisfy the purpose of #1.
3. The normative requirement that a compliant architecture description provide a set of views of the architecture of the system using the required viewpoints.
4. Tests for consistency and completeness among the views produced.
5. As a practical matter, an architecture framework may contain other advisory information, like guidelines on its use.

All of the frameworks discussed below are roughly consist with the five points above. We will use this simple formalism for what an architecture framework is to better compare some of the existing standards.

Current Architecture Frameworks

Several standards explicitly labeled architecture frameworks have emerged from the 1990s to the 2000s. The four standards we consider here are the U.S. Department of Defense Architecture Framework (DODAF), the Ministry of Defence Architecture Framework (MODAF), the International Standards Organization's RM-ODP standard, and the ANSI/IEEE 1471 Recommended Practice for Architectural Description for Software-Intensive Systems (now ISO 42010). All four use the basic concepts given above but take different approaches to the selection of views, the models specified, and the overall approach to formalization and rigor. We also discuss current research problems and issues that commonly arise in application of the current architecture frameworks.

U.S. DODAF

In the early 1990s, the U.S. Department of Defense (DoD) undertook to develop an architecture framework for Command, Control, Communications, Computing, Intelligence, Surveillance, and Reconnaissance (C4ISR) systems. The stated goal for this project was to improve interoperability across commands, services, and agencies by standardizing how architectures of C4ISR systems are represented. It also became a response to U.S. Congressional requirements for reform in how information technology systems are acquired.

The Architecture Working Group (AWG) published a version 1.0 of the framework (which became known as the C4ISR Architecture Framework) in June 1996. This was followed by a version 2.0 document in December 1997. The version 2.0 document was widely published and is available through the U.S. DoD Web sites, although it would now be considered obsolete. Subsequent to the publishing of the C4ISR Framework, it was further extended and designated the DOD Architecture Framework,[1] now applicable to a much wider array of systems. The DODAF reached version 1.0 status in October 2003. A 1.5 version was released in August 2007.

The DODAF requires that the architecture description be organized into summary information (also referred to as a "view") and three additional "architecture views." The three views are called the "Operational Architecture View," the "System Architecture View," and the "Technical Architecture View." These are often contracted in discussion to the "Operational Architecture" or "Operational View," the "System Architecture" or "System View," and the "Technical Architecture" or "Technical View." Speaking of the "operational view of the architecture" is more consistent with the notion of view than any of the common contractions.

The DODAF is a blueprint standard in that it defines how to represent a system's architecture, but it does not restrict the nature of the architecture of the underlying system. It is possible to embed the equivalent of "building codes" using the mechanisms of the DODAF, however. For example, the Joint Technical Architecture (JTA) was a particular instance of a standards profile that could be incorporated as the technical architecture view. Although there was at one time an intent to do exactly that, to drive compliance via inclusion of broader standards documents within system-specific architecture descriptions, it has not been successful. The JTA effort still exists, but it is not included within DODAF compliant documents by mandate.

In describing the contents of the DODAF, we draw directly from it in some cases, and interpolate commentary and guidelines from the point of view of this book.

Summary Information

The required summary information is contained in the "all-view" and is denoted AV-1 Overview and Summary Information and AV-2 Integrated Dictionary. Both are simple, textual objects. The first is information on scope, purpose, intended users, findings, and so forth. The second is definitions of all terms used in the description.

Operational View

The operational view shows how military operations are carried out through the exchange of information. It is defined as follows:

> *Operational View*: A description of task and activities, operation elements, and information flows integrated to accomplish support military operations.

There are nine individual models defined within the DODAF operational view of the architecture. Each has a specified modeling language, although none of the languages is defined very formally. Some are entirely informal, as in the required High Level Concept Graphic (OV-1), while others (such as the Logical Data Model, OV-7) suggest the use of more formalized notations, though they do not require it. The defined elements are as follows:

High-Level Operational Concept Graphic (OV-1): A relatively unstructured graphical description of all aspects of the systems operation, including organizations, missions, geographic configuration, and connectivity. The rules for composing this are loose with no real requirements.

Operational Node Connectivity Description (OV-2): Defines the operational nodes, and activities at each node, and the information flows between nodes. The rules for composing this are more structured than for OV-1 but are still loose.

Operational Information Exchange Matrix (OV-3): A matrix description of the information flows among nodes. This is normally done as an augmented form of data dictionary table.

Command Relationships Model (OV-4): A modestly structured model of command relationships.

Activity Model (OV-5): Similar to a data flow diagram for operational activities.

Operational Rules Model (OV-6a): Defines the sequencing and timing of activities and information exchange through textual rules.

Operational State Transition Model (OV-6b): Defines the sequencing and timing of activities and information exchange through a state transition model, which is usually quite formal.

Operational Event/Trace Description (OV-6a): Defines the sequencing and timing of activities and information exchange through scenarios or use-cases. This is behavioral specification by example, as discussed in Chapter 8.

Logical Data Model (OV-7): Usually a class-object model or other type of relational data model. No specific notation is required, but most of the popular notations used are fairly formal. The intent of the OV-7 is to define the data requirements and relationships.

As a guideline, it is suggested that the OV-1, OV-2, OV-3, and OV-5 should always be provided. It is not mandated that they be, but it is typically done.

System View

The system view is defined as follows:

> *System View*: Description, including graphics, of a system and interconnections providing for, and supporting, warfighting functions.

There are 16 individually defined elements within the DODAF under the system view, although several are just small variations on each other. The most important is as follows:

System Interface Description (SV-1): This model identifies the systems physical nodes and their interconnections. It is similar to an architecture interconnection diagram in the Hatley-Pirbhai sense, described in Chapters 8 and 10. A graphic representation method is called out but is not formally defined.

The 15 other elements are mostly concerned with more detailed descriptions of system-level data interchange or operation. However, some of the supporting products wander very far afield from these concerns. For example, the System/Services Technology Forecast (SV-9) is a tabular compilation of the technologies expected to be available, broken out by time frame, for the system.

Technical View

The technical view is concerned with standards. The technical sets out the required or forecast standards (typically information technology standards) that are to be used in the construction of the system. It has two elements:

Technical Standards Profile (TV-1): A listing of the standards mandatory for the system being described.

Technical Standards Forecast (TV-2): A projection of how standards and products compliant with standards will emerge during the time the system is developed and operated.

Evaluation and Issues with the Use of the DODAF

The DODAF has been available and in wide use long enough for a body of experience to be generated. One clear issue is that it is often being misused for purposes for which it was not intended. Recall that the purpose was primarily to facilitate interoperability through commonality of description. Interoperability specification, analysis, and improvement are highlighted as primary goals. There is no stated intention, within the DODAF, for it to be considered as a framework for acquisition documentation. So there is no place in the views for performance models, cost models, acquisition requirement documents, or other management models. Yet all of those are clearly necessary when the client is an acquirer and must make acquisition decisions.

Likewise, the DODAF does not contain the elements necessary to cover all the architectural concerns of a builder of, for example, software-intensive systems. We have a fairly good understanding of best practices in software architecture description, and those practices are not mirrored in the DODAF. This cannot be considered a fault of the DODAF developers, as it was not part of their original purpose. It is a fault of those who specify use of the DODAF for purposes for which it is not intended.

Just as an architecture must be fit for purpose, so must an architecture description framework. If the DODAF is misused, the fault is much more in the misuser than in the framework. Nevertheless, it can be cited as a weakness of the DODAF that its parts are very loosely related. Very disparate concerns and models are lumped together into the views. Neither intraview nor interview consistency is addressed at all. The individual models are so loosely defined, especially in some of the required elements, that ostensibly compliant descriptions can be produced that will not come close to guaranteeing interoperability. Because the DODAF adopts such a neutral stance to methodology, it cannot enforce stronger consistency and completeness checks. It is probably not possible to strengthen consistency and completeness properties without adopting much more formal modeling methods, which would negatively impact ongoing programs on which the DODAF might be mandated.

Aside from this large issue, there are numerous practical and conceptual problems that must be resolved by individual user groups:

1. The notion of what constitutes an "operational node" in the operational view is unclear and inconsistently used. In the activity

modeling sections of an operational view, an operational node is purely functional. In the higher-level diagrams, like OV-1s and OV-2s, common usage, and even examples in the defining documents, equate operational nodes to specific physical entities (for example, an AWACS aircraft, a command center).

2. The high-level operational view diagrams are often, in practice, elaborately produced professional graphics with little technical content. Yet, they are held up in examples as centerpieces.

3. The hierarchy in the system view does not incorporate the widely accepted layering concepts from computer networks. At least five layers of the network stack (physical, data-link, network, transport, and application) are firmly established in theory and practice, but the concept is absent from the system view definitions. This is a barrier as systems are built that incorporate the widely available off-the-shelf network components.

4. The models in the system view do not make clear distinctions between node and connector types by layer. Within a particular layer (for example, physical or network), the identity of the physical nodes and the nature of connection channel and data exchange is usually clear. But, the DODAF models do not directly support that information.

5. The definitions of the elements within a view focus on diagrams rather than graphics independent models, which could have various visualizations. As a result, users fixate on the diagrams rather than the model content.

6. There is relatively little in the way of explicit consistency and completeness checks, especially between views. There has been some improvement on this point, but it is still immature.

MODAF[2]

The Ministry of Defence Architecture Framework (MODAF) is a United Kingdom Ministry of Defence extension to and modification of the DODAF. In many ways, it can be seen as a reaction to the issues encountered in use of the DODAF. The primary differences between the MODAF and the DODAF are as follows:

1. Terminology has been adjusted, and in some cases sharpened. The concepts and terminology are generally close to those in ANSI/IEEE 1471, now ISO 42010 (see later in this chapter for additional detail). The terminology associated with what is an operational versus a system node has been sharpened.

2. Various models are broken out into more formal pieces. For example, the high-level operational depiction, OV-1 in DODAF, is broken into a purely pictorial element and other tabular and even quantitative parts.

3. A "Strategic Viewpoint" has been added. This viewpoint specifies models of policy, capability deployment, and related trade-offs for larger-scale planning. Its intended audience is mainly higher-level planners and staffers.

4. An "Acquisition Viewpoint" has been added. This is largely in response to the practice of mandating Framework-compliant documents for acquisition programs. If compliant documents are going to be required for an acquisition to go forward, then it would be desirable that the standard incorporate acquisition concerns. Compared to the full range of models usually used by project managers, the specified set here is rather thin. However, the intent in the MODAF is mainly to support planning and visibility between projects, so the models focus on dependencies and the clustering of projects.

The DODAF and MODAF address descriptions where the objects of interest are themselves significant systems and programs instead of the component-level elements that were typical in our discussion of integrated modeling methods. A description standard that reaches much further down the hierarchy is ISO RM-ODP.

ISO RM-ODP

The International Standards Organization (ISO) has also developed an architecture description framework known as Reference Model for Open Distributed Processing (RM-ODP).[3] As the name implies, RM-ODP is computation and software-centric. It addresses open distributed processing — that is, multivendor, multiorganization, heterogeneous computing systems whose processing nodes are spatially distributed. As defined in RM-ODP, a distributed system is generally characterized by one that is spatially distributed, does not have a global state or clock, may have individual node failures, and operates concurrently.

The scope of RM-ODP is larger than just architectural description. RM-ODP makes extensive normative statements about how systems should be described but also goes on to specify functions they should provide, and structuring rules to provide those functions. The architecture concerns of RM-ODP include both description (through viewpoints) and the provision of what are considered critical functions, called "transparencies" in the RM-ODP model.

The RM-ODP defines the following*:

1. A division of an ODP system specification into viewpoints in order to simplify the description of complex systems.

* ISO/IEC 10746-1: 1995 DIS (E), p. 8.

2. A set of general concepts for the expression of those viewpoint specifications.
3. A model for an infrastructure supporting, through the provision of distribution transparencies, the general concepts it offers as specification tools.
4. Principles for assessing conformance for ODP systems.

This is certainly larger than just description of architectures. Points one and two of RM-ODP are our concern here. RM-ODP is much more strongly normative than the other architecture frameworks discussed in this chapter. It takes a more normative approach both because of the inclinations of the authors (and their beliefs about best practices) and because the domain of application is narrower. RM-ODP applies to a particular class of computing systems (albeit a broad class), and it seeks to be both a consistent and complete approach to describing such systems.

The heart of RM-ODP in regard to descriptions is its five normative viewpoints. RM-ODP uses viewpoint to mean essentially what view means here, although it also carries the meaning of a generic specification method to be applied to any system. The RM-ODP notion of viewpoint is really a mixture of the language specification, the concerns covered, and the actual model instances for a particular system. The five ODP viewpoints are enterprise, information, computational, engineering, and technology. ODP adopts the notion that each viewpoint is a "projection" of the system's whole specification onto some set of concerns (using a specific language). The five viewpoints are chosen to be complete with respect to the concerns assumed to be relevant for an open distributed processing system. The definitions of the five viewpoints are as follows*:

1. An *enterprise* specification of an ODP system is a model of the system and the environment with which the system interacts. It covers the role of the system in the business, and the human user roles and business policies related to the system. The enterprise viewpoint is a viewpoint on the system and its environment that focuses on the purpose, scope, and policies for the system.
2. The *information* specification of an ODP system is a model of the information that it holds and of the information processing that it carries out. The information model is extracted from the individual components and provides a consistent common view that can be referenced by the specifications of information sources and sinks, and the information flows between them. The information viewpoint on the system and its environment focuses on the semantics of the information and information processing performed.

* Ibid, pp. 8–9, 16.

3. The *computational* specification of an ODP system is a model of the system in terms of the individual, logical components that are sources and sinks of information. Using the computational language, computational specifications can express the requirements of the full range of distributed systems, providing the maximum potential for portability and interworking and enabling the definition of constraints on distribution, while not specifying the detailed mechanisms involved. The computational viewpoint is a viewpoint on the system and its environment that enables distribution through functional decomposition of the system into objects that interact at interfaces.

4. The *engineering* specification of an ODP system defines a networked computing infrastructure that supports the system structure defined in the computational specification and provides the distribution transparencies that it identifies. It describes mechanisms corresponding to the elements of the programming model, effectively defining an abstract machine that can carry out the computational actions and the provision of the various transparencies needed to support distribution. The engineering viewpoint is a viewpoint on the system and its environment that focuses on the mechanisms and functions required to support distributed interaction between objects in the system.

5. The *technology* specification defines how a system is structured in terms of hardware and software components. The technology viewpoint is a viewpoint on the system and its environment that focuses on the choice of technology in that system.

Each viewpoint has a language associated with it, defined in the RM-ODP standard. The language specification in the standard is less specific than a typical programming language. The language specification consists of the definitions of the terms used to compose the language and constraints on constructing statements. All terms and constructions are in-built on object modeling concepts. The RM-ODP standard uses OMT conventions, although they could as easily be transferred to Unified Modeling Language (UML). Because RM-ODP is a component of the Object Management Group (OMG) of standards (UML is also a prominent component), such a transfer is already under way. ISO/IEC 10746-4 has mappings between the viewpoint language concepts and mathematically based formal languages from computer science.

RM-ODP recognizes the problem of interview and interview consistency. A conformant description must perform a number of cross-view checks for consistency. These checks are not a true guarantee, and the models involved do not have a precise notion of consistency built in, but the checks do serve as an explicit attempt to look for inconsistencies.

Proprietary and Semi-Open Information Technology Standards

Architecture has been widely addressed through proprietary and semi-open standards in information technology. Many firms have architectural standards, and many have developed their own description standards, typically as part of a development process standardization activity. The architectural description standard is typically tied to making specific go-ahead decisions about system development. Standardization of description products helps make those go-ahead decisions more consistent and facilitates process standardization.

One of the more widely known architecture description frameworks in information technology is usually called the Zachman framework after the name of the author. The Zachman framework is not fixed as it has evolved with Zachman's writings. There are a number of similarities between the various Zachman frameworks and the RM-ODP standard as some of Zachman's early work popularized some notions of viewpoints and viewpoint languages. More recent published works by Zachman have added many more views than five, and have particularly emphasized the enterprise and business management aspects of choosing information technology architectures.

ANSI/IEEE 1471, ISO 42010

In April 1995, the IEEE Software Engineering Standards Committee (SESC) convened an Architecture Planning Group (APG) to study the development of an architecture standard for software-intensive systems. After publication of their report, the APG upgraded to the Architecture Working Group and was charged with the development of a Recommended Practice for architectural description, a particular type of standard. A Recommended Practice is one form of standard, commonly used for relatively immature fields as it provides more general guidance rather than normative requirements. After extended debate and community consensus building, a Recommended Practice for architecture description was published.* Subsequently, the standard was accepted by the American National Standards Institute as ANSI/IEEE 1471. In 2006 and 2007 the standard was submitted to ISO and adopted as ISO 42010, though with the proviso that it enter a revision cycle with the IEEE. At the time of this writing, that revision cycle is ongoing.

The 1471 project was intended to codify the areas of community consensus on architecture description. Originally it was envisioned that the

---

* IEEE 1471 Recommended Practice for Architectural Description of Software-Intensive Systems, 2000. 1471 is part of the Computer Society's software engineering standards set. This standard subsequently became ANSI/IEEE 1471 and ISO 42010.

standard would codify the notion of view and prescribe the use of particular views. In the end, consensus developed around a framework of views and viewpoints and an organizing structure for architecture descriptions, but there was no prescription of any particular views. As a recommended practice, it is assumed that community experience will eventually lead to greater detail within the standard. In practice, 1471 has become useful for other standards groups in organizing more domain-specific architecture description frameworks. Some application in this fashion was already seen in the review of how the MODAF extends and clarifies the DODAF.

1471 Concepts

Because the ontology of 1471 is independent of a specific framework of views and viewpoints, its ideas have been threaded into the discussion of this book. Thus, much of what 1471 contains will already be familiar to the reader. 1471 codifies the structure of an architecture description and the definitions of its parts. The terminology of 1471 is shown in Figure 11.1. The diagram is written in UML, but it can be easily interpreted even without knowledge of UML. In the 1471 ontology, every system has one architecture. That architecture can have several architecture descriptions. This expresses the idea that an architecture is a conceptual property of a thing,

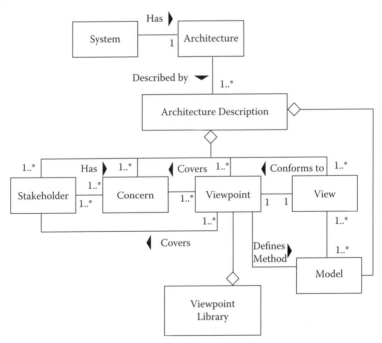

Figure 11.1 Information model of ANSI/IEEE 1471 concepts.

but an architecture description is a representation of the conceptual object. Again, we should be familiar with this concept from the earliest chapters of this book. Several mutually consistent representations of a thing can exist, so we need not specify that there be a single representation. 1471 does not make a distinction between types of system, so the relationship of architecture and architecture description could hold for an individual system, a family-of-systems, a system-of-systems, or a subsystem.

Returning to Figure 11.1, an architecture description is composed of stakeholders, concerns, viewpoints, views, and models. Stakeholders have concerns. Viewpoints cover stakeholders and concerns by their choice of language with which to represent the system. Views are groups of models, which must conform to exactly one viewpoint by using its language and rules.

Viewpoints may be drawn from a viewpoint library. A viewpoint library is not required by 1471, but it is expected in organizations that frequently develop architectural descriptions.

1471 makes an explicit distinction between the concepts of viewpoint and view, a distinction adopted in the MODAF, although not made in the DODAF and RM-ODP. If our goal is simply to write an architecture description, or to form a single standard, it is not necessary to separate the concepts. It is necessary to separate the concepts in 1471 because 1471 may be used to form many standards. Viewpoints are the vehicle for forming a standard. Indeed, viewpoints may be placed into a library to be drawn from at the discretion of the architect and the specific stakeholders involved in a particular project. Thus, this organizing element must be separately named and defined to allow them to be separately assembled for the needs of different sets of clients.

The viewpoints of RM-ODP are examples of 1471 compliant viewpoints. The only distinction between the viewpoint concept in RM-ODP and 1471 is that the RM-ODP version combines the abstraction (the 1471 viewpoint) and the actual instance of a representation of a particular system (the 1471 view).

1471 Normative Requirements

The normative requirements of 1471 are limited, particularly compared to RM-ODP and even the DODAF. An architecture description conformant to 1471[4] must meet the following main requirements:

1. The stakeholders identified must include users, acquirers, developers, and maintainers of the system.
2. The architecture description must define its viewpoints, with some specific elements required.
3. The system's architecture must be documented in a set of views in one-to-one correspondence with the selected viewpoints, and each view must be conformant to the requirements of its associated viewpoint.

4. The architecture description document must include any known interview inconsistencies and a rationale for the selection of the described architecture.

There are a variety of other relatively minor normative requirements, along with various recommendations. Many of these are to make 1471 consistent with other IEEE software engineering standards, notably the overarching software engineering standard 12207.

Research Directions

The current state of the art and practice in architecture description leaves much work undone. As the RM-ODP example shows, the basic architectural concepts of viewpoint, view, stakeholder, and concern can be extensively refined and tied to modeling formalisms if the domain of application is narrowed. A cost is the intellectual complexity of the resulting methods. RM-ODP is a complex standard. Its conceptual makeup is not complex in comparison to other areas in computer science, but it is quite complex compared to common practice in information technology. There may be strong benefits in mastering the complexity, but it acts as a barrier to the adoption of this technology. To make it more widely used, we need better tools and better explanation and training mechanisms to pass these ideas on to the professional community.

As we move to more general systems, the range of formalized models drops off dramatically. It seems very unlikely that we can develop a really general architecture framework that will simultaneously be formalizable. It seems more likely that we must continue to work up from the engineering disciplines to create more general notations. One problem will be dealing the disjunction between models common in the hardware (and some of the systems engineering) world and those coming from computer science and software. The hardware models are typically performance-centric and physics based. They work from physical objects. The computer science models are now commonly based on object-orientation and encapsulation of functionality within data models. It is not obvious how these will be reconciled or to what extent it is necessary. It may be a better approach to leave the modeling techniques as they are, taking the modeling techniques as they have been validated within the engineering disciplines. The higher-level challenge will then be to develop interview consistency checking techniques that do not require the disciplinary modeling methods to be altered but can work with them as they are.

There are two notable, relatively structured areas of research for frameworks: data relationships in subsystem flowdown and choosing models to match analytical needs.

Subsystem Architecture Flowdown

Architecture frameworks are commonly employed in situations of high complexity, where the elements of the main system of interest are reasonably complex systems. We might ascribe the term "system-of-systems" here, except, as argued in the chapter on collaborative systems, the concept is not well formed until we introduce the intentionality of the integration. Nevertheless, it is clear in normal usage that the standard architecture frameworks are typically employed only in cases where subsystems are reasonably complex. So, a natural question is, if we write an architecture view of a given system, and that system has subsystems, what is the relationship between the subsystems corresponding architecture views?

To be more formal, suppose we are interested in a system S, which may or may not exist at the time we write an architecture description. System S is composed of subsystems S_1 through S_N. For design purposes, we form architecture views of S, each composed of various models. Let those views be denoted $V_1(S)$ through $V_M(S)$. For the moment, we need not be concerned with the identity of the views. Some subsequent design groups may also be concerned with forming views of one or more subsystem. The direction of formation need not be downward in the hierarchy. We might start with a set of subsystems and views of those subsystems and want to form the architecture views of some integrated supersystem of those subsystems. Indeed, this upward assembly might be the usual case.

Given the terminology, some representative questions are as follows:

1. What is (or should be) the relationship between $V_k(S)$ and $V_k(S_n)$?
2. What is the relationship between $V_k(S)$ and the collection $V_k(S1)$ through $V_k(S_N)$? What views other than V_k need to be interrogated to make the determination?
3. Given $V_k(S)$, what is the relationship between $V_k(S_i)$ and $V_k(S_j)$?
4. If S_1 through S_N are to be integrated into S, and they have a given set of views, can you derive the corresponding views of the integrated whole? Is the corresponding view of the integrated whole completely specified?

Some of the integrated modeling methods discussed earlier provide some interesting examples of these issues. Consider a data flow, functional decomposition view of a system. If that system is divided into subsystems, there are clearly some required relationships between the corresponding data flow, functional decomposition views. The decompositions must match across corresponding partitions of the functions. Methods like Hatley-Pirbhai have a set of formal consistency and completeness checks based on how those models decompose. Physical models from system to subsystem produce requirements for correspondence on interfaces.

It seems unlikely that there are very general rules for relating views. In any particular instance, various strong constraints can be worked out, but it does not appear that the rules carry over in detail from one case to another. On the other hand, viewpoints, the rules and templates for constructing views, do carry over strongly from one case to another.

Matching Analytical Needs

We do not (or should not) build complex architecture description documents in compliance to frameworks without clear purposes in mind. The main reason for having an architecture framework is so that the uses of the document can be consistently performed. For example, a stakeholder group may want to evaluate all systems being constructed within their purview for mutual interoperability. In order to facilitate that analysis, they could specify in an architecture framework that a set of models needed in interoperability analysis be provided. In this way, the framework would be linked to a specific set of purposes — in this case interoperability analysis.

In principle, today's architecture frameworks are supposed to be linked to particular analyses. In practice, the linkage is often disappointing. For example, the DOD Architecture Framework is intended to facilitate interoperability analysis. Compliant documents have been used for interoperability analysis. However, there have been many problems with such use,* and it cannot be claimed that the analysis tools are firmly linked to architecture products, especially as a matter of consistent practice. This is an area where there could be clear improvement through research on linking analysis tools and framework compliant products.

Adapting Processes to Frameworks

At their best, architecture frameworks are still only the equivalent of blueprint standards. There remains the problem of how to organize the architecting process to reliably produce both useful documents and corresponding systems fit for use. Unfortunately, many groups, when faced with the challenge of building a framework compliant architecture description, simply default to producing a set of models that "check off the boxes" in the framework document. This is very unlikely to lead to a useful document.

---

* The U.S. military Joint Staff (specifically J8) was tasked to critique architecture-based planning. A variety of shortcomings, some reflected in the discussion here, were identified in the report from that study. The results of that study, although not formally published, have been generally available. See SPG-Directed Planning Task, Integrated Architectures, from the Joint Staff, 17-May-04. A paper from roughly the same time, Mara, M., and J. Grobman, A Capability-Based Planning Methodology for Networked Systems-of-Systems, The Phalanx, Military Operations Research Society-Military Applications Society of INFORMS, Vol. 36, Number 4, 2003, contains several of the major points from the longer briefing report.

We can identify a variety of heuristics that are useful for guiding the development of framework compliant (or framework-driven) architecture description documents.

Focus on Architecture Decisions, and Let Descriptions Flow from the Decisions

Having an architecture description document, no matter its size or the expense of construction, is no guarantee that any architectural decisions have been made well. The value of the system to its stakeholders is determined by the structure and content of that system (determined by architectural decisions), not by the architecture description document. So, architecture work should focus squarely on the decisions foremost, and let the documents flow from those decisions.

This leads to a rather simple test for managers or evaluators of an architecture description. When faced with a large architecture description document (enterprise architecture projects are especially notorious), ask "What decisions about changing our systems are contained in this document?" If the answer is "none," or something equivocal, it is clear that the volume of the document is no guide to its value. If there are crisp answers, then the rationale for those answers can be pursued and evaluated. If the rationale is sound, the document is probably sound (although an edit down to the crisp decisions might be worthwhile).

Always Use an Iterative Process to Do Architecting

Architecting is something you can virtually never get right at the beginning. It requires pursuit of multiple paths and the presentation of disparate alternatives to stakeholders. Any architecting process that requires a year or more to execute, and does not result in any clear decisions before its end, is almost certain to be ineffective. Aside from the substantial risk that issues will change over the year, the most likely failure path is that the initial understanding of the problem will be poor, resulting in poor alternatives. It is easy to advocate "just get the problem right at the beginning," but the reality is that we are unlikely to know whether the problem is "right" until multiple solutions have been examined.

Do Not Overstaff Early; the Best Architectures Come from Small Teams

Another common failure path is that once a project at the architecting phase is initiated, it quickly comes under pressure to spend at a rate commensurate with its full budget later in the development cycle. Even worse, a large staff may be assigned to the architect very early on. The consequence is that the architects and managers are quickly consumed with getting people working on anything that results in a stream of visible deliverables, whether or not they are directly relevant to the architecting-specific goals. The important work (the decision-centric, strategic work) is swamped by visible document production. The only cure is to strictly avoid early over-staffing, and for management to make go-ahead decisions on the basis of strategic issues and not on the basis of unread document delivery.

A particular form of this problem is the project that gets mired in as-is data collection and documentation. Very often a project that requires significant architecting needs to begin by evaluating the state of existing systems (the "as-is architecture"). While this is often necessary, an as-is documentation effort has the potential to expand to fill all available funds and staff. The as-is documentation work is in many ways tailor-made for project management. It is well structured, relatively easy to measure prog-ress, and requires the coordination of a potentially large staff. It is easy to see how staff can become enamored with endless pursuit of broader and deeper as-is documentation, and want to avoid the hard choices, complex decision making, and political difficulties of prescribing future change.

Avoid Underreach and Overreach

Effective architectures must constrain enough, but not too much. A defect in some architecture efforts is to either draw their scope too large (result-ing in attempts to control things the team has no hope of controlling) or drawing the scope too narrow (resulting in a failure of the constraints to ensure that enough joint functionality is present to be useful). Architecture framework standards do not themselves help define an appropriate scope; the scope definition (an essential architectural decision) must come from those directing the project.

A famous example of good scoping and effective decision making is the Motion Pictures Expert Group (MPEG) standard.[5] The MPEG standard was a multicompany effort to create a standard for encoding compressed digital video. The motivations of the players were to build the largest possible market, while avoiding a destructive standards war (as in the case of VHS versus Betamax). With regard to underreach and overreach, the key is the scope of the standard. The standard as structured defines the structure of a compressed stream and defines a free reference decom-pressor. The standard leaves open the exact compression algorithm.* As a result, end-to-end quality is a free variable; a given piece of source mate-

---

* It is more correct to say that the standards leave open a large number of algorithmic parameters rather than the algorithm itself, although pre- and postprocessing algorithms are left undefined, but the details are immaterial to the discussion here.

rial at a given bit rate has a huge number of compliant encodings. This particular scope works effectively for all of the stakeholders concerned, though a larger or small scope likely would not have:

1. The scope implemented guarantees interoperability between all content providers and all equipment providers. This is the primary concern of the consumer.
2. The scope of the standard is narrow enough to minimize the need of the participants to contribute intellectual property to the licensing pool. This is important to those stakeholders concerned about poststandard competitiveness.
3. The undefined elements of the standard encourage further technology development. Because consumer content is typically compressed once and decompressed millions of times, there are great advantages into putting a lot of intelligence into the compression side. A significant market has grown up for boutique compression providers who know how to optimize algorithm parameters and preprocessing to specific types of material (for example, sports programming, movies, animation).

Orient the Project Effectively

To orient an architecture project means to define its context and effective attributes as it begins. Many architecture projects that are driven to use frameworks are not oriented effectively, perhaps because the orientation questions are not embedded in the framework. For an architecture project to be oriented effectively, the following questions should be answered:

1. What is the "system-of-interest" to the project? Typically this is a new, discrete system to be produced; a family-of-systems or product-line; a collaborative system; or a document.
2. What is the basis for the project, or the driving reason that we are pursuing it? Typical answers are purpose driven (it is being done for a specific user-client), technology driven (we wish to exploit a technology without client-user available), or bureaucracy driven (it is a mandate to produce an architecture description document).
3. What is the scope of the effort? Is the scope restricted to a controlled system-of-interest or does it range outside to systems controlled by others outside the control of the sponsor?
4. What will be done with the product of the architecture effort when it is completed?
5. Can we disentangle the purposes of the system-of-interest, the architecture project, and any architecture description document being produced? In many cases, these three properly have different purposes, but they need to be known.

Conclusion

The problem of "blueprint" standards for complex systems architectures has yielded a number of architecture frameworks that are true or de facto standards. None is an ideal solution, but all contain important ideas. The architect faced with a normative requirement to use one of the frameworks must keep in mind their limitations and the architect's core role. The architect's core role is to assist the system's client in making the key technical decisions, particularly what system concept to go-ahead on construction with or how to constrain a larger assemblage like a family of systems or a collaborative system. This places a premium on models and methods that communicate effectively with the client, regardless of their correspondence (or lack thereof) to engineering models. Only as the architect's role evolves to transitioning the system to development and maintaining conceptual integrity during development does that correspondence to engineering methods become foremost.

A number of common problems with employing frameworks have been identified, and some important mitigations and relevant heuristics have been presented. The most important is to concentrate on the architect's core role in facilitating effective decision making. Architecture documentation plays a key role in the architect's work but is never a substitute for decision making. Architecture frameworks work best when their role is understood — to establish a common language among stakeholders within which architectural decision making can be conducted.

Notes and References

1. The DODAF documentation is distributed electronically by the U.S. Department of Defense. See www.defenselink.mil/cio-nii/docs/DoDAF_Volume_I.pdf for the current main volume.
2. See www.modaf.org.uk for online documentation on the MODAF and a detailed discussion of its concepts.
3. ISO/IEC JTC1/SC21/WG7 Reference Model for Open Distributed Processing officially titled ITU-T X.901 ISO/IEC 10746 Reference Model, Parts 1–4.
4. The complete details are in the standard, ANSI/IEEE 1471 Recommended Practice for Architectural Description of Software-Intensive Systems, issued in 2000.
5. The architect within the early MPEG standards was clearly Leonardo Chiariglione, a classic example of an architect within a collaborative system environment. A particularly good article on this is Leonardo's Art in the now defunct *Brill's Content* magazine. There is an interview with Chiariglione at www.eetimes.com/disruption/interviews/chiariglione retrieved 15-May-2008, as well as numerous other interviews and articles.

part IV

The Systems Architecting Profession

The first three parts of this book have been about systems architecting as an activity or as a role in systems development. This fourth part is about systems architecting as a profession — that is, as a recognized vocation in a specialized field. Three factors are addressed here. The first is the embedding of architecting in the context of commercial or government systems developments, with primary attention to how architecting and organizational strategy overlap and interrelate. This is vital because architecting can only happen in a supportive organizational environment, whether in business or government. The second, the political process,* is important because it interacts strongly with the architecting process, directly affecting the missions and designs of large-scale complex systems. The third, the professionalization of systems architecting, is important because it affects how the government, industry, and academia perceive systems architects as a group.

Chapter 12 covers the situating of architecting in business and government in general, but its major focus is on how strategy relates to architecture. Organizations have strategies in the sense of objectives, selected means for achieving those objectives, and patterns for changing as their environment changes. Of course, the dominant means of executing strategy is the conduct of operations by an organization's personnel. But, organizations also build systems, create programs to build systems, and structure themselves as organizations. Building is, at least in part, an architectural activity. Because the architecting of systems is already the subject of most of this book, Chapter 12 focuses primarily on the architecting of programs, and how the architecture of systems, programs, and organizational strategy relate.

---

* By "political process" is meant the art and science of government, especially the process by which it acquires large-scale, complex systems.

Chapter 13 is based on a course originated and taught at the University of Southern California by Brenda Forman of the University of Southern California and the Lockheed Martin Corporation. The chapter describes the political process of the American government and the heuristics that characterize it. The federal process, instead of company politics or executive decision making,* was chosen for the course and for this book on architecting for three reasons.

First, federal governments are major sponsors *and clients* of complex systems and their development. Second, the American federal political process is a well-documented, readily available, open source for case studies to support the heuristics given here. And third, the process is assumed by far too many technologists to be uninformed, unprofessional, and self-serving. Nothing could be worse, less true, or more damaging to a publicly supported system and its creators than acting under such assumptions. In actuality, the political process is the properly constituted and legal mechanism by which the general public expresses its judgments on the value to it of the goods and services that it needs. The fact that the process is time consuming, messy, litigious, not always fair to all, and certainly not always logical in a technical sense, is far more a consequence of inherent conflicts of interests and values of the general public than of base motives or intrigue of its representatives in government.

The point that has been made many times in this book is that value judgments must be made by the client — the individual or authority that pays the bills — and not by the architect. For public services in representative democratic countries, that client is represented by the legislative, and occasionally the judicial, branch of the government.† Chapter 13 states a number of heuristics, the "facts of life," if you will, describing how that client operates. In the political domain, those rules are as strong as any in the engineering world. The architect should treat them with at least as much respect as any engineering principle or heuristic. For example, one of the facts of life states:

> *The best engineering solutions are not necessarily the best political solutions.*

Ignoring such a fact is as great a risk as ignoring a principle of mathematics or physics — one can make the wrong moves and get the wrong answers.

---

* Company politics were felt to be too company specific, too little documented, and too arguable for credible heuristics. Readers with experience in corporate politics will have little difficulty extending the heuristics of the chapter to other political settings. Executive decisions are the province of decision theory and are best considered in that context.

† In the United States, the executive branch *implements* the value judgments made by the Congress unless the Congress expressly delegates certain ones to the executive branch.

Chapter 14 addresses the challenge in the Preface to this book to *professionalize* the field — that is, to establish it as a profession* recognized by its peers and its clients. In university terms, this means at least a graduate-level, specialized education, successful graduates, peer-reviewed publications, and university-level research. In industry terms, it means the existence of acknowledged experts and specialized techniques. Elliott Axelband† reports on progress toward such professionalization by tracing the evolution of systems standards toward architectural guidelines, by describing architecture-related programs in the universities, and by indicating professional societies and publications in the field. Axelband concludes the book with an assessment of the profession and its likely future.

---

* "Any occupation or vocation requiring training in the liberal arts or the sciences and advanced study in a specialized field." *Webster's II, New Riverside University Dictionary.* Boston, MA: Riverside, 1984, p. 939.

† Formerly, at the time of the original writing, Associate Dean, School of Engineering, University of Southern California, and the director of the Systems Architecting and Engineering program. Axelband previously was an executive of the Hughes Aircraft Company until his retirement in early 1994. He is currently on staff at the RAND Corporation.

chapter 12

Architecting in Business and Government

Architecture is the technical embodiment of strategy.

Most engineering disciplines continue to make sense when divorced from the context of their application. You do not have to know that someone is working for a builder or a government department to judge the application of aerodynamics or circuit design. Aerodynamics and circuit design (and most other methods from the established engineering disciplines) are application neutral. The equations are the same no matter who applies them. They are grounded primarily in physics and mathematics, and we can judge much of their work by the standards of science.

Architecting is much more deeply embedded in the context of its practice. Although many techniques will remain constant from one context to another, the architect's practice is heavily influenced by where it is carried out. Moreover, architecting is not just about the technical nature of the system of interest. It is about the structure of the program that builds and operates the system and the organization that either buys or conducts architecting.

This chapter explores that linkage between architecting and the business or governmental organization in which it is embedded. The focus, as suggested by the opening quote, is on strategy. There are many occasions when the architect may feel more like a strategic consultant than an engineer. Sometimes this is a sign of healthy practice, and sometimes it is a sign of looming trouble. This chapter will examine how we might tell the conditions apart.

Problem-System-Program-Organization

We can identify many different scopes of interest, but architecture and strategy are most clearly understood with four scopes: problem, system, program, and organization. Problem is that we are trying to solve or achieve by way of building a system. A system is a technical object we build to solve a problem. The program is the means by which the system

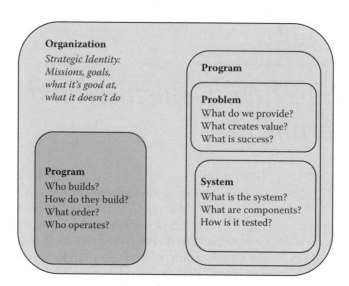

Figure 12.1 Organizations have programs that build systems in response to problems. Each exists in its own context, with issues unique to each context.

is developed, produced, and deployed. The organization, really the organizations, is the human construct that carries out the program. This is schematically illustrated in Figure 12.1 (refer also to Figure 1.6).

First-level systems architecting is about the relationship between problem and system. The architect seeks a consistent and harmonious linkage between a problem to be solved and a system to do it. Architecting is a problem-seeking activity and not solely a problem-solving activity. Good architecting examines the problem scope in parallel with solutions. The best architectural solution often involves reformulating the problem.

> *DC-3 Example*: The DC-3 was a success because it allowed the restructuring of the airline business. The Boeing 247, using nearly identical technology, was optimized for the operational environment of the time, where profits came from carrying airmail. The DC-3 had the size, capacity, range, and safety margin to allow profitable operation without subsidized airmail. To some extent, this was a happy accident, as the predecessors of the DC-3 (the DC-1 and DC-2) failed to pass the revolutionary threshold, although they were excellent airplanes. But, the architects of the DC-3 knew they were aiming at a new problem, space, as well as a new exploitation of system technology.

> *GPS Example:* The revolutionary aspects of the Global Positioning System (GPS) have come from exploitation into new problems. The original problem formulation was to navigate platforms to improve weapons delivery; thus the slogan "five bombs in the same hole." But, the most effective exploitations have been in placing guidance on weapons, in surveying, in network synchronization, and other civilian applications. Those applications represent not only the application of GPS technology but the reformulation of concepts of operation for both military and civilian activities.

In the scope of problem-system, we talk about the fundamental structure of the system, its architecture. But, the system has to be brought into being. Beyond the architecture design, it has to be fully developed, produced (in quantities from one to millions), deployed to users, and supported over a life cycle. We refer to these activities as the "program." A program also has a fundamental or organizing structure or an architecture. We can identify a number of basic forms or architectural styles, which we discuss in detail in a subsequent section:

- Single object, waterfall construction, as in buildings and occasional one-of-a-kind systems.
- Prototype development followed by serial production, with parallel manufacturing system development (discussed in Chapter 4).
- Breadboard-Brassboard-Flight incremental development, a typically hardware-centric process where functionality remains constant while environmental fitness is improved.
- Risk-spiral incremental development, where increments represent case-specific steepest-descent reduction of risk.
- Incremental delivery, where multiple systems are delivered with increasing functionality.

Program structure may play an equal role with the architecture of the system in realizing stakeholder value. A fine system may be crippled by poor execution or doomed by a program structure that is inappropriate to the surrounding circumstances. Conversely, a well-chosen program structure may allow successful adaptation to errors in execution and surprises in technology or operational conditions.

Layered software is a response to rapidly changing technology and uncertainties in user demands. Well-chosen and implemented layers isolate areas of rapid change from each other and allow change in those isolated areas to proceed as quickly as technology or market changes

demand. The Internet Protocol (IP) layer in the Internet effectively separated the very high rates of change in physical communication technologies and network applications from each other, and allowed both
to repeatedly abandon existing legacies independently. In negative
contrast, layered architectures can introduce broad dependencies that
may damage an organization's ability to deliver. A change to a deeper
layer may have rippling effects in all of the higher-layer applications that
use the shared, deep layer. If the deeper layers are pushed to change in
response to user demands falling on applications, and the surrounding
organization and technological infrastructure is unable to make changes
without risk of affecting all applications, the layered structure may lead
to development paralysis.

A program is carried out by an organization, which may be a single
company or government division or a consortium of many. By organization we simply mean an organized grouping of humans whose purpose
here is to carry out programs to build systems. Programmatic structures
should be chosen to best fit the programmatically related objectives of a
given development. In practice, the structures will also be influenced by
the standing concerns of the organization. So, if the overall identity of the
organization is well aligned with the programmatic and system mission,
things are likely to go well. If the strategic identity of the organization
is in conflict with stakeholder concerns, programmatic imperatives, and
system objectives, things are likely to go badly. This rising scope, and
changing nature of concerns, is illustrated in Figure 12.2.

The strategic identity of an organization is the basic representation
of what it does. The strategic identity should be a shared understanding

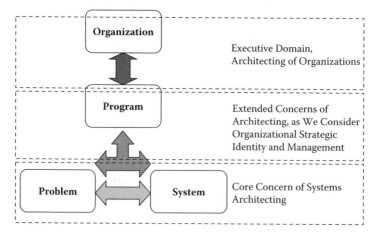

Figure 12.2 Systems architecting is primarily concerned with the relationships
between elements, whether at the level of system-problem, program and management, or (occasionally) at the level of organizations.

among the organization's members. A strategic identity specifies what an organization's mission is, how that mission relates to other organization's missions, and how the organization's members can take action in service of that mission. Organization's can be viewed productively as systems, and so have their own architectures. The architecture of an organization is its basic structure, not just in organizational chart terms, but in the fuller terms of expertise, experiences, working relationships internal and external, shared objectives among its members, and resources available to it. The architecture of an organization is not a principal concern of this book,[1] but we cannot understand the architecting of systems without considering how the hierarchy of contexts from problem to organization relate, through the system and program.

Strategy and Architecture in Business and Government

In the classic model of architecting, the paradigm derived directly from classic civil architecting, the architect is an executor of the client's strategy. The client has a strategy. Perhaps it is to build a house well suited to his or her family's life, to build a profit-making facility in a given business, to combine business return with brand identity, or to build a long-term educational institution. Whatever that strategy is, the job of the architect is to understand it well enough to be able to produce a fit physical, technical embodiment of that strategy.

The architect does not create the strategy, although he may need to elicit the client well beyond just asking "what is your strategy?" Architecting accepts that the problem is unlikely to be presented in well-structured form, and is probably fundamentally ill structured. With ill-structured problems, the process of forming a solution influences the client's understanding of his or her problem, and not just the architect's. Thus, to some extent, the process of working with an architect may help a client formulate his or her own strategy, even though it is not the architect's role to formulate the client's strategy.

As we talk about unprecedented systems, the sharp border between the architect and client in strategy becomes unclear. Who formulated the strategy of moving to operations and capabilities for global positioning beyond the concept in the original program? It was not entirely the client or the architect. In the case of GPS, the long-term revolution was driven by organizations beyond the client, and that in many cases did not even exist when the program was formed. These lateral exploitation applications, which are growing to the point they now drive global positioning well beyond one program, were inside the original architect's visions, but were not the original architect's responsibility. Many others had to become participants, typically independent participants in what became a collaborative system, for the revolution to happen.

When the border between architecting and strategy formation becomes fuzzy, the architect may find himself or herself acting more like a strategy consultant than an architect. Even though it is not impossible for this to be effective, it is fraught with difficulty and danger. Architecting requires technical depth, and good architects have that technical depth. Effective strategy formulation requires much more knowledge and insight into the operational situation faced by an organization (whether business, military, or diplomatic) than is necessary for architecting, and requires much less technical depth. Both architects and strategists are bridging the engineering to operations gap, but they approach the gap from opposite sides. As such, they can be extremely effective partners but are less likely to be effective substitutes.

A basic embodiment of strategy produced by architecture is in the structure of the program. Architecture may also embody strategy in other ways, but program structure is the most common. To explore this, we need to consider different forms of strategy as related to technical system development, specifically static and dynamic strategies.

Static Strategies

A static strategy is unchanging, or slowly changing. Static strategies seek to understand the world and to determine a set of objectives for systems that will yield a superior position. Typical static strategies include the following:

- Be the low-cost producer or supplier. This involves leaning down design, production, and delivery systems relative to other competitors.
- Be feature superior to the competition. Deliver systems with superior quality, cost, and delivery, measured on the same scales as the competition.
- Bring superior firepower and concentration to the battlefield.
- Use concepts of operation similar to one's opponent, but with longer range, greater accuracy, and larger effect.

Architecting in a static strategy environment is relatively similar to classic systems engineering. The problem may still be ill structured, because we do not know where in the feature space we will find suitable problem–solution combinations, but the feature space is assumed to be knowable. The objective of architecting is to elaborate on the problem and solution spaces and find excellent combinations.

Dynamic Strategies

A dynamic strategic approach assumes that the playing field is continuously changing, or even better, and that our actions can force the playing field to change.[2] Instead of trying to beat competitors at an established game, we seek to create new games. We try to avoid head-to-head competition on features, cost, or delivery and instead choose actions that are unexpected by the competition and that the competition is incapable of imitating. A truly dynamic approach to strategy is to further assume that any dominant move we make is temporary and must be followed by a succession of game-changing moves.

Architecting in a dynamic strategy is a twofold process. First, the architect is driven to come up with unprecedented system concepts. A dynamic strategy has a continuous appetite for the new and unexpected. Second, the organizational processes and supporting system must be architected to support continuous and rapid change. An organization very good at executing a static strategy is unlikely to be good at executing a dynamic strategy, and vice versa.

The real world is not cleanly divided between exponents of static and dynamic strategy. Even if one believes that a dynamic strategy is inherently superior, in considerable measure much of economic and military life is dominated by mature operational concepts where fierce competition in static strategy prevails. Where the investment in legacy capabilities is very large, it is very uncommon for dynamic shifts to upset the entire operational picture quickly. Even when the operational picture can be changed quickly, it may settle down to maturity with time. A useful metric for understanding where static or dynamic strategies are likely to play a larger role is system obsolescence time or depreciation rate. How quickly after introduction does a system lose most of its value? In how much time will the original system owner be willing to throw away the system and find it is not worth the upkeep?

- In aircraft, from the 1920s to the early 1960s, aircraft depreciated in 5 years, lengthening to 10 years. Military aircraft built at the end of World War II were of low value within 5 years and scrap before 10 years elapsed. This time lengthened considerably from the 1960s to the present, when both military and civilian aircraft are still flying usefully 25 years or longer.
- At the beginning of the space era, a given satellite design was useful for a few years at most. By the 1970s, satellite architectures settled down. Lifetimes of 5 to 10 years are now not unusual, and designs can be valuable longer.
- A 5-year-old computer is, with few exceptions, something to sell at a flea market.

- Cell phones and related personal electronics are disposable on a 2-year timeline.

An organization should have an explicit strategic position about which it pursues its mission. Some organizations, those supporting stable mission areas or markets, logically pursue strategies that are mostly static. This is not a bad thing. If a static strategy that carefully focuses on stable sources of value and stable means of delivering value can achieve a competitive advantage against static strategy measures, that is very hard to match or overcome. As an example, various automakers, both U.S. and Japanese, established long-term competitive advantages that endured for decades (in different eras). But, static strategies can be overcome by frontal competition, and by "end-runs" when strategic conditions change.

Organizations that pursue pure dynamic strategies can likewise be very successful, and can fail abjectly. An organization that solely pursues the unprecedented is vulnerable every few years. Even for the most capable, the business of producing unprecedented capabilities is very uncertain. Luck is required, and runs of luck always come to an end. If the organization cannot weather a string of failures, it will disappear.

Architecture of Programs

As discussed above, the next step of context above Problem-System is Program. At the program level we are concerned with the structure of the effort to develop, produce, deploy, and maintain the system of interest. Obviously, as with systems, there are countless possible such structures. We cannot enumerate them all. However, we can identify a number of program styles or repeating patterns of program organization, and the heuristics for their application. These program styles relate directly to an organization's pursuit of a dominantly static or dynamic strategy.

Single Pass, Waterfall Construction

The paradigm for this case is constructing a house or other building. The process normally proceeds very linearly: A design is developed and approved, contractors are hired, the building is constructed on the site, and approved for occupancy and delivery after completion. There are few or no intermediates. Modeling is conducted during design and may involve the construction of scale models, but we do not build trial buildings as part of the process. On occasion, some subsystem elements might be built early for testing. An example is building a unit of windows for a major skyscraper to test their weather integrity if they use an innovative method of holding the window glass.

In practice, there may be some level of incrementalism. For example, in a building complex, we may build all of the infrastructure but only some of the buildings in an initial phase, with the remaining buildings deferred for a later phase after the first set are occupied and in use. Sometimes a building is designed with options for remodeling or extension in mind. However, the basic pattern is simple; we directly design and build the final system we intend to deliver in one pass.

This programmatic pattern is most applicable where:

- Only a single system is to be built and delivered.
- Risks are low. There is high certainty that a satisfactory system can be built and delivered at predictable cost from a design.
- The strategy is static. We can build a system in response to the strategy and believe it will be fit for the natural lifetime of the system.

Serial Production

The basic pattern here is that we build one or more prototype systems, probably using the one-shot waterfall pattern, freeze a final design from analysis of the prototype, and then produce many copies of that prototype design. Alternatively, we may use one of the other patterns for prototype developments before freezing the design and proceeding to production. This pattern is most applicable when:

- Many copies of the system are required.
- The cost of production is high relative to the cost of design and development. The overall cost is dominated by the costs incurred in production (typical for hardware-centric systems).
- Risks can be resolved by a prototype. Once we have the prototype, and have worked with it, we can have confidence that the produced system will be fully acceptable.

Breadboard-Brassboard-Flight

This pattern is an incremental pattern, in that we build a series of systems that are less capable than the final system and that lead to the final system. In the breadboard-brassboard-flight pattern, the series of systems that we build should all be functionally equivalent to the final delivered system, but are not all environmentally suitable for operation in the delivered systems environment. In the classic version of this pattern, the breadboard system is spread out over laboratory benches. In electronics, it consists of large boards with many parts and no effort of design shrink. In optics, the components are spread over an optical bench. In chemical engineering, the early versions are physically much smaller than the target version, with far lower capacity (the direction of improvement is increasing scale instead of decreasing size). The brassboard version has classically been shrunk to an operationally suitable size and form factor, but is not yet fully hardened or reliable enough for operational use. The "Flight" version is the final version to be delivered to operational use.

This pattern is particularly applicable when:

- Functional risks are low. We have high confidence that we can identify all of the desired functional characteristics early in the design process.
- Technology and implementation risks are (relatively) high. We have low confidence that we can build and package the desired functional characteristics in an environmentally suitable unit.
- Production numbers can either be very low (a single flight system), or this can be combined with the serial production pattern.
- The strategy allows for a static functional aim point.

Because of the second bullet, the pattern is mostly seen in hardware-centric systems. In most software-centric systems, the technology and implementation risks are relatively low. We know that if we can write functionally acceptable software we can probably package it in an acceptable way. Obviously, many exceptions exist, but the point is that in software-centric systems we are typically driven by functional risks rather than technology and packaging risks. In contrast, in many sensor, aircraft, and spacecraft systems, we have mature knowledge of how to build a functionally acceptable system but not how to make it operate in the environmentally constrained environment of the operational target. Consider the problem of sending remote sensing instruments to Jupiter. In most cases, the instruments we wish to send are well understood and widely used already in terrestrial or even earth-orbiting environments. But, packaging the instrument in a size, weight, power, reliability, radiation-resistant form factor usable in an outer planets mission is always a great challenge.

Incremental Delivery

The incremental delivery pattern can be thought of as the converse of the breadboard-brassboard-flight pattern. In the incremental delivery pattern, we again build a series of systems, each different from the previous, but the sequence grows in functional capability and not in environmental suitability. Each member of the sequence is fully usable. In the classic version of this, each member of the sequence is not just fully usable, but each is delivered and used operationally. In commercial market terms, this is a series of incrementally developed products.

This pattern is particularly applicable where:

- The risks and uncertainty about what is functionally acceptable are high and can only be resolved by operational experience. No amount of requirements elicitation in the absence of a real system can be expected to resolve the questions about what functional capabilities the users really want.
- Risks in developing an environmentally suitable system are low. It is not difficult to meet user expectations of size, weight, power, reliability, or other physical quality characteristics.
- The cost, price, and revenue issues are such that multiple replacements of a delivered system are acceptable (or even desired).
- The strategy is dynamic, and we realize value substantially by adapting to change and new knowledge with different system configurations.

The third bullet is characteristic of software-centric systems, because software production deployment costs can be very low. The third bullet may also apply to systems with significant hardware content where market forces lead to rapid turnover. As an example, consider many consumer electronic segments where people rarely keep a device for more than 2 to 3 years and are willing to pay for replacements (as long as they offer new features).

Risk Spiral

The risk spiral is an integrated combination of breadboard-brassboard incrementalism and incremental delivery. In the risk spiral (the concept original to Boehm), each cycle through development yields a system. The objectives of each cycle are driven by an overall assessment of risk. If the assessment is that currently the risks and uncertainties about what functions have value to users, then the next spiral cycle will emphasize a user delivered system that can assess the value of functions. If the assessment is that the highest risk is engineering and packaging, then the next spiral cycle will emphasize the breadboard-brassboard-type of development.

Each spiral cycle consists of all of the conventional activities of the waterfall: requirements analysis, design, build, integration, and test, as illustrated in Figure 1.5 and Figure 6.1. Architecting in all spiral or incremental situations differs in two basic ways from one-shot or waterfall architecting. First, architecting becomes episodic. We do architecting every time we go around the spiral. Each cycle around the spiral requires decisions about the problem and solution content of that cycle around the spiral. Each cycle is a complete pass through development and requires a set of architectural decisions on the concept to be developed (at the beginning of the cycle), and decisions on acceptance for use (at the end of the cycle). Second, we architect the invariants, or the things that do not change as we spiral.

In addition, the choice of an incremental development approach, versus a one-shot development or some other pattern, is an architectural choice. It is the choice of program architecture. That choice, of program architecture, is rooted in an understanding of the overall strategy and how architectural decisions embody that strategy.

Collaborative Formation

In Chapter 7 we examined the concept of a collaborative system, a system formed by the partially or whole voluntary interaction of autonomous systems. We can undertake the formation of a collaborative system as a deliberate effort, although the fact that we cannot control all aspects means that there is an element of uncertainty. In creating a program whose goal is the collaboration formation of a system, we are deliberately choosing to orchestrate a process whose end point we cannot precisely predict. We must accept the uncertainty of the collaborative assemblage process in return for the benefits that it brings. A collaborative formation approach is especially appropriate when:

- The strategy is dynamic, and we believe our power to shape is greater than our power to actually implement.
- The environment inherently contains multiple autonomous players, and it is neither sensible nor feasible to replace them.
- The risks associated with the strong preexisting players are more significant than the risks of a particular configuration being achieved or not.

Strategic Architecting of Programs

Given that programs have architectures, and that the architecture of the program needs to be considered in parallel with the architecture of the system, how does the architect go about it? Architectural thinking in business and government should consist of all of the following:

- Understand the organizational context in which architecting takes place.
    - Who are the competitors?
    - Who are the opponents (not the same as competitors)?
    - Is the organization involved with architecting a constant or variable? Is a new organization logically an outcome of architecting?

- Understand the overall strategy of the organization involved with architecting. What are the static and dynamic aspects of the strategy? What is the strategic identity of the organization?
- Use the context in Problem-System aspects of architecting. Begin exploration of the program architecture as system concepts emerge.
- Select system and program styles consistently with the strategic identity of the organization.

Consider how these factors interrelated in the DC-3 example. The example involved several organizations with different positions relative to each other, different strategies, and different architecting responsibilities. Boeing was responsible for architecting (and building) airplanes to satisfy a commercial mission (make money by being sold to airlines to be operated commercially by those airlines). Their strategy, at the level of the 247, was a static strategy — perform existing missions with better performance and cost. At the level of the whole company, the strategy was much more dynamic because of their parallel pursuit of much larger, military aircraft with technology that overlaps with commercial applications. The strategic identity of the 247 group was the pursuit of the static strategy. The strategic identity of the corporation as a whole was a dynamic one of shaping the aircraft market. The architects of the 247 dutifully pursued the static strategy of performance and cost improvements and were successful within that context. The program style was prototyping followed by serial production. There was no incremental development. When the DC-3 overtook the 247, Boeing's response was to make another architectural jump (the corporate-level dynamic strategy), but that was cut off by the beginning of World War II.

Douglas and Boeing were competitors, and at the corporate level they were pursuing similar dynamic strategies. At the local level of the aircraft program, Douglas pursued a more dynamic strategy. Considering the DC-1, DC-2, and DC-3 as a series, we see an incremental development strategy. The architectural jump on the problem side was to move away from the known airmail market. As a result of the uncertainties, this was pursued with incremental development, with each subsequent aircraft a bigger step into the unknown in size and performance. The program style was incremental development because each of the models was a fully usable, customer deliverable system. Indeed, all three models were customer delivered, although the total production of the DC-3 swamps the other two.

Jump and Exploit

The DC-3 and its numerous models are an illustration of a larger heuristic applicable at the program and organization level, the pattern of "jump and exploit." Jump and exploit describes the strategic approach of seeking unprecedented systems (the jump) followed by extensive lateral exploitation of the unprecedented jump. This combines the notions of static and dynamic strategy. We make jumps in creating new systems *and* coupled new concepts of operation or markets. When the jump is successful, we vigorously pursue the static strategy of improving performance and cost for the newly revealed operational concept or market.

The interplay of architectural jumps and long-term steady improvement is a strategic challenge. Leaders must be able to evaluate when the time for a jump is ripe, invest when the time is ripe, and stop focusing on incremental improvement. Conversely, failing to run a strategy of focused incremental improvement can easily cede competitive advantage to another player who does focus on continuous improvement. While recognizing when each situation pertains is inherently hard, one heuristic has been found useful.

> *An architectural leap can rarely be justified when the consequence of a successful leap is a drop in revenue. Markets must expand to make cost reductions justifiable.*

This is a hard heuristic to swallow in many cases, but it is important to examine. The simplest case is where an essentially fixed number of systems will be produced. Consider the case of space launch vehicles. Imagine that the government buys an average of five of a particular type per year. If each launch vehicle costs $100 million, the government expends an average of $500 million a year with this particular supplier. Now suppose there is a proposal to develop a new launch vehicle with the same performance but an estimated per launch cost of $50 million. Is this likely to be a workable proposal? The heuristics suggest it would not be, and the heuristic is developed from past experience with space systems and other limited market systems where demand is inelastic with price.

Why might this be so, and when might it not be true? When the supplier base is relatively fixed, we can imagine that existing suppliers would be less than enthusiastic about a program that promises to cut their revenue in half. Even if the government was to pay for the development, the overall situation is unlikely to be favorable. Financial capital is not the only capital of importance. Human capital is attracted to growing markets and is an essential fuel. In most cases, the program to cut costs in half is likely to be successful only if price elasticity is such that volume will likewise increase by at least a factor of two. Our hypothetical launch vehicle cost reduction program might be successful if a price cut by a half would more than double the launch rate. We see this effect in play in the electronics industry. While the price per transistor drops steadily by factors of two, total production and sales of transistors goes up even more quickly. The total revenues of the electronics industry have risen rapidly. Were that not so, they would be unlikely to have been able to attract the capital (both financial and human) that was necessary to fuel the growth.

Enterprise Architecture

A natural conclusion to this discussion of architecting in organizations is enterprise architecture. Enterprise architecture is big business. It exists as an established practice with numerous books, consultants, service providers, methods, and courses. The purpose of this book is not to replace any of the large body of work, or even to engage in a detailed critique. Nevertheless, the concepts of this book, and especially this chapter, can be used to usefully inform the practice of enterprise architecture and understand some of the most commonly encountered difficulties.

Given that the field is large and the companies are so diverse, it is perhaps not surprising that there is a good deal of disagreement on what enterprise architecture is. If we take an enterprise to be an organization with a defined mission (a company and a government department would both normally qualify), then a "natural" definition of the architecture of an enterprise would be the fundamental and unifying structure of the enterprise. Then the practice of enterprise architecture would concern itself largely with business strategy and business processes and how the enterprise might be best organized to carry out its mission. In the case of a company, this would mean long-term value creation in particular markets. In the case of a government department, it would depend on the case (human services versus environment versus research versus security). But in reality, enterprise architecture as actually practiced almost always is largely concerned with information technology, either substantially or solely.

A good definition of *enterprise architecture* comes from Peter Weill of the Massachusetts Institute of Technology (MIT):

> The organizing logic for key business processes and
> IT capabilities reflecting the integration and standard-
> ization requirements of the firm's operating model.[3]

This book is concerned primarily with the architecting of systems. The information technology of a firm is certainly a system. The structure of that system should support the overall mission of the firm. As stated at the beginning of this chapter, the architecture of the firm's information technology (IT) should embody the strategy for the firm. From a simple insight, we can glean some important lessons.

The strategy an IT system embodies should be that of the organization it belongs to as a whole, not that of just the IT controlling organization.

Several times one of the present authors has encountered the situation of how a research and development (R&D) group's IT is arranged within a larger organization of which R&D is a small part. Consider a hypothetical large specialty chemical manufacturer. The firm will undoubtedly have a chief information officer (CIO) and a corporate-wide information system. That corporate-wide information system needs to support internal functions (such as time and attendance, human resources, corporate-wide e-mail, and so forth) as well as core business activities. The core business activities would include sales and marketing, customer interaction, production and transportation planning, finance and reporting, and many others. Because these core functions represent virtually all of the firm's revenue, the CIO is likely to be very concerned with how they are supported. The CIO's priorities are likely to be dominated by system and application stability, availability, security, regulatory compliance, and cost control. When the CIO's office conducts an enterprise architecture exercise, it is likely to focus on central control and standardization. The ideal will change slowly, be carefully controlled, and provide a well-chosen set of common services with high availability.

Within this large manufacturer, there is likely to be a R&D group. The R&D group may have dominant responsibility for new products and production methods. At any given time, they are not a revenue source; they are likely a sink for money. However, the long-term future of the company (5 and 10 years out) depends almost totally on the success of the R&D group. In an environment where products age out in 5 to 10 years, failure to have a full pipeline of new products will spell the end of the company and its value. How do the information technology needs of the R&D group align with the priorities of the CIO? In many cases, they align very poorly.

On the one hand, the employees of the R&D group have many of the same needs for centralized information services as everybody else. They need access to corporate-wide e-mail, human resources applications, and other centralized tools just like other groups. But, today, the R&D group in a specialty chemical company is likely to be trying to rapidly exploit computational chemistry, cheminformatics, a myriad of tools for chemical engineering, biology-based products and production methods, and collaborations with groups around the world. This environment changes quickly with tools being updated monthly, tools coming from all over the world, and all being run on a diversity of platforms. There is often a serious collision of strategies between the R&D group and the CIO.

Good enterprise architecture recognizes the diversity of business strategies within a firm and tries to appropriately accommodate them all. It realizes that the strategy the whole firm's IT should embody is the strategy of the whole firm, not that of a narrow segment. The IT in a firm should be there to execute the purpose of the firm, over both the long and the short term. Good architecting goes back, again and again if necessary, to root purpose. The purpose of a firm's IT is not to cost less than it did last year (even if that makes somebody's metrics look good); it is to support the business strategy of the whole firm. This is the holistic view of architecting, a system that we embraced from Chapter 1.

> *An architecture description is not an architecture, and neither is an architecture framework.*

This is simply a reiteration of a point made early in this book to not confuse architecture with architecture description. It is, unfortunately, not uncommon for a group to point at a large binder and say "This is our enterprise architecture." It is not and cannot be. At best, the binder will contain a description of the decisions that define the architecture of the enterprise. At best, those decisions will be good ones and will have captured the firm's business strategy effectively. Unfortunately, the best may not be true. Quite possibly the key decisions contained in that binder are buried from view and unwise to boot. If the decisions are clear, the architect should be able to highlight and explain them without many pages of description in the binder. If the decisions are wise, the reasoning should be clear and explainable and not buried in an opaque trade study where an answer is touted as the best simply because it scored the best on an evaluation function but nobody can clearly articulate why and with what sensitivity.

The frameworks commonly cited in the enterprise architecture literature are all architecture description frameworks. That is, they define how to write a document about a system or systems. They do not define the decisions, and in most cases they provide little guidance on how to go about the decision making. By itself this is not a great problem. Standardization of description methods can be quite useful in promoting wide communication. Where it becomes a problem is when framework adherence and unthinking artifact production take the place of critical thinking about an organization's missions, the diversity of missions that make up the overall mission, and the sources of value from what is being architected. Rote application of frameworks typically yields large documents that are then either applied inflexibly or ignored (which may be better than inflexible application).

In thinking about the example of the large chemical company above, no amount of framework application will resolve the essential tensions. The essential tensions are how to balance the need for diversity, change, and local control within the R&D group with the need for stability and standardization in the firm as a whole. Technology or business process adaptation may provide ways to relieve that tension. The tension is unlikely to ever be fully resolved, and there will inevitably be problems between the R&D group and other groups over how IT is selected and used. But, the absence of perfection is no excuse to avoid deep thinking about how best to resolve the tension. There is almost certainly a great deal of long-term business value to allowing R&D to fully utilize the rapidly growing technology in accelerating product and production development and a great threat in the possibility that competitors will resolve it better, sooner.

> *Program structure may be as important, or more important, than product structure.*

In a large firm, the way they select, procure, deploy, and operate their IT is likely to be more important than the precise components chosen. In the terms of the chapter, the structure of the program is likely to be as important as or more important than the Problem-System pair. The structure of the program must be inside the scope of architectural consideration, not outside it.

> *Good architecting thinks as much or more about the problem as about the solution. Architecting teams need the skills relevant to the problem scope they are engaged in.*

It often seems natural that an IT-centric enterprise architecture job should be done by IT specialists. But, if the scope of consideration includes how we might change business processes in concert with IT deployments in order to better carry out the firm's mission, an all-IT-specialist team will be wholly inadequate. This is reflective of the basic nature of systems architecting. As discussed here, the lowest scope that is "architectural" is Problem-System. That is, the problem is inside the scope of investigation, not outside it. As discussed in the beginning chapters of the book, architecting addresses ill-structured problems where the statement of the problem is in-play. If all the requirements can be readily determined, it is not architecting. If the nature of the problem is "Find the best solution to this precisely stated and well-structured problem," it might be a very worthy and difficult thing to do, but it is not architecting. If it is not architecting, it does not need the machinery of architecting, and we can rely instead on the established machinery of engineering science.

In the enterprise case, the situation is often more difficult because the problem includes basic strategic issues for the firm. A team constructed to do what is viewed as a technical architecting job rarely contains the expertise and authority to challenge enterprise-level strategic decisions. They may be unable to command sufficient attention from senior executives whose purview definitely includes strategy. Lacking that attention, they may default to extracting a well-structured problem they can solve, whether or not that is really relevant to the organization's greatest need.

Many darkly humorous tales can be told of the technologist supposedly empowered to make a revolution in a firm running after executives trying to have a strategic discussion,[4] or giving up, pursuing a technological solution, and failing because of an inability of the firm to connect technological success to business strategy. The DC-3 and GPS stories are stories of success, albeit with all of the fits and starts and blind alleys of the real world. One of the most famous stories of failure to connect technological architecting to business strategy execution is Xerox PARC[5] in the 1970s. The story is lengthy and well told in the published literature, but certain points bear repeating.

Xerox executives had a clear strategic vision of the need to make a change from the copier business by the late 1960s. They took tremendous advantage of the availability of a whole cadre of the best computer scientists and engineers who became available as a result of U.S. Defense Advanced Research Projects Agency (DARPA) funding cuts. They set their recruits up in a new organization that produced an unprecedented series of technical innovations (laser printers, object-oriented programming languages, and window-mouse-graphics displays famously among them), with many of those innovations taken to the point of prototype products. But, they were then unable to convert those innovations and products into revenue. The failure was largely one of mismatched strategies and a lack of coupled change in business models (that is, operational concepts) to go with the innovative products. To turn the new products into value would have required new business concepts to go with them. As it turned out, other firms were much faster to find those altered operational concepts and implement them.

Ironically, it was the ability and willingness of Xerox to discover and use coupled technological and business model change in the 1950s, replacing a purchase model for copiers with a lease and per-page-charge model that made the company such an enormous success originally. But what was possible in the start-up days became impossible in the days of maturity. The lesson is only reinforced from the other case studies. Dealing with change at the right scope is critical. The biggest successes come from coupled change in technology, system, and operational concept.

Of course, this story about Xerox is only loosely about a firm's IT architecture. But, that is exactly the point. Good architecting knows its scope. Focusing on just a firm's IT architecture is a narrow focus, one unlikely to lead to changes in strategy or overall approach, but likely to lead to efficiency and improvements within a mission area. It leads to small solutions within preexisting mission areas instead of large solutions in new mission areas. Focusing on efficiencies can create strong competitive advantage when missions or markets are stable. When missions or markets are unstable, and revolution is in the air, focusing on efficiencies is a distraction, and possibly a fatal distraction. Of course, the opposite is also possible when missions and markets are stable, and one can be distracted by a futile search for revolutions where none can exist and fail to develop the efficiencies that others will use to win the competition. An organization has to be wise enough to know the difference.

An Enterprise Thought Experiment

Let us return briefly to Chapter 6 on software and the case discussion on layered systems. Many exercises in enterprise architecture turn, one way or another, on how to provide common services across a large enterprise. A popular buzz-word is "services-oriented architecture" (SOA). As a thought experiment, what are issues in the program or strategic sense for providing a common infrastructure of software services? Two programmatic alternatives (of course, there are others) are to license a commercial enterprise service bus (ESB) and have new software written on top of it, and to license an open-source ESB equivalent and write new software on top of it using a continuing open-source rule (all components written become enterprise shared property). Some of the issues are as follows:

- What is the impact on which developers can be used? Will some developers refuse to contract to write code that is shared with other enterprise developers? With a commercial ESB, will the cost of developer licenses inhibit how many developers can be used and when (for example, cannot afford experimental programming)?
- In either case, will making a transition away from current practice devalue current developers? Can the organization afford the costs involved in building a new developer community? As an aside, that cost might run from very large to negative depending on the nature of the market and the skills of the current developers.
- Does business with other enterprises on the part of the ESB vendor bring economies of scale? Versus, is there a significant open-source development community beyond the local enterprise for an open-source alternative?
- How are third-party developers impacted? If you want a particular tool of high importance brought in to the enterprise service environment, how will that tool vendor integrate with the ESB (commercial from another party versus open source)?
- Does the choice of implementation strategy affect how the larger business strategy will be achieved?

Conclusion

Architecting must be situated in its business or government operational context. When viewed in its context, the relationship to strategy becomes evident. Where strategies are clear, good architecting can follow. Where strategies are unclear, good architecting will be very difficult. As we consider not just the classic architecting relationship of problem-system but expand the focus to the structure of development programs, there becomes a synergy of architectural and strategic thinking.

The most successful, unprecedented systems involve changes to business or operational models in parallel with new systems and technology. Just introducing a new system is not enough; when the system is revolutionary, the context has to change as well for the greatest success to be realized. Coupled business or operational change must be enabled by the organization's strategy.

The leading guidelines in this chapter were as follows:

- Remember problem-system-program-organization.
- Understand static and dynamic strategies, and how they map into program styles.
- Have a catalog of development program styles and understand where each is best suited.
- Know when to jump, and know when to settle for continuous improvement.
- Architect holistically for strategy, not just for local stakeholders.
- Do not confuse architectures, architecture descriptions, and architecture description frameworks.
- Think of architecture as the technical embodiment of strategy.

Notes and References

1. Although see Rechtin, E., *Systems Architecting of Organizations: Why Eagles Can't Swim.* Boca Raton, FL: CRC Press, 1999.
2. This discussion of static and dynamic strategy is heavily influenced by the strategy concepts of Colonel John Boyd. Boyd's concepts have become well known in spite of there being no regularly published reference. The briefings that capture Boyd's concepts are generally available on the Internet, see www.d-n-i.net/dni/john-r-boyd/ for a collection. The most relevant references here are the Boyd briefing entitled Patterns of Conflict and the book *Certain to Win* by Boyd's long-time colleague Chet Richards. In Boyd's terms, one can regard a static strategy as "attrition warfare" and a dynamic strategy as "maneuver warfare." Boyd comes down strongly on the side of the superiority of maneuver warfare, but it must be admitted that attrition warfare armies win wars too, albeit often at terrible cost.

3. Weill, Peter, Innovating with Information Systems: What Do the Most Agile Firms in the World Do? Sixth e-business conference PwC and IESE Barcelona, Spain, March 27, 2007. Retrieved from www.iese.edu/en/files/6_29338.pdf 14-May-2008.
4. A favorite is My Year at a Big High Tech Company by Joe Kay, appearing from August to November 2000 in *Forbes ASAP*.
5. See Hiltzik, M. A., *Dealers of Lightning*. New York: Harper Business, 2000, for a detailed recounting.

chapter 13

The Political Process and Systems Architecting*

Brenda Forman

Introduction: The Political Challenge

The process of systems architecting requires two things above all others — value judgments by the client and technical choices by the architect. The political process is the way that the general public, when it is the end client, expresses its value judgments. High-tech, high-budget, high-visibility, publicly supported programs are therefore far more than engineering challenges; they are *political* challenges of the first magnitude. A program may have the technological potential of producing the most revolutionary weapon system since gunpowder, elegantly engineered and technologically superb, but if it is to have any real life-expectancy or even birth, its managers must take its political element as seriously as any other element. It is not only possible but likely that the political process will not only drive such design factors as safety, security, producibility, quantity, and reliability, but may even influence the choice of technologies to be employed.† The bottom line is:

If the politics don't fly, the system never will.

---

* This chapter is based on a course originated and taught at the University of Southern California by Dr. Forman, now with the Lockheed Martin Corporation in Washington, DC. As indicated in the Introduction to Part IV, the political process of the American Federal Government was chosen for the course for three reasons: it is the process by which the general public expresses its value judgments as a customer, it is well documented and publicized, and it is seriously misunderstood by the engineering community, to the detriment of effective architecting.

† Outside of the government sphere, it may seem that politics disappears, but it does not. Large corporations are also political entities, and the heuristics of politics operate in organizations on many levels. Of course, as an organization becomes commercially focused, its objectives are different and the motivations of its leaders are likewise different, but many of the same rules will apply. Thus, this chapter can be seen as a guide to the political demands on architects, even outside of the purely political sphere.

Politics as a Design Factor

Politics is a determining design factor in today's high-tech engineering. Its rules and variables must be understood as clearly as stress analysis, electronics, or support requirements. However, its rules differ profoundly from those of Aristotelian logic. Its many variables can be bewildering in their complexity and often downright orneriness.

In addition to the formal political institutions of the Congress and the White House, a program must deal with a political process that includes interagency rivalries, intra-agency tensions, dozens of lobbying groups, influential external technical review groups, powerful individuals both within and outside government, and always and everywhere, the media.

These groups, organizations, institutions, and individuals interact in a process of great complexity. This confusing and at times chaotic activity, however, determines the budgetary funding levels that either enable the engineering design process to go forward or threaten outright cancellation. More often of late, it directly affects the design in the form of detailed budget allocations, assignments of work, environmental impact statements, and the reporting of risks or threats.

Understanding the political process and dealing successfully with it are therefore crucial to program success.

> *Example*: Perhaps no major program has seen as many cuts, stretch outs, reviews, mandated designs, and risk of cancellation as the planetary exploration program of the 1970s and 1980s. Much of the cause was the need to fund the much larger Shuttle program. For more than a decade, there were no planetary launches and virtually no new starts. From year to year changes were mandated in spacecraft design, the launch vehicles to be used, and even the planets and asteroids to be explored. The collateral damage to the planetary program of the Shuttle Challenger loss was enormous in delayed opportunities, redesigns, and wasted energy. Yet the program was so engineered that it still produced a long series of dramatic successes, widely publicized and applauded, using spacecraft designed and launched many years before.*

---

* As a follow-up, in the 1990s, JPL made an aggressive move toward shorter missions with more rapid turnover. This can be seen as emphasizing the longer term importance of learning and adapting in scientific exploration over immediate cost efficiency. A consistent, year-to-year stream of results also helps build constituency. But, ironically, in the mid-2000's planetary exploration is once again under pressure, this time from the expense of *terminating* the Shuttle program.

Begin by understanding that *power is very widely distributed in Washington*. There is no single, clear-cut locus of authority to which to turn for support for long-term, expensive programs. Instead support must be continuously and repeatedly generated from widely varying groups, each of which may perceive the program's expected benefits in quite different ways and many of whose interests may diverge rather sharply when the pressure is on.

> *Example*: The nation's space program is confronted with extraordinary tensions, none of which are resolvable by any single authority, agency, branch, or individual. There are tensions between civilian and military, between science and application, between manned and unmanned flight, between complete openness and the tightest security, between the military services, between NASA centers and between the commercial and government sectors, to name a few. Typical of the contested issues are launch vehicle development, acquisition and use; allocations of work to different sections of the country and the rest of the world; and of the future direction of every program. No one, anywhere, has sufficient authority to resolve any of these tensions and issues, much less to resolve them all simultaneously.

This broad dispersion of power repeatedly confuses anyone expecting that somebody will really be in charge there. Rather the opposite is true: anything that happens in Washington is resultant of dozens of political vectors, all pulling in different directions. Everything is the product of maneuver and compromise. When those fail, the result is policy paralysis — and all too possibly, program cancellation by default or failure to act.

There are no clear-cut chains of command in the government. It is nothing like the military or even like a corporation. The process gets even more complicated because *power does not stay put* in Washington. Power relationships are constantly changing, sometimes quietly and gradually, at other times suddenly, under the impact of a major election or a domestic or international crisis. These shifts can alter the policy agenda — and therefore funding priorities — abruptly and with little advance warning. A prime example is the ever-changing contest over future defense spending levels in the wake of the welcomed end of the Cold War.

The entire process is far better understood in dynamic than static terms. There is a continuous ebb and flow of power and influence between the Congress and the White House, among and within the rival agencies, and among ambitious individuals. And through it all, everyone is playing to the media, particularly to television, in efforts to change public perceptions, value judgments, and support.

The First Skill to Master

To deal effectively with this process, *the first skill to master is the ability to think in political terms.* And that requires understanding that the political process functions in terms of an entirely different logic system than the one in which scientists, engineers, and military officers are trained. Washington functions in terms of the logic of politics. It is a system every bit as rigorous in its way as any other, but its premises and rules are profoundly different. *It will therefore repeatedly arrive at conclusions quite different from those of engineering logic, based on the same data.*

Scientists and engineers are trained to marshal their facts and proceed from them to proof: for them, proof is a matter of firm assumptions, accurate data, and logical deduction. Political thinking is structured entirely differently. It depends not on logical proof but on past experiences, negotiation, compromise, and perceptions. Proof is a matter of "having the votes." If a majority of votes can be mustered in Congress to pass a program budget, then — by definition — the program has been judged to be worthy, useful, and beneficial to the nation. If the program cannot, then no matter what its technological merits, the program will lose out to other programs that can.

Mustering the votes depends only in part on engineering or technological merit. These are always important — but getting the votes frequently depends as much or even more on a quite different value judgment, the program's benefits in terms of jobs and revenues among the Congressional districts.

> *Example*: After the Lockheed Corporation won NASA's Advanced Solid Rocket Motor (ASRM) program, the program found strong support in the Congress because Lockheed located its plant in the Mississippi district of the then Chairman of the House Appropriations Committee. Lockheed's factory was only partially built when the chairman suffered a crippling stroke and was forced to retire from his Congressional duties. Shortly thereafter, the Congress, no longer obliged to the chairman, reevaluated and then cancelled the program.

In addition to the highest engineering skills, therefore, the successful architect-engineer must have at least a basic understanding of this political process. The alternative is to be repeatedly blindsided by political events — and worse yet, not even to comprehend why.

Heuristics in the Political Process: "The Facts of Life"

Following are some basic concepts for navigating these rocky rapids — "The Facts of Life." They are often unpleasant for the dedicated engineer, but they are perilous to ignore. Understanding them, on the other hand, will go far to help anticipate problems and cope more effectively with them. They are as follows and will be discussed in turn:

- Politics, not technology, sets the limits of what technology is allowed to achieve.
- Cost rules.
- A strong, coherent constituency is essential.
- Technical problems become political problems.
- The best engineering solutions are not necessarily the best political solutions.

> FACT OF LIFE # 1: *Politics, not technology, sets the limits of what technology is allowed to achieve.*

If funding is unavailable for it, any program will die, and getting the funding — not to mention keeping it over time — is a political undertaking. Furthermore, funding — or rather, the lack of it — sets limits that are considerably narrower than what our technological and engineering capabilities could accomplish in a world without budgetary constraints. *Our technological reach increasingly exceeds our budgetary grasp.* This can be intensely frustrating to the creative engineer working on a good and promising program.

> *Example*: The space station program can trace its origins to the mid-1950s. By the early 1960s it was a preferred way station for traveling to and from the moon. But when, for reasons of launch vehicle size and schedule, the Apollo program chose a flight profile that bypassed any space station and elected instead a direct flight to lunar orbit, the space station concept went into limbo until the Apollo had successfully accomplished its mission. The question then was, what next in manned spaceflight? A favored concept was a manned space station as a waypoint

to the moon and planets, built and supported by a
shuttle vehicle to and from orbit. Technologically,
the concept was feasible; some argued that it was
easier than the lunar mission. Congress balked.
The President was otherwise occupied. Finally,
in 1972, the Shuttle was born as an overpromised,
underbudgeted fleet, without a space station to
serve. Architecturally speaking, major commitments and decisions were made before feasibility
and desirability had been brought together in a
consistent whole.

FACT OF LIFE #2: *Cost rules.*

High technology gets more expensive by the year. As a result, the only
pockets deep enough to afford it are increasingly those of the government.*
The fundamental equation to remember is *Money = Politics*. Reviews and
hearings will spend much time on presumably technical issues, but the
fundamental and absolutely determining consideration is always affordability — and *affordability is decided by whichever side has the most votes.*

Funding won in one year, moreover, does not stay won. Instead it
must be fought for afresh every year. With exceedingly few exceptions,
no program in the entire federal budget is funded for more than one year
at a time. Every year is therefore a new struggle to head off attackers who
want the program's money spent somewhere else, to rally constituents, to
persuade the waverers, and, if possible, to add new supporters.

This is an intense, continuous, and demanding process requiring huge
amounts of time and energy. And after one year's budget is finally passed,
the process starts all over again. There is always next year. Keeping a
program "sold," in short, is a continuous political exercise, and like the
heroine in the old movie serial, "The Perils of Pauline," some programs at
the ragged edge will have to be rescued from sudden death on a regular
basis. Rescue, if and when, may be only partial — not every feature can or
will be sustained. If one of the lost features is a *system* function, the end
may be near.

---

* Although the economic expansion through the end of the 1990s sets an interesting counter-
point. The government is less and less able to influence technology in certain areas, for
example, computing, simply because the commercial market has become so large relative
to the federal market. Similarly, some of the most ambitious space and launch ventures
in the 1990s were privately funded. Although many of those did not come to fruition,
the continuing development of some such efforts is testimony to the role of return-on-
investment thinking over cash-flow thinking.

Example: After the Shuttle had become operational, the question again was, what next in manned spaceflight? Although a modestly capable space station had been successfully launched by a Saturn launch vehicle, the space station program had otherwise been shelved once the Shuttle began its resource-consuming development. With developmental skill again available, the space station concept was again brought forward. However, order-of-magnitude life-cycle cost estimates of the proposed program placed the cost at approximately that of the Apollo, which in 1990-decade dollars would have been about $100 billion — clearly too much for the size of constituency it could command. The result has been an almost interminable series of designs and redesigns, all unaffordable as judged by Congressional votes. Even more serious, the cost requirement has resulted in a spiraling loss of system functions, users, and supporters. Microgravity experiments, drug testing, on-board repair of on-orbit satellites, zero-g manufacturing, optical telescopes, animal experiments, military research and development — one after another had to be reduced to the point of lack of interest by potential users. A clearly implied initial purpose of the space station, to build one because the Soviet Union had one, was finally put to rest with the U.S. government's decision to bring Russia into a joint program with the United States, Japan, Canada, and the European Space Agency. One apparent certainty: the U.S. Congress made the value judgment that a yearly cap of $2.1 billion is all that a space station program is worth. The design must comply or risk cancellation. Cost rules.

Example: Now that the decision to terminate the Shuttle has been made, a new program (Constellation) is required to maintain U.S. human access to space. All agree that the Shuttle should not be terminated until the new system is flying to avoid a break in U.S. human access to space. However, that can be accomplished only by an expensive overlap of the program of several years that requires a large (several billion U.S. dollars) increase in the NASA top-line budget.

NASA has instead structured the programs with a multiyear gap in space access, but avoids the funding hiccup (using savings from Shuttle termination to ramp up production funding for Constellation). Although there is widespread unhappiness, in Congress and NASA, about the resulting gap, there is no willingness to raise budgets enough to close it. Again, cost rules.

FACT OF LIFE #3: *A strong, coherent constituency is essential.*

No program ever gets funded solely — or even primarily — on the basis of its technological merit or its engineering elegance. By and large, the Congress is not concerned with its technological or engineering content (unless, of course, those run into problems — see Fact of Life #4). Instead, program funding depends directly on the strength and staying power of its supporters — that is, its *constituency.*

Constituents support programs for any number of reasons, from the concrete to the idealistic. At times, the reasons given by different supporters will even seem contradictory. From the direct experience of one of the authors, some advocates may support defense research programs because they are building capability; others because research in promising better systems in the future permits reduction if not cancellation of present production programs.

Example: The astonishing success of the V-22 tilt-rotor Osprey aircraft program in surviving 4 years of hostility during the 1988–1992 period, and several fatal accidents, is directly attributable to the strength of its constituency, one that embraced not merely its original Marine Corps constituency but other Armed Services as well — plus groups that see it as benefiting the environment (by diminishing airport congestion), as improving the balance of trade (by tapping a large export market), and as maintaining U.S. technological leadership in the aerospace arena.

Assembling the right constituency can be a delicate challenge because a constituency broad enough to win the necessary votes in Congress can also easily fall prey to internal divisions and conflicts. Such was the case for the Shuttle and is the case for the Space Station. The scientific community proved to be a poor constituency for major programs; the more fields that were brought in, the less the ability to agree on mission priorities. On the other hand, a tight homogeneous constituency is probably too small to win the necessary votes. The superconducting supercollider proved to be such. The art of politics is to knit these diverse motivations together firmly enough to survive successive budget battles and keep the selected program funded. Generally speaking, satellites for national security purposes have succeeded in this political art. It can require the patience of a saint coupled with the wiliness of a Metternich, but such are the survival skills of politics.

> FACT OF LIFE #4: *Technical problems become political problems.*

In a high-budget, high-technology, high-visibility program, *there is no such thing as a purely technical problem.* Program opponents will be constantly on the lookout for ammunition with which to attack the program, and technical problems are tailor-made to that end.

And the problems will normally be reported in a timely fashion. As many programs have learned, *mistakes are understandable; failing to report them is inexcusable.* In any case, reviews are mandated by the Congress as a natural part of the program's funding legislation. Any program that is stretching the technological envelope will inevitably encounter technical difficulties at one stage or another. The political result is that "technical" reports inevitably become political documents as opponents berate and advocates defend the program for its real or perceived shortcomings.

Judicious damage prevention and control, therefore, are constantly required. Reports from prestigious scientific groups such as the Nuclear Regulatory Commission (NRC) or Defense Science Board (DSB) will routinely precipitate Congressional hearings in which hostile and friendly Congressmen will pit their respective expert witnesses against one another and the program's fate may then depend not only on the expertise, but on the political agility and articulateness of the supporting witnesses. Furthermore, although such hearings will spend much time on ostensibly technical issues, the fundamental and absolutely determining consideration is always affordability — and affordability is decided by whichever side has the most votes.

> *Examples*: Decades-long developments are particularly prone to have their technical problems become political. Large investments have to be made every year before any useful systems appear. The widely reported technical difficulties of the B-1 and B-2 bombers, the C-17 cargo carrier, the Hubble telescope, and the Galileo Jupiter spacecraft became matters of public as well as legislative concern. The futures of

long-distance air cargo transport, of space exploration, and even of NASA are all brought up for
debate and reconsideration every year. Architects,
engineers, and program managers have good reason to be concerned.

FACT OF LIFE #5: *The best engineering solutions are
not necessarily the best political solutions.*

Remember that we are dealing with two radically different logic
systems here. *The requirements of political logic repeatedly run counter of those
of engineering logic.* Take construction schedules: in engineering terms,
an optimum construction schedule is one that makes the best and most
economical use of resources and time and yields the lowest unit cost.
In political terms, the optimum construction schedule is the one that the
political process decides is affordable in the current fiscal year. These two
definitions routinely collide; the political definition always wins.

Example: NASA and other agencies often refer to
what is called the program cost curve. It plots total
cost of development and manufacture as a function
of its duration (Figure 13.1).

The foregoing example leads to another provisional heuristic:

*With few exceptions, schedule delays and life-cycle cost
increases are accepted grudgingly; annual cost overruns
are not, and for good reason.*

The reason is basic. A cost overrun — that is, an increase over budget
in a given year — will force the client to take the excess from some other
program, and that is not only difficult to do, it is hard to explain to the
blameless loser and to that program's supporters. Schedule delays mean
postponing benefits at some future cost — neither of which affect anyone
today. At the worst end of this heuristic, it leads to managers "kicking
problems down the road" knowing they will have moved on before the
problem comes due.

Example: Shuttle cost overruns cost the unmanned
space program and its scientific constituency two
decades of unpostponable opportunities, timely
mission analyses, and individual careers based on
presidentially supported, wide-consensus planning.

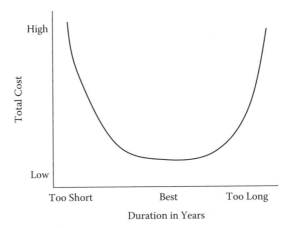

Figure 13.1 The curve is logical and almost always true. But, it is irrelevant because the government functions on a cash-flow basis. Long-term savings will almost always be foregone in favor of minimizing immediate outlays. Overall life-cycle economies of scale will repeatedly be sacrificed in favor of slower appropriations, even if they cause higher unit costs and greater overall program expense. There is also the contradictory perception that if a given program is held to a tight, short schedule it will cost less, facts notwithstanding. (See Chapter 5, "Social Systems," Facts versus Perceptions: An Added Tension.)

By the same token, a well-run program that sticks to budget can encounter very difficult technical problems and survive.

> *Examples*: Communication and surveillance satellite programs.

As an aside, this heuristic leads us directly back to observations on program and organization architecture and the role of feedback loops at the enterprise level. Consider what happens when an organization caps the duration of programs within its sphere of responsibility to, for example, 5 years. If they do so honestly they will, of course, push some long-term, ambitious endeavors out of feasibility. But, they will have created a situation where program managers and architects can be expected to be assigned to the same program for its full lifetime. They can expect to be personally accountable for the end-point consequences of their start-point actions. Moreover, when selecting program managers and architects for the next program, executives can use the actual results of past programs in evaluations, not the results reported before the responsible parties have moved onto other projects. In the language of economists, the "moral hazard" issue of people undertaking risks that others will have to suffer from would be largely eliminated. The feedback loop at the enterprise level now synchronizes programmatic and human responses.

The political process can be bewildering and intimidating to the uninitiated. But it need not be so. Because in addition to being confusing and chaotic, this is a profoundly interesting and engrossing process, every bit as challenging as the knottiest engineering problem. Indeed it *is* an engineering challenge because it molds the context in which systems architecting and engineering must function.*

The reader may well find the craziness of the political process distasteful — *but it will not go away.* The politically naive architect may experience more than a fair share of disillusion, bitterness, and failure. The politically astute program manager, on the other hand, will understand the political process, will have a strategy to accommodate it, and will counsel the architect accordingly. Some suggestions for the architect: It helps to document accurately when and why less-than-ideal technical decisions were made — and how to mitigate them later, if necessary. It helps to budget for contingencies and reasonably foreseeable risks. It helps to have stable and operationally useful interim configurations and fallback positions. It helps to acknowledge the client's right to have a change of mind or to make difficult choices without complaint from the supplier. Above all, it helps to acknowledge that living in the client's world can be painful, too. And finally, select a kit of prescriptions for the pain such as the following from Appendix A:

- The Triage, when there is only so much that can be done: Let the dying die. Ignore those who will recover on their own. And treat only those who would die without help.
- The most important single element of success is to listen closely to what the customer [in this case, the Congress] perceives as his requirements and to have the will and ability to be responsive. (J. E. Steiner, The Boeing Company, 1978)
- Look out for hidden agendas.

That does not mean that architects and engineers have to become expert lobbyists — but it does mean having an understanding of the political context within which programs must function, the budget battle's rules of engagement, and of those factors that are conducive to success or failure. The political process is not outside, it is *an essential element of,* the process of creating and building systems.

---

* This is one area where commercial and government politics are often dramatically different. Commercial enterprises tend to be return-on-investment driven rather than cash-flow driven. However, firms that are driven by quarterly results will tend to resemble the cash-flow-driven government budgeting process.

A Few More Skills to Master

Following are a few more basic coping skills for the successful systems architect. Foremost, *understand that the Congress and the political process are the owners of your project. They are the ultimate clients.* It is absolutely essential to deal with them accordingly by making sure they understand what you are trying to do, why it is important, and why it makes political sense for them to support you.

Be informed. This is your life, so be active. Learn the political process for yourself and keep track of what is going on. Figure out what information the political system needs in order to understand what the program needs — and arrange to supply it to them. A chief engineer has utterly different information requirements from a Congressional oversight committee. Learn what sort of information furthers your program's fortunes in Washington and then get it to your program managers so they can get it to the political decision makers who determine your program's funding. Maybe your program has a great job-multiplier effect in some crucial lawmaker's district. Maybe its technology has some great potential commercial applications in areas where the United States is losing a competitive battle with another country.

The point is that the political process bases its decisions on very different information than does the engineering process. Learn to satisfy both those sets of requirements by plan.

Conclusion

The political process is a necessary element of the process of creating and building systems. It is not incomprehensible; it is different. Only when they are not understood do the political Facts of Life instill cynicism or a sense of powerlessness. Once understood, they become tools like any others in the hands of an astute architect. It is a compliment to the client to use them well.

chapter 14

The Professionalization of Systems Architecting

Elliott Axelband

> Profession: Any occupation or vocation requiring training in the liberal arts or the sciences and advanced study in a specialized field.[1]

Introduction

To readers who have progressed this far, the existence of systems architecting as a *process,* regardless of who performs it, can be taken for granted. Functions and forms have to be matched, system integrity has to be maintained throughout development, and systems have to be certified for use. Visions have to be created, realized, and demonstrated.

This chapter, in contrast, covers the evolution of the systems architecting *profession.* An appropriate place to begin is with the history of the closely related profession of systems engineering, the field from which systems architecting evolved.

The Profession of Systems Engineering

Systems engineering as a process began in the early 1900s in the communication and aircraft industries. It evolved rapidly during and after World War II as an important contributor to the development of large, innovative, and increasingly complex systems. By the early 1950s, systems engineering had reached the status of a recognized, valued profession. Communication networks, ballistic missiles, radars, computers, and satellites were all recognized as systems. The "systems approach" entered into everyday language in many fields, social as well as technical. Government regulations and standards explicitly addressed systems issues and techniques. Thousands of engineers called systems engineering their vocation. Professional societies formed sections with journals devoted to systems and their development.[2] Universities established systems engineering departments or systems-oriented programs.* Books addressing the process, or aspects of it, started to appear.[3] Most recently, the profession became formally represented with the establishment of the International Council on Systems Engineering (INCOSE).[4]

The core of the systems approach from its beginnings has been the definition and management of system interfaces and trade-offs, of which there can be hundreds in any one system. Systems analysis, systems integration, systems test, and computer-aided system design were progressively developed as powerful and successful problem-solving techniques. Some have become self-standing professions of their own under the rubric of systems engineering. Their academic, industrial, and governmental credentials are now well established.

All are science based — that is, based on measurables and a set of assumptions. In brief, these are that requirements and risks can be quantified, solutions can be optimized, and compliance specified. But these same assumptions are also constraints on the kinds of problems that can be solved. In particular, science-based systems engineering does not do well in problems that are abstract, data deficient, perceptual, or for which the criteria are immeasurable.

For example, the meanings of such words as safe, survivable, affordable, reliable, acceptable, good, and bad, are either outside the scope of systems engineering — "ask the client for some numbers" — or are force-fitted to it by subjective estimates. Yet these words are the language of the clients. Quantifying them can distort their inherent vagueness into an unintended constraint.

There is no group of professionals that better understands these difficulties than systems engineers and executives — nor who wish more to convert immeasurable factors to quantitatively statable problems by whatever techniques can help. The first step they made was to recognize the nature of the problems. The second was to realize that almost all of them occur at the front (and back) ends of the engineering cycle. Consider the following descriptive heuristics, developed long ago from systems engineering experience:

- All the serious mistakes are made in the first day.
- Regardless of what has gone before, the acceptance criteria determine what is actually built.
- Reliability has to be designed in, not tested in.

* Among the best known are the University of Arizona at Tucson, Boston University, Carnegie Tech, Case Western Reserve, the University of Florida, Georgia Tech, the University of Maryland, George Mason University, Ohio State, MIT (Aerospace), New York Polytech, the University of Tel Aviv, the University of Southern California, Virginia Polytechnic Institute, and the University of Washington.

It is no coincidence that many systems engineers, logically enough, now consider systems architecting to be "the front end of systems engineering" and that architectures are "established structures." More precisely, systems architecting can be seen as setting up the necessary conditions for systems engineering and certifying its results. In short, systems architecting provides concepts for analysis and criteria for success. In evolving systems, the functions of systems architecting, systems engineering, and disciplinary engineering are all more episodic. Concepts for analysis and criteria for success are established in early phases, but are revised with each new spiral through the development process. Systems engineers must control interfaces through many cycles of design, development, and integration, not just through one. In addition to conducting classical architecting episodically, the systems architect must also consider the issue of stable forms. The evolving system should not change everything on each cycle; it needs to retain stable substructures to evolve effectively. The definition of these substructures is part of the architect's role.

The immediate incentive for making architecting an explicit process, the necessary precursor to establishing it as a self-standing profession complementary to systems engineering, was the recognition in the late 1980s by systems executives that "something was missing" in systems development and acquisition. And the omission was causing serious trouble: system rejection by users, loss of competitive bids to innovators, products stuck in unprofitable niches, military failures in joint operations, system overruns in cost and schedule, and so on — all traceable to their beginnings. Yet there was a strong clue to the answer. Retrospective examinations of earlier, highly successful systems showed the existence in each case of an overarching vision, created and maintained by a very small group, that characterized the program from start to finish.[5]

Software engineers and their clients were among the first to recognize that the source of many of their software system problems was structural — that is, architectural.* Research in software architecture followed in such universities as Carnegie Mellon, the University of North Carolina Chapel Hill, the University of California at Irvine, and the University of Southern California. Practitioners began identifying themselves as software architects and forming architectural teams. Communication, electronics, and aerospace systems architects followed shortly thereafter.

---

* One of the earliest and most famous books on systems architecting is *The Mythical Man-Month, Essays on Software Engineering* by Frederick P. Brooks, Jr. (1974, Addison-Wesley, Reading, MA), which not only recognized the structural problems in software but explicitly, on p. 37, calls for a systems architect, a select team, conceptual integrity, and for the architect to "confine himself scrupulously to architecture" and to stay clear of implementation. Brook's analogy for the architectural team was a surgical team. He credits a 1971 Harlan Mills proposal as the source of these precepts.

Societies established architecture working groups, notably the INCOSE Systems Architecting Working Group[6] and the IEEE Software Engineering Standards Committee's Architecture Working Group,[7] to formulate standard definitions of terms and descriptions for systems and software architectures. These activities are essential to the development both of a common internal language for systems architecting and for the integration of software architecture models and overall systems architectures in complex, software-intensive systems.

At the scale of the profession of engineering, the recognition that something was missing led to identifying it, by direct analogy with the processes of the classical architectural profession, as systems architecting.[8] Not surprisingly, the evolution of systems architecting tools was found to be already underway in model building, discussed in Part III, and systems standards,* discussed in the next section.

Systems Architecting and Systems Standards

Earlier chapters have pointed out that the abstract problems of the conceptual and certification phases require different tools from the analytic ones of system development, production, and test. One of the most important sets of tools is that of systems standards. Chapter 11 discussed one type of architecture standard, standards for architecture description. Here we discuss a different category of standards, those that define development processes. For historical reasons, *architectural* process standards were not developed as a separate set. Instead, general systems process standards were developed that included systems architecting elements and principles understood at the time, most of them induced from lessons learned in individual programs. As will be seen, some key elements appeared as early as in the 1950s.

Driven by much the same needs, the recognition of systems architecting in the late 1980s was paralleled, independently, by a recognition that existing systems standards needed to be modified or supplemented to respond to long-standing systems-level structuring problems. Bringing the two tracks, architecting and standards, together should soon help both. Architecting can improve systems standards. Systems standards can provide valuable tools for the systems architecting profession.

Some of the earliest systems standards in which elements of systems architecting appeared were those of system specification, interface description, and interface management. They proliferated rapidly. A system specification can beget 10 subsystem specifications, each of which is supported

---

* Systems standards, for the purposes of this book, are those engineering standards having impact on the system as a whole, whether explicitly identified as such in their titles or not. They are a relatively small part of the totality of engineering standards. Many, if not most, are interface and test standards.

by 10 lower-level subsystem specifications, and so on. All of these had to be knitted together by a system of 100 or so interface specifications.

Even though modern computer tools (computer-assisted system engineering [CASE] tools) have been developed to help keep track of the systems engineering process, extraordinarily disciplined efforts are still required to maintain the systems integrity.[9]

As systems complexity increased, systems engineers were faced with increasingly difficult tasks of assuring that the evolving form of the system met client needs, guaranteeing that trade-offs maintained system intent in the face of complications arising during development, and finally assuring that the system was properly tested and certified for use. In due course, the proliferation of detailed specifications led to a need for overarching guidelines, an overview mechanism for "structuring" the complexity that had begun to obscure system intent and integrity.

Before continuing, it should be pointed out that overarching guidelines are not, and cannot be, a *replacement* for quantitative system standards and specifications. The latter represent decades of corporate memory, measurable acceptance criteria, and certified practices. Guidelines — performance specifications, tailorable limits, heuristics, and the like — have a fundamental limitation. They cannot be certified by measurables. They are too "soft" and too prone to subjective perceptions to determine to the nearest few percentage points whether a system performs, or costs, what it should. At some point, the system has to be measured if it is to be judged objectively.

From the standpoint of an architecting profession, the most important fact about system standards is that they are changing. To understand the trend, their development will be reviewed in some detail, recognizing that some of them are continuing to be updated and revised.

The Origins of Systems Standards

The Ballistic Missile Program of the 1950s

Urgent needs induce change, and, eventually, improvement. The U.S./Soviet ballistic missile race begun in the mid-to-late 1950s brought about significant change, as it led to the development and fielding of innovative and complex systems in an environment where national survival was threatened. To its credit, the U.S. Air Force recognized the urgent need to develop and manage the process of complex system evolution, and did so.* The

---

* Those responsible for this development, Simon Ramo and General Bernard Schriever, in particular, from time to time referred to their respective organizations as architects as well as system integrators. "Architecture," as a formalism, was largely bypassed in the urgency to build ballistic missiles as credible deterrents. Nonetheless, the essential "architectural" step of certification of readiness for launch was incorporated from the beginning and executed by all successor organizations. It became a centerpiece for the space launch programs of the 1960s and thereafter.

response in the area of standards was the "375" System Standard, subsequently applied to the development of all new complex Air Force equipment and systems.

"375" required several things that are now commonplace in systems architecting and engineering. Timelines depicting the time-sequenced flow of system operation were to be used as a first step in system analysis.* From these, system functional block diagrams and functional requirements were to be derived as a basis for subsequent functional analysis and decomposition. The functional decomposition process in turn generated the subsystems that with their connections and constraints comprised the system, and allowed the generation of subsystem requirements via trade-off processes.

"375" was displaced in 1969 by a MILSTD 499 (Military Standard — 499),† which was applied throughout the Department of Defense. MILSTD 499A, an upgrade, was released in 1974 and was in effect for 20 years. MILSTD 499B, a later upgrade, was unofficially released in 1994, and was almost immediately replaced by EIA/IS 632 Interim Systems Engineering Standard.[10]

The Beginning of a New Era of Standards

The era of MILSTDs 499/499A/499B was an era in which military standards became increasingly detailed. It was not only these documents that governed system architecting and engineering, but they in turn referenced numerous other DoD (Department of Defense) MILSTDs that addressed aspects of system engineering, and which were imposed on the military system engineering process as a consequence. To cite a few: MILSTD 490, Specification Practices, 1972; MILSTD 481A, Configuration Control — Engineering Changes, Deviations and Waivers (Short Form), 1972; MILSTD 1519, Test Requirements Document, 1977; and MILSTD 1543, Reliability Program Requirements for Space and Missile Systems, 1977. See Eisner (1994)[11] for additional examples.

This mindset changed with the end of the Cold War in the late 1980s. Cost became an increasingly important decisive factor in competitions for military programs, supplanting performance, which had been the dominant factor in the prior era. Lowest cost, it was argued, could only be achieved if the restrictions of the military standards were muted. The detailed process ("how to") standards of the past, which specified how to conduct systems engineering and other program operations, needed to be

---

* This is not necessarily appropriate for all systems, but it was well suited to the missile, airplane, and weapon systems the Air Force had in mind at the time.
† The official form is "Mil. Std. - 499," but for ease of reading in a text, an alternate form, "MILSTD 499," will be used here.

replaced by standards that only provided guidelines,* leaving the engineering specifics to the proposing companies that would select these so as to be able to offer a low-cost product.[12] Further supporting this reasoning was the reality that the most sophisticated components and systems in fields such as electronic computer chips and computers were now available at low cost from commercial sources, whereas in the past the state of the art was available only from MILSTD-qualified sources. It was in this environment that EIA/IS 632 was born.

EIA/IS 632, an Architectural Perspective

EIA/IS 632 is short by comparison with other military standards. Its main body is 36 pages. Including appendices, its total length is 60 pages, and these include several which have been left intentionally blank. And most significantly, no other standards are referenced.

The scope and intent of the document is best conveyed by the following quotes from its contents:

- "The scope ... of systems engineering (activities) are defined in terms of what should be done, not how to do ... (them)." (p. i)
- "(EIA/IS-632) identifies and defines the systems engineering tasks that are generally applicable throughout the system life cycle for any program." (p. 7)

From a systems architecting perspective, it is clear that the scope of the life-cycle perspective includes the modern understanding of systems architecting.

One of the major activities of the systems architect, that of giving form to function, is addressed in pages 9 through 11. These pages summarize, in their own words and style, the client/architect relationship, the establishment of the defining system functions, the development of the system's architecture, and the process of allocating system functions to architectural elements via trade-offs. By implication, the trade-offs continue, with varying degrees of concentration, throughout the life cycle.

Curiously, test and validation are deferred to a later section entitled "4.0 Detailed Requirements." This is consistent with the historical organization of the preceding military standards, wherein section 4 was dedicated to product assurance, a term that included system test. It is, however, a significant departure from the systems architecting point of view. A basic tenet of systems architecting is that certification for use is one of its most important functions, and that this should be developed

---

* As noted earlier, *replacement* is a questionable motivation for guidelines. Nonetheless, the establishment of a high-level guideline document — a key architecting technique — was a milestone.

in parallel with, and as a part of, the development of a system's architecture. Consider, for example, some of the architecting heuristics that could apply:

- To be tested, a system must be designed to be tested.
- Regardless of what has gone before, the acceptance (and test) criteria determine what is actually built.

There are other sound and basic architecting principles that, suitably explained, could and should have been included as historically validated guiding principles in EIA/IS 632 which, in its own words, "provides guidance for the conduct of a systems engineering effort." Some applicable heuristics would include the following:

- Simplify. Simplify. Simplify.
- The greatest leverage in systems architecting is at the interfaces.
- Except for good and sufficient reasons, functional and physical structuring should match.
- In partitioning a system into subsystems, choose a configuration with minimal communications between subsystems.
- It is easier to match a system to the human one that supports it than the reverse.

Beyond these, the need for an unbiased agent — the systems architect — to represent the client and technically guide the process is absent and a serious omission.

Commercial Standards

Even though EIA/IS 632 applies only to military systems engineering, that was not its original intent. The objective was to develop a universal standard for systems engineering that would apply to both the military and commercial worlds and be ratified by all of industry. However, there was an urgency to publish a new military standard and in the 4-month schedule that was assigned, only it could be developed. This led to two consequences. First, IEEE 1220,[13] a commercial systems engineering standard, was separately published. Second, the merging of EIA/IS 632 with IEEE 1220 to create the first universal standard for system engineering was planned for publication in 1997. The development of this universal systems engineering standard involves personnel from several organizations including ANSI (American National Standards Institute) and EIA.

At the international level, ISO (The International Standards Organization) has issued an overall standard covering the development and engineering of systems (15288). This standard is not a systems engineering standard per se, but it lays out a set of activities, including systems engineering activities, associated with developing a system. 15288 is intended to be a standard at a level above that of the other cited standards.

IEEE 1220, An Architectural Perspective

The IEEE working group that generated 1220 was sponsored by the IEEE Computer Society and included representatives from INCOSE, the EIA, and the IEEE AES Society. It is the first commercial standard to formally address systems engineering. 1220's similarity with EIA/IS 632 derives from a fair degree of common authorship plus a deliberate effort to (1) coordinate efforts in order to present a common view of systems engineering and (2) anticipate the eventual merger of the two documents. The similarities are therefore not surprising, but there are significant differences that are worthy of mention.

To begin with similarities, both standards are guides and not "how to" instruction manuals. Both address the entire life cycle of a product. Both share a common architecture, addressing in similar ways things that are becoming similar — the processes of system engineering in the military and commercial environments. This extends to a fair degree of common vocabulary, although mercifully 1220 is freer of acronyms.

Compared to EIA/IS 632, 1220 is more complex and longer (58 pages in the body of the report versus 36, and 66 pages overall versus 60). It is much more rigorous in its definitions and use of system hierarchical structures. It has several significant differences that tend to favor the recognition and processes of systems architecting.

- The subsystems that comprise a system are understood and treated as systems. (pp. 2, 4, 13)
- The customer (client) is explicitly identified along with a need to determine and quantify his or her expectations. (page 35)
- External constraints including public and political constraints are recognized as part of the process. (page 35)
- The role of system boundaries and constraints in system evolution is considered. (page 36).
- The need to evolve test plans with product evolution is expressed. (pp. 18, 19)
- The explicit need to generate functional and physical architectures is recognized, unfortunately (from a systems architect's view), in the same section of the document which through usage defines systems architecture as the sum of the product and its defining data package. (page A-3)

In summary, 1220 better recognizes the systems architecting process than does EIA/IS 632. It does, however, have significant systems architecting shortfalls and would better serve as a systems engineering guide if the role of the systems architect were included and if the architecting heuristics given in this chapter were added.

A continuing problem in all of these systems standards, highlighted in Chapter 6, is the difference in system/subsystem hierarchies across hardware and software. Both can be thought of hierarchically, but the hierarchical model for software is often changed to become layered, and the hierarchy of software units in a distributed system often does not match the associated hardware. This often leads to significant problems in development and contributes to poorly structured software in systems where software development cost dominates total development cost. Standards for distributed system development, such as Reference Model for Open Distributed Processing (RM-ODP) and Unified Modeling Language (UML), recognize the disjunction and allow the software and hardware elements to be represented in their own hierarchies. This frees the software architects from an imposed, and often damaging, hardware-based hierarchy, but introduces new problems in reconciling the two models to assure consistency. Engineering process standards have only begun to address this issue.

Company Standards

Each company has its own set of standards and practices that incorporate unique core competencies, practices, and policies. These need to evolve for a company to improve its performance and competitive posture. Company standards serve two other functions: instructing its initiates and relating to its customers. The latter function is stimulated whenever customers change their standards, and it is from this perspective that the systems engineering standards of several companies were examined. This was not an easy task because systems architecting and engineering are viewed by those companies engaged in them as an enabler of efficient product generation, and as such, applicable practices providing a competitive advantage are considered trade secrets.

Several generalizations are possible. Today's competitive pressures have caused self-examination and particularly reengineering to become a regular way of life. This has also been encouraged by popularized business literature.[14] Process, as opposed to product, is the focus of such institutionalized activity. In reviewing the process of product generation and support, systems architecting and, in some cases, systems architects are gaining recognition, although not always in a way clearly separated from systems engineering.

The Harris Corporation Information Systems Division culminated 4 years of activity by publishing their revised *Systems Engineering Guide Book*. A generalized description is provided in Honour 1993.[15] The 128-page book is company proprietary. Discussions with its author, Eric Honour, indicated that although systems architecting is not delineated per se, the *processes* that constitute systems architecting account for approximately 25% of its pages.

Sarah Sheard and Elliot Margolis reported on the evolving systems engineering process within The Loral Federal Systems Organization.[16] Their conclusion is that there is an important relationship between the nature of a product and the team developing it, and that as such there is no one best organization for product development. However, their recent experience indicates success with a team structure that includes a distinct architecture team, with a clearly identified chief architect, working in conjunction with a management team and both software and hardware development teams. Their use of the terms architect, architectures, and architecting are consistent with those of this book.

Hughes Aircraft published its 3-inch thick *Systems Engineering Handbook* in 1994.[17] Its objective, stated on page P-1, is to "improve both the quality and efficiency of systems engineering at Hughes." The *Handbook* describes the then-applicable MILSTD 499B and the Hughes systems engineering processes for both DoD and non-DoD programs, including Hughes' organizational and other resources available to implement these. It is a very comprehensive user-friendly book, clearly adapted from MILSTD 499B, and includes the activities of the systems architect — who is never identified by that name — within the framework of systems engineering. The systems engineering function and organization are identified as the technical lead organization for product development and provided a unique identity in all forms of organization discussed: functional, projectized, and matrix. In that the *Handbook* is patterned after MILSTD 499B, which has a strong resemblance to EIA/IS 632, the comments made earlier with respect to EIA/IS 632 apply.

A Summary of Standards Developments, 1950–1995

For a variety of reasons and by a number of routes, system standards and specifications are evolving consistent with the principles and techniques of systems architecting. The next step is the use of systems architects to help improve systems standards, particularly in system conception, test, and certification. At the same time, improved systems standards can provide powerful tools for the systems architecting profession.

A cautionary note: the recent and understandable enthusiasm of the Department of Defense to streamline standards and eliminate all references to prior MILSTDs could make systems architecting considerably more difficult. Useful as guidelines are, they are no substitute for quantitative standards for bidding purposes, for certifying a system for use, or for establishing responsibility and liability.* MILSTDs in many instances incorporate specific philosophical and quantitative requirements based on lessons dearly learned in the real world. They reduce uncertainty in areas that should not or need not be uncertain. To ignore these by omission is to run the risk of learning them all over again, at great cost. To the extent that the lessons relearned are architectural, the risks can be enormous. As the heuristic states, all the serious mistakes are made on the first day.

Systems Architecting Graduate Education

Systems Engineering Universities and Systems Architecting

Graduate education, advanced study, and research give a profession its character. They distinguish it from routine work by making it a vocation, a calling of the particularly qualified.

The first university to offer masters and doctorate degrees in systems engineering was the University of Arizona, beginning in 1961. The program began as a graduate program; an undergraduate program and the addition of Industrial Engineering to the department title came later. Still in existence, the graduate department has well over 1,000 alumni.

The next to offer advanced degrees was the Virginia Institute of Technology in 1971, but not until after 1984 did additional universities join the systems engineering ranks. They included Boston University, George Mason University, the Massachusetts Institute of Technology (MIT), the University of Maryland, the University of Southern California (USC), the University of Tel Aviv, and the University of Washington. It is worth noting that all are located at major centers of industry or government, the principal clients and users of systems engineering.

To the best knowledge of the authors of this book, the University of Southern California was the first to offer a graduate degree in Systems Architecting and Engineering with the focus on systems architecting. However, of the universities offering graduate degrees in systems engineering, some half dozen now include systems achitecting within their curricula. Notable among them is the MIT Systems Design and Management (SDM) program. This program, which is intended as a new

---

* In this connection, the Department of Defense has explicitly retained interface and certification standards as essential, not to be considered as candidates for elimination.

kind of graduate education program for technical professionals, is built on three core subjects: Systems Engineering, Systems Architecture, and Project Management. Although the degree is not focused on systems architecting, that subject forms a major part of the curriculum's core. The MIT SDM curriculum is becoming more of a national model as it is spread through the Product Development in the 21st century (PD21) program. PD21 is creating programs that are similar to MIT's SDM program in universities across the country. The current universities involved are the Rochester Institute of Technology, the University of Detroit–Mercy, and the Naval Postgraduate School.

Architecting is also becoming a strong interest in universities offering advanced degrees in computer science with specializations in software and computer architectures; notably, Carnegie Mellon University, the Universities of California at Berkeley and Irvine, and USC. At USC, the systems architecture and engineering degree began with an experimental course in 1989, and formally became a master's degree program in 1993 following its strong acceptance by students and industry.

In the last 10 years, there has been growing recognition of the value of interdisciplinary programs, which of itself would favor systems architecting and engineering. These have been soul-searching years for industry, and the value of systems architecting and engineering has become appreciated as a factor in achieving a competitive advantage. Also, the restructuring of industry has caused a rethinking of the university as a place to provide industry-specific education. These trends, augmented by the success of the systems architecting and engineering education programs, have caused university architect-engineering programs to prosper.

The success of these programs can be measured in several ways. First, the direction is one of growth. Seven out of the eight existing masters programs were started in the last 20 years. And the Universities of Maryland, Tel Aviv, and Southern California are all considering expanding their programs to include a Ph.D. MIT has formed a cross-cutting Engineering Systems Division (ESD) that draws from all of the traditional departments, and offers a Ph.D. in Engineering Systems. Second, systems architecting and systems architecting education are making a positive difference in industry, as supported by industry surveys. In point of fact, company-sponsored systems architecting enrollments have increased even during the part of this period where there was industrial contraction.

Curriculum Design

It is not enough in establishing a profession to show that universities are interested in the subject. The practical question is what is actually taught; that is, the curriculum. Because USC was apparently the first university to offer an advanced degree specifically in systems architecture and engineering, its curriculum is described. It should be pointed out that this curriculum is at the graduate level. To date, no undergraduate degree is offered or planned.

The USC master's program admits students satisfying the School of Engineering's academic requirements and having a minimum of 3 years applicable industrial experience. Students propose a 10-course curriculum that is reviewed, modified if required, and accepted as part of their admission. The curriculum requires graduate-level courses as follows:

- An anchor systems architecting course.
- An advanced engineering economics course.
- One of several specified engineering design courses.
- Two elective courses in technical management from a list of eleven that are offered.
- One of eight general technical area elective courses.
- Four courses from one of eleven identified technical specialization areas, each of which has six or more courses offered.

The structure of this M.S. in Systems Architecture and Engineering curriculum has been designed based on both industrial and academic advice. Systems architecture is better taught in context. It is too much to generally expect a student to appreciate the subtleties of the subject without some experience. And the material is best understood through a familiar specialty area in which the student already practices. The 3-year minimum experience requirement and the requirement of four courses in a technical specialty area derive from this reasoning.

The need for an anchor course is self-evident. Systems architecture derives from inductive and heuristic reasoning, unlike the deductive reasoning used in most other engineering courses. To fully appreciate this difference, the anchor course is taken early, if not first, in the sequence. The course contains no exams as such, but requires two professional-quality reports so that the student can best experience the challenges of systems architecting and architecture by applying his or her knowledge in a dedicated and concentrated way.

Experience has shown that a design experience course, the advanced economics course, and courses in technical management are valuable to the systems architect, and therefore they are curriculum requirements. The additional course in a general technical area allows the student to select a course that most rounds out the student's academic experience. Possibilities include a systems architecting seminar, a course on decision support systems, and a course on the political process in systems architecture design.

Advanced Study in Systems Architecting

A major component of advanced study in any profession is graduate-level research and refereed publications at major universities. In systems architecting, advanced study can be divided into two relatively distinct parts: that of its science, closely related to that of systems engineering, and of its art. The universities committed to systems engineering education were given earlier. Advanced study in its art, though often illustrated by engineering examples, has many facets, including research in the following:

- *Complexity*, by Flood and Carson[18] at City University London, England
- *Problem solving*, by Klir[19] at the State University of New York at Binghamton and by Rubinstein[20] at the University of California at Los Angeles
- *Systems and their modeling*, by Churchman[21] at the University of California at Berkeley (UCB) and Nadler[22] and Rechtin[23] at the University of Southern California (USC)
- *The behavioral theory of architecting*, by Lang[24] at the University of Pennsylvania, Rowe[25] at Harvard, and Losk, Pieronek, Cureton, Geis, and Carpenter[26] at USC
- *The practice of architecture*, by Alexander[27] and Kostof[28] at UCB
- *Machine (artificial) intelligence and computer science*, by Genesereth and Nilsson[29] at Stanford, Newell[30] and Simon[31] at Carnegie Mellon University and Brooks[32] at the University of North Carolina at Chapel Hill
- *Software architecting*, by Garlan and Shaw[33] at Carnegie Mellon University and Barry Boehm at USC

All have contributed basic architectural ideas to the field. Many are standard references for an increasing number of professional articles by a growing number of authors. Most deal explicitly with systems, architectures, and architects, although the practical art of systems architecting was seldom the primary motivation for the work. That situation predictably will change rapidly as both industry and government face international competition in a new era.

Professional Societies and Publications

Existing journals and societies were the initial professional media for the new fields of systems architecting and engineering. Because much early work was done in aerospace and defense, it is understandable that the IEEE Society on Systems Man and Cybernetics, the IEEE Aerospace Electronics Society, and the American Institute of Aeronautics and Astronautics, and their journals, along with others, became the professional outlets for these fields. One excellent sample paper from this period (Booton and Ramo 1984[34]) explained the contributions that systems engineering had made to the U.S. ballistic missile program.

The situation changed in 1990 when the first International Council on Systems Engineering conference was held and attracted 100 engineers. INCOSE became the first professional society dedicated to systems engineering and soon established a Systems Architecture Working Group.

The society, with a current membership of 3,500 (one-third of which are outside the United States), publishes a quarterly newsletter and a journal. The journal first appeared in 1994, and it published jointly with the IEEE AES Society in 1996. Since then, it has become a stand-alone, quarterly publication.

Conclusion: An Assessment of the Profession

The profession of systems architecting has come a long way — and its journey has just begun. Its present body of professionals in industry and academia, beginning most often in electronics, control, and software systems, soon broadened into systems engineering, formed the core of small design teams, and now consider themselves as architects. The profession has been nurtured within the framework of systems engineering, and no doubt will maintain a tight relationship with it. A masters-level university curricula now exists, and the material and ideas are suffusing into many other systems-oriented programs. Applicable research is underway in universities. Applicable standards and tools are being developed at the national level. It has an acknowledged home within INCOSE as well as other professional societies that, together with their publications, provide a medium for professional expression and development.

It is interesting to speculate on where the profession might be going and how it might get there. The cornerstone thought is that the future of a profession of systems architecting will be largely determined by the perceptions of its utility by its clients. If a profession is useful, it will be sponsored by them and prosper. To date, all indicators are positive.

Judging by the events that have led to its status today, and by comparable developments in the history of classical architecture, systems architecting could well evolve as a separate business entity. The future could hold more systems architecting firms that bid for the business of acting as the technical representative or agent of clients with their builders. There are related precedents today in Federally Funded Research and Development Centers (FFRDCs) and Systems Engineering and Test Assistance Contractors (SETACs), independent entities selected by the Department of Defense to represent it with defense contractors that build end products. Similar precedents exist in NASA and the Department of Energy.

The role of graduate education is likely to grow and spread. Today's products and processes are more netted and interrelated than those of 10 years ago, and tomorrow's will be even more so. System thinking is proving to be fundamental to commercial success, and systems architecting will increasingly become a crucial part of new product development. It is incumbent upon universities to capture the intellectual content of this phenomenon and embody it in their curricula. This will require a tight coupling with industry to be aware of important real-world problems, a dedication to research to provide some of the solutions, and an education program that trains students in relevant architectural thinking.

Published peer-reviewed research has stood the test of time, providing the best medium for the rapid dissemination of state-of-the-art thinking. Today INCOSE's Systems Architecture Working Group provides one such outlet. Still others will be needed for further growth.

In summary, all the indicators point to a future of high promise and value to all stakeholders.

Notes and References

1. *Webster's II, New Riverside University Dictionary*, Boston, MA: Riverside, 1984, p. 939.
2. *Transactions on Systems, Man and Cybernetics,* Institute of Electrical and Electronic Engineers (IEEE), New York; *IEEE Transactions on Aerospace and Electronic Systems,* Institute of Electrical and Electronic Engineers (IEEE), New York; *Journal of the American Institute of Aeronautics and Astronautics,* American Institute of Aeronautics and Astronautics, Washington, DC.
3. Machol, R. E., *Systems Engineering Handbook.* New York: McGraw-Hill, 1965; Chestnut, Harold, *Systems Engineering Methods.* New York: John Wiley, 1967; Blanchard, Benjamin S., and W. J. Fabrycky, *Systems Engineering and Analysis.* Upper Saddle River, NJ: Prentice Hall, 1981.
4. *Proceedings of the First Annual Symposium of the National Council on Systems Engineering,* National Council on Systems Engineering, Seattle, WA, 1990; *The Journal of the National Council on Systems Engineering,* Inaugural Issue, National Council on Systems Engineering, Seattle, WA, 1994; *Tools for System Engineering,* A Brochure. Ascent Logic Corporation, San Jose, CA, 1995.
5. Rechtin, E., *Systems Architecting, Creating and Building Complex Systems.* Englewood Cliffs, NJ: Prentice Hall, 1991, pp. 299–301. Note that throughout the rest of the chapter, this reference will be referred to as Rechtin 1991.
6. See the Inaugural Issue of *Systems Engineering, the Journal of the National Council on Systems Engineering,* Vol. 1, Number 1, July/September 1994.
7. See Toward a Recommended Practice for Architectural Description of the IEEE SESC Architecture Planning Group, April 9, 1996.
8. Rechtin 1991.
9. *Doors.* A Brochure. Zycad Corporation, Freemont, CA, 1995; *SEDA — System Engineering and Design Automation.* A Brochure. Nu Thena Systems, Inc., McLean, VA, 1995.

10. *EIA Standard IS-632.* Electronic Industries Association Publication, Washington, DC, 1994.
11. Eisner, Howard, *Computer Aided Systems Engineering.* Upper Saddle River, NJ: Prentice Hall, 1988.
12. Perry, William J., *Specifications and Standards — A New Way of Doing Business.* DODI 5000.2, Part 6, Section I. Department of Defense. Washington, DC, 1995.
13. *IEEE P1220 Trail Use Standard for Application and Management of the Systems Engineering Process.* IEEE Standards Department, New York, 1994.
14. Hammer, Michael, and James Champy, *Reengineering the Corporation.* New York: Harper, 1994; Hamel, Gary, and C. K. Prahalad, *Competing for the Future.* Boston, MA: Harvard Business School Press, 1994.
15. Honour, Eric C., TQM Development of a Systems Engineering Process. *Proceedings of the Third Annual Symposium of the National Council on Systems Engineering.* Sunnyvale, CA: National Council on Systems Engineering, 1993.
16. Sheard, Sarah A., and Elliot M. Margolis, Team Structures for Systems Engineering in an IPT Environment. *Proceedings of the Fifth Annual Symposium of the National Council on Systems Engineering.* Sunnyvale, CA: National Council on Systems Engineering, 1995.
17. *Systems Engineering Handbook.* Culver City, CA: Hughes Aircraft Company, 1994.
18. Flood, Robert L., and Ewart R. Carson, *Dealing with Complexity, An Introduction to the Theory and Application of Systems Science.* New York: Plenum Press, 1988.
19. Klir, George J., *Architecture of Problem Solving.* New York: Plenum Press, 1988.
20. Rubinstein, Moshe F., *Patterns of Problem Solving.* Englewood Cliffs, NJ: Prentice Hall, 1975.
21. Churchman, C. W., *The Design of Inquiring Systems.* New York: Basic Books, 1971.
22. Nadler, Gerald, *The Planning and Design Approach,* New York: John Wiley & Sons, 1981.
23. Rechtin 1991.
24. Lang, Jon, *Creating Architectural Theory, The Role of the Behavioral Sciences in Environmental Design.* New York: Van Nostrand Reinhold, 1987.
25. Rowe, Peter G., *Design Thinking.* Cambridge, MA: MIT Press, 1987.
26. Losk, Pieronek, Cureton, Geis, and Carpenter graduate reports are unpublished but available through the USC School of Engineering, Los Angeles, CA, 0089-1450.
27. Alexander, Christopher, *Notes on the Synthesis of Form.* Cambridge, MA: Harvard University Press, 1964. The first in a series of books on the subject.
28. Kostof, Spiro, *The Architect, Chapters in the History of the Profession.* New York: Oxford University Press, 1977.
29. Genesereth, Michael R., and Nils J. Nilsson, *Logical Foundations of Artificial Intelligence.* Los Altos, CA: Morgan Kaufmann, 1987.
30. Newell, Allen, *Unified Theories of Cognition.* Cambridge, MA: Harvard University Press, 1990.
31. Simon, Herbert A., *The Sciences of the Artificial.* Cambridge, MA: MIT Press, 1988.
32. Brooks, Frederick P., Jr., *The Mythical Man-Month.* Reading, MA: Addison-Wesley, 1982.

33. Garlan, David, and Mary Shaw, *An Introduction to Software Architecture.* Pittsburgh, PA: Carnegie Mellon University, 1993.
34. Booton, Richard C., Jr., and S. Ramo, The Development of Systems Engineering. *The IEEE Transactions on Aerospace and Electronic Systems* AES–20, Vol. 4, pp. 306–309, July 1984.

Appendix A:
Heuristics for Systems-Level Architecting

Experience is the hardest kind of teacher.
It gives you the test first and the lesson afterward.

Susan Ruth, 1993

Introduction: Organizing the List

The heuristics to follow were selected from Rechtin 1991,[1] the *Collection of Student Heuristics in Systems Architecting, 1988–1993,*[2] and from subsequent studies in accordance with the selection criteria of Chapter 2. The list is intended as a tool store for top-level systems architecting. Heuristics continue to be developed and refined not only for this level, but for domain-specific applications as well, often migrating from domain-specific to system level and vice versa.*

---

* The manufacturing, social, communication, software, management, business, and economics fields are particularly active in proposing and generating heuristics — though they usually are called principles, laws, rules, or axioms.

For easy search and use, the heuristics are grouped by architectural task and categorized by being either descriptive or prescriptive — that is, by whether they describe an encountered situation or prescribe an architectural approach to it, respectively.

There are over 180 heuristics in the listing to follow, far too many to study at any one time. Nor were they intended to be. The listing is intended to be scanned as one would scan software tools on software store shelves, looking for ones that can be useful immediately, but remembering that others are also there. Although some are variations of other heuristics, the vast majority stand on their own, related primarily to others in the near vicinity on the list. Odds are that the reader will find the most interesting heuristics in clusters, the location of which will depend on the reader's interests at the time. The section headings are by architecting task. A "D" signifies a descriptive heuristic; a "P" signifies a prescriptive one. When readily apparent, prescriptions are grouped by insetting under appropriate descriptions or alternate prescriptions; otherwise not. In the interests of brevity, an individual heuristic is listed in the task where it is most likely to be used most often. As noted in Chapter 2, some 20% can be tied to related ones in other tasks.

A major difference between a heuristic and an unsupported assertion is the credibility of the source. To the extent possible, the heuristics are credited to the individuals who, to the authors' knowledge, first suggested them. To further aid the reader in judging credibility or in finding the sources, the heuristics to follow are given symbols. These symbols indicate the following:

[ ] *An informal discussion* with the individual indicated, unpublished.
( ) A formal, dated source, with examples, located in the University of Southern California (USC) Master of Science in Systems Architecture and Engineering (MS-SAE) program archive, especially in the Collection of Student Heuristics in Systems Architecting, 1988–1993. For further information, contact the Master of Science Program in Systems Architecture and Engineering, USC School of Engineering, University Park, Los Angeles, California 90089-1450.
* Rechtin 1991, where it is sourced more formally. By permission of Prentice Hall Inc., Englewood Cliffs, New Jersey 07632.
Bold Key words useful for quick search. Otherwise, heuristics to follow are in plain type to make page reading easier. Real-world examples of each can be found in the references indicated.

The authors apologize in advance for any miscrediting of sources. Corrections are welcome. The readers are reminded that not all heuristics apply to all circumstances, just most to most.

Heuristic Tool List

Multitask Heuristics

D **Performance, cost, and schedule** cannot be specified independently. At least one of the three must depend on the others.**

D With few exceptions, schedule **delays** will be **accepted** grudgingly; cost **overruns** will **not**, and for good reason.

D The **time to completion** is proportional to the ratio of the time spent to the time planned to date. The greater the ratio is, the longer the time to go.

D **Relationships** among the elements are what give systems their added value.*

D **Efficiency** is inversely proportional to **universality**. (Douglas R. King 1992)

D **Murphy's Law**, "If anything can go wrong, it will."*

P **Simplify**. Simplify. Simplify.*

P The first line of defense against complexity is simplicity of design.

P Simplify, combine and eliminate. (Suzaki 1987)

P Simplify with smarter elements. (N. P. Geiss 1991)

P The most **reliable part** on an airplane is the one that isn't there — because it isn't needed. [DC-9 Chief Engineer 1989]

D One person's **architecture** is another person's **detail**. One person's system is another's component. [Robert Spinrad 1989]*

P In order to **understand anything**, you must not try to understand everything. (Aristotle, 4th century B.C.)

P Don't confuse the functioning of the parts for the functioning of the system. (Jerry Olivieri 1992)

D In general, each **system level** provides a context for the level(s) below. (G. G. Lendaris 1986)

P Leave the **specialties** to the specialist. The level of detail required by the architect is only to the depth of an element or component critical to the system as a whole. (Robert Spinrad 1990) But the architect must have **access** to that level and know, or be informed, about its criticality and status. (Rechtin 1990)

P Complex systems will develop and **evolve** within an overall architecture much more rapidly if there are **stable intermediate** forms than if there are not. (Simon 1969)*

D Particularly for social systems, it's the **perceptions**, not the facts, that count.

---

* As indicated in the introduction to this appendix, an asterisk indicates that this heuristic is taken from Rechtin 1991. (With permission of Prentice Hall, Englewood Cliffs, New Jersey.)

D In introducing technological and **social** change, **how** you do it is often more important than **what** you do.*

   P   If social cooperation is required, the **way** in which a system is **implemented** and introduced must be an **integral part** of its architecture.*

D If the **politics** don't fly, the hardware never will. (Brenda Forman 1990)

   D   Politics, not technology, sets the limits of what technology is allowed to achieve.

   D   Cost rules.

   D   A strong, coherent constituency is essential.

   D   Technical problems become political problems.

   D   There is no such thing as a purely technical problem.

   D   The best engineering solutions are not necessarily the best political solutions.

D **Good products** are not enough. Implementations matter. (Morris and Ferguson 1993)

   P   To remain competitive, determine and **control the keys** to the architecture from the very beginning.

Scoping and Planning

> The beginning is the most important part of the work.
>
> **Plato, 4th century B.C.**

> **Scope!** Scope! Scope!
>
> **William C. Burkett, 1992**

D **Success** is defined by the **beholder**, not by the architect.*

   P   The most important single element of success is to **listen** closely to what the customer perceives as his requirements and to have the will and ability to be responsive. (J. E. Steiner 1978)*

   P   **Ask early** about how you will **evaluate** the success of your efforts. (F. Hayes-Roth et al., 1983)

   P   For a system to meet its **acceptance criteria** to the satisfaction of all parties, it must be architected, designed, and built to do so — no more and no less.*

   P   Define how an **acceptance** criterion is to be certified at the same time the criterion is established.*

   D   Given a **successful** organization or system with valid criteria for success, there are some things it **cannot do** — or at least not do well. Don't force it!

P The **strengths** of an organization or system in one context can be its **weaknesses** in another. Know when and where!*

D There's nothing like being the **first success.***

P If at first you don't succeed, but the architecture is sound, try, try again. Success sometimes is where you find it. Sometimes it finds you.*

D A system is successful when the **natural intersection** of technology, politics, and economics is found. (A. D. Wheelon 1986)*

D Four questions, **the Four Who's**, need to be answered as a self-consistent set if a system is to succeed economically; namely, who benefits? who pays? and, as appropriate, who loses?

D **Risk** is (also) defined by the **beholder**, not the architect.

P If being absolute is impossible in estimating system risks, then be relative.*

D **No** complex system can be **optimum** to all parties concerned, nor all functions optimized.*

P Look out for hidden agendas.*

P It is sometimes more important to know **who** the customer is than to know what the customer **wants**. (Whankuk Je 1993)

D The phrase, "**I hate it**," is direction. (Lori I. Gradous 1993)

P Sometimes, but not always, the best way to solve a difficult problem is to **expand** the problem, itself.*

P Moving to a **larger purpose** widens the range of solutions. (Gerald Nadler 1990)

P Sometimes it is necessary to **expand the** *concept* in order to simplify the problem. (Michael Forte 1993)

P [If in difficulty,] **reformulate** the problem and re-allocate the system functions. (Norman P. Geis 1991)

P Use **open** architectures. You will need them once the market starts to respond.

P Plan to **throw one away**. You will anyway. (F. P. Brooks, Jr. 1982)

P You can't avoid **redesign**. It's a natural part of design.*

P Don't make an architecture **too smart** for its own good.*

D Amid a **wash of paper**, a small number of documents become critical pivots around which every project's management revolves. (F. P. Brooks, Jr. 1982)*

P Just because it's written, doesn't make it so. (Susan Ruth 1993)

D In architecting a new [software] program, all the **serious mistakes** are made in the **first day**. [Spinrad 1988]

P The most **dangerous** assumptions are the **unstated** ones. (Douglas R. King 1991)

D Some of the **worst** failures are **systems** failures.

D In architecting a new [aerospace] system, by the time of the first **design review**, performance, cost, and schedule have been **predetermined**.

One might not know what they are yet, but to first order all the critical assumptions and choices have been made which will determine those key parameters.*

P **Don't assume** that the original statement of the problem is necessarily the best, or even the right, one.*

    P **Extreme** requirements, expectations, and predictions should remain under challenge. throughout system design, implementation, and operation.

    P Any extreme requirement must be intrinsic to the system's design philosophy and must validate its selection. "Everything must pay its way on to the airplane." [Harry Hillaker 1993]

    P **Don't assume** that previous **studies** are necessarily complete, current, or even correct. (James Kaplan 1992)

    P Challenge the process and solution, for surely someone else will do so. (Kenneth L. Cureton 1991)

    P Just because it worked in the past there's no guarantee that it will work now or in the future. (Kenneth L. Cureton 1991)

    P Explore the situation from more than one point of view. A seemingly impossible situation might suddenly become transparently simple. (Christopher Abts 1988)

P **Work forward and backward**. (A set of heuristics from Rubinstein 1975)*

Generalize or specialize.

Explore multiple directions based on partial evidence.

Form stable substructures.

Use analogies and metaphors.

Follow your emotions.

P Try to hit a solution that, at worst, won't put you **out of business**. (Bill Butterworth as reported by Laura Noel 1991)

P The **order** in which decisions are made can change the architecture as much as the decisions themselves. (Rechtin 1975, IEEE SPECTRUM)

P Build in and maintain **options** as long as possible in the design and build of complex systems. You will need them. OR ... Hang on to the agony of decision as long as possible. [Robert Spinrad 1988]*

    P Successful architectures are **proprietary, but open**. [Morrison and Ferguson 1993]

D Once the architecture begins to take shape, the sooner contextual constraints and **sanity checks** are made on assumptions and requirements, the better.*

D Concept **formulation** is complete when the **builder** thinks the system can be built to the **client's** satisfaction.*

D The **realities** at the end of the conceptual phase are not the models but the **acceptance criteria**.*

P Do the **hard parts** first.

P Firm **commitments** are best made after the **prototype works**.

Modeling*

P If you can't analyze it, don't build it.
D Modeling is a **craft** and at times an art. (William C. Burkett 1994)
D A **vision** is an **imaginary architecture** ... no better, no worse than the rest of the models. (M. B. Renton Spring 1995)
D From psychology: If the concepts in the mind of one person are very different from those in the mind of the other, there is no **common model** of the topic and no communication. [Taylor 1975] OR ... From telecommunications: The **best receiver** is one that contains an internal model of the transmitter and the channel. [Robert Parks & Frank Lehan 1954]*
D A model is not **reality.***
  D The **map** is not the territory. (Douglas R. King 1991)*
  P Build **reality checks** into model-driven development. [Larry Dumas 1989]*
  P Don't believe **nth order consequences** of a first order [cost] model. [R. W. Jensen circa 1989]
D Constants aren't and variables don't. (William C. Burkett 1992)
D One **insight** is worth a thousand **analyses**. (Charles W. Sooter 1993)
  P Any war game, systems analysis, or study whose results can't easily be explained on the back of an envelope is not just worthless, it is probably dangerous. [Brookner-Fowler circa 1988]
D Users develop **mental models** of systems based [primarily] upon the user-to-system interface. (Jeffrey H. Schmidt)
D If you can't explain it in **five minutes**, either you don't understand it or it doesn't work. (Darcy McGinn 1992 from David Jones)
P The **eye** is a fine **architect**. Believe it. [Wernher von Braun 1950]
D A good solution somehow **looks nice**. (Robert Spinrad 1991)
  P **Taste:** an aesthetic feeling that will accept a solution as right only when no more direct or simple approach can be envisaged. [Robert Spinrad 1994]
  P Regarding **intuition**, trust but **verify**. (Jonathan Losk 1989)

Prioritizing (Trades, Options, and Choices)

D In any resource-limited situation, the **true value** of a given service or product is determined by what one is willing to **give up** to obtain it.
P When choices must be made with unavoidably inadequate information, **choose** the best available and then **watch** to see whether future

---

* See also Chapters 3 and 4.

solutions appear faster than future problems. If so, the choice was at least adequate. If not, go back and **choose** again.*

P When a decision makes sense through several different **frames,** it's probably a good decision. (J. E. Russo 1989)

D The **choice** between architectures may well depend upon which set of **drawbacks** the client can handle best.*

P If trade results are inconclusive, then the wrong selection criteria were used. Find out [again] what the customer wants and why they want it, then repeat the trade using those factors as the [new] selection criteria. (Kenneth Cureton 1991)

P The triage: Let the dying die. Ignore those who will recover on their own. And treat only those who would die without help.*

P Every once in a while you have to go back and see what the real world is telling you. [Harry Hillaker 1993]

Aggregating ("Chunking")

P **Group** elements that are strongly **related** to each other, **separate** elements that are unrelated.

D Many of the **requirements** can be **brought together** to complement each other in the total design solution. Obviously the more the design is put together in this manner, the more probable the overall success. (J. E. Steiner 1978)

P Subsystem interfaces should be drawn so that each subsystem can be implemented independently of the specific implementation of the subsystems to which it interfaces. (Mark Maier 1988)

P Choose a configuration with **minimal communications** between the subsystems. (computer networks)*

    P Choose the elements so that they are as independent as possible; that is, elements with low external complexity (low coupling) and high internal complexity (high cohesion). (Christopher Alexander 1964 modified by Jeff Gold 1991)*

    P Choose a configuration in which local activity is high speed and global activity is slow change. (P. J. Courtois 1985) *

P Poor aggregation results in **gray** boundaries and **red** performance. (M. B. Renton Spring 1995)

    P **Never aggregate** systems that have a **conflict** of interest; partition them to ensure checks and balances. (Aubrey Bout 1993)

    P **Aggregate** around "testable" subunits of the product; **partition** around logical **subassemblies.** (Ray Cavola 1993)

    P Iterate the partition/aggregation procedure until a model consisting of **7 ± 2 chunks** emerge. (Moshe F. Rubinstein 1975)

P The **optimum number** of architectural elements is the amount that leads to distinct **action**, not general planning. (M. B. Renton Spring 1995)

P System structure should resemble functional structure.*

  P Except for good and sufficient reasons, **functional and physical** structuring should match.*

  P The architecture of a **support** element must **fit** that of the system which it supports. It is easier to match a support system to the human it supports than the reverse.*

P **Unbounded limits** on element behavior may be a **trap** in unexpected scenarios. [Bernard Kuchta 1989]*

Partitioning (Decompositioning)

P Do not **slice** through regions where **high rates** of information exchange are required. (computer design)*

D The greatest **leverage** in architecting is at the **interfaces.***

  P **Guidelines** for a good quality interface specification: They must be simple, unambiguous, complete, concise, and focus on substance. Working documents should be the same as customer deliverables; that is, always use the customer's language, not engineering jargon. [Harry Hillaker 1993]

  P The **efficient architect**, using contextual sense, continually looks for likely **misfits** and redesigns the architecture so as to eliminate or minimize them. (Christopher Alexander 1964)* It is inadequate to architect up to the boundaries or **interfaces** of a system; one must architect **across** them. (Robert Spinrad as reported by Susan Ruth 1993)

  P Since boundaries are inherently limiting, look for solutions outside the boundaries. (Steven Wolf 1992)

  P Be prepared for **reality** to add a few interfaces of its own.*

P Design the structure with **good "bones."***

P Organize personnel tasks to **minimize** the **time** individuals spend interfacing. (R. C. Tausworthe 1988)*

Integrating

D **Relationships** among the elements are what give systems their **added value.***

  P The greatest leverage in system architecting is at the interfaces.*

  P The greatest **dangers** are also at the interfaces. [Raymond 1988]

  P Be sure to ask the question, "What is the worst thing that other elements could do to you across the interface?" [Kuchta 1989]

D Just as a piece and its template must match, so must a system and the resources which make, test, and operate it. Or, more briefly, the **product and process** must match. Or, by extension, a system architecture cannot be considered complete lacking a suitable match with the process architecture.*

    P When confronted with a particularly difficult interface, try changing its **characterization**.*

P Contain **excess energy** as close to the source as possible.*

    P Place barriers in the paths between energy sources and the elements the energy can damage. (Kjos 1988)*

Certifying (System Integrity, Quality, and Vision)

D As **time to delivery** decreases, the **threat** to functionality increases. (Steven Wolf 1992)

    P If it is a good design, **insure** that it stays **sold**. (Dianna Sammons 1991)

D Regardless of what has gone before, the **acceptance criteria** determine what is actually built.*

    D The number of **defects remaining** in a (software) system after a given level of test or review (design review, unit test, system test, etc.) is proportional to the **number found** during that test or review.

    P **Tally** the defects, analyze them, **trace** them to the source, make corrections, keep a **record** of what happens afterwards and keep **repeating** it. [Deming]

    P **Discipline.** Discipline. Discipline. (Douglas R. King 1991)

    P The principles of **minimum communications** and proper partitioning are key to system testability and **fault isolation**. (Daniel Ley 1991)*

    P **The five whys** of Toyota's lean manufacturing. (To find the basic cause of a defect, keep asking "why" from effect to cause to cause five times.)

D The test **setup** for a system is itself a system.*

    P The test system should always allow a part to pass or fail on its own merit. [James Liston 1991]*

    P To be tested, a system must be designed to be tested.*

D An element **"good enough"** in a small system is unlikely to be good enough in a more complex one.*

D Within the same class of products and processes, the **failure rate** of a product is linearly proportional to its **cost**.*

D The **cost** to find and **fix** an inadequate or failed part increases by an **order of magnitude** as it is successively incorporated into higher levels in the system.

    P The least expensive and most effective place to find and fix a problem is at its source.

D Knowing a **failure has occurred** is more important than the actual failure. (Kjos 1988)

D **Mistakes** are **understandable**, failing to report them is **inexcusable**.

D Recovery from failure or flaw is not complete until a specific mechanism, **and no other**, has been shown to be the cause.*

D Reducing **failure rate** by each **factor of two** takes as much effort as the original development.*

D **Quality** can't be tested in, it has to be **built in.***

    D You can't achieve quality ... unless you specify it. (Deutsch 1988)

    P Verify the quality close to the source. (Jim Burruss 1993)

    P The five why's of Japan's lean manufacturing. (Hayes et al. 1988)[3]

    D High-**quality**, reliable systems are produced by high-quality architecting, engineering, design and manufacture, **not by inspection,** test, and rework.*

    P Everyone in the development and production line is both a customer and a supplier.

D Next to interfaces, the greatest **leverage** in architecting is in aiding the recovery from, or exploitation of, **deviations** in system performance, cost or schedule.*

Assessing Performance, Cost, Schedule, and Risk

D A good design has benefits in more than one area. (Trudy Benjamin 1993)

D System quality is defined in terms of customer satisfaction, not requirements satisfaction. (Jeffrey Schmidt 1993)

D If you think your **design** is perfect, it's only because you haven't shown it to **someone else**. [Harry Hillaker, 1993]

    P Before proceeding too far, **pause and reflect!** Cool off periodically and seek an independent review. (Douglas R. King 1991)

D Qualification and **acceptance** tests must be both definitive and **passable.***

    P High **confidence**, not test completion, is the **goal** of successful qualification. (Daniel Gaudet 1991)

    P Before ordering a **test** decide what you will do if it is 1) **positive** or if 2) it is negative. If both answers are the same, **don't do** the test. (R. Matz, M.D. 1977)

D "**Proven**" and "**state of the art**" are mutually **exclusive** qualities. (Lori I. Gradous 1993)

D The **bitterness** of **poor performance** remains long after the sweetness of low prices and prompt delivery are forgotten. (Jerry Lim 1994)

D The **reverse of diagnostic** techniques are good architectures. (M. B. Renton 1995)

D Unless everyone who **needs to know** does know, somebody, somewhere will foul up.

   P   Because there's no such thing as immaculate communication,
       don't ever stop **talking** about the system. (Losk 1989)*

D Before it's tried, it's **opinion**. After it's tried, it's **obvious**. (Wm.
C. Burkett 1992)

D Before the **war**, it's opinion. After the war, it's too late! (Anthony
Cerveny 1991)

D The first **quick look** analyses are often **wrong**.*

D In correcting system deviations and failures, it is important that all
the participants know not only **what** happened and how it happened,
but **why** as well.*

   P   Failure reporting without a **close out** system is meaningless.
       (April Gillam 1989)

   P   Common, if undesirable, responses to **indeterminate outcomes**
       or failures:*
       If it **ain't broke**, don't fix it.
       Let's **wait and see** if it goes away or happens again.
       It was just a **random** failure. One of those things.
       Just treat the **symptom**. Worry about the cause later.
       Fix everything that might have caused the problem.
       Your **guess** is as good as mine.

D Chances for recovery from a **single failure** or flaw, even with
complex consequences, are fairly good. Recovery from **two or
more** independent failures is unlikely in real time and uncertain
in any case.*

Re-Architecting, Evolving, Modifying, and Adapting

> The test of a good architecture is that it will last.
> The sound architecture is an enduring pattern.

[Robert Spinrad 1988]

P The team that created and built a presently successful product is often the
best one for its evolution — but seldom for creating its replacement.

D If you **don't understand** the existing system, you can't be sure you're
rearchitecting a **better** one. (Susan Ruth 1993)

P When implementing a change, keep some elements constant to provide an **anchor** point for people to cling to. (Jeffrey H. Schmidt 1993)

   P   In large, mature systems, **evolution** should be a process of
       ingress and **egress**. (IEEE 1992, Jeffrey Schmidt 1992)

P   Before the change, it is your opinion. After the change it is your problem. (Jeffrey Schmidt 1992)

D   Unless constrained, **rearchitecting** has a natural tendency to proceed unchecked until it results in a substantial transformation of the system. (Charles W. Sooter 1993)

D   Given a change, if the anticipated actions don't occur, then there is probably an invisible barrier to be identified and overcome. (Susan Ruth 1993)

Exercises

Exercise: What favorite heuristics, rules of thumb, facts of life, or just plain common sense do you apply to your own day-to-day living — at work, at home, at play? What heuristics have you heard on TV or the radio (for example, on talk radio, action TV, children's programs)? Which ones would you trust?

Exercise: Choose a system, product, or process with which you are familiar and assess it using the appropriate foregoing heuristics. What was the result? Which heuristics are or were particularly applicable? What further heuristics were suggested by the system chosen?

Were any of the heuristics clearly incorrect for this system? If so, why?

Exercise: Try to spot heuristics and insights in the technical literature. Some are easy; they are often listed as principles or rules. The more difficult ones are buried in the text but contain the essence of the article or state something of far broader application than the subject of the piece.

Exercise: Try to create a heuristic of your own — a guide to action, decision making, or to instruction of others.

Notes and References

1. Rechtin, E., *Systems Architecting, Creating and Building Complex Systems.* Englewood Cliffs, NJ: Prentice Hall, 1991. Note that throughout chapter, this reference will be referred to as Rechtin 1991.

2. Rechtin, E., ed., *Collection of Student Heuristics in Systems Architecting, 1988–1993.* Los Angeles, CA: University of Southern California, March 15, 1994 (unpublished but available to students and researchers on request).
3. Hayes, Robert H., S. C. Wheelwright, and Kim B. Clark, *Dynamic Manufacturing, Creating the Learning Organization.* New York: The Free Press, 1988.

Appendix B: Reference Texts Suggested for Institutional Libraries

The following list of texts is offered as a brief guide to books that would be particularly appropriate to an architecting library.

Architecting Background

Alexander, C., *A Pattern Language: Towns, Buildings, Construction*, Oxford University Press, New York, 1977.

Alexander, C., *Notes on the Synthesis of Form*, Harvard University Press, Cambridge, MA, 1964.

Alexander, C., *The Timeless Way of Building*, Oxford University Press, New York, 1979.

Kostoff, Spiro, *The Architect*, Oxford University Press, New York, 1977 (paperback).

Lang, Jon, *Creating Architectural Theory*, Van Nostrand Reinhold, New York, 1987.

Rowe, P. G., *Design Theory*, MIT Press, Cambridge, MA, 1987.

Vitruvius, *The Ten Books on Architecture*, Dover Publications, Mineola, New York, 1960 (paperback). Translated by Morris Hicky Morgan.

Management

Augustine, N. R., *Augustine's Laws*, AIAA, Inc., Reston, VA, 1982.

Deal, Terrence E., and A. A. Kennedy, *Corporate Cultures, The Rites and Rituals of Corporate Life*, Addison-Wesley, Reading, MA, 1988.

DeMarco, Tom, and Timothy Lister, *Peopleware: Productive Projects and Teams*, Dorset House, New York, 1987.

Juran, J. M., *Juran on Planning for Quality*, The Free Press, New York, 1988.

Modeling

Eisner, H., *Computer Aided Systems Engineering*, Prentice Hall, Upper Saddle River, NJ, 1988.

Hatley, D. J., and I. Pirbhai, *Strategies for Real-Time System Specification*, Dorset House, New York, 1988.

Rumbaugh, J., M. Blaha, W. Premerlani, F. Eddy, and W. Lorensen, *Object-Oriented Modeling and Design*, Prentice Hall, Upper Saddle River, NJ, 1991.

Ward, P. T., and S. J. Mellor, *Structured Development for Real-Time Systems, Volume 1: Introduction and Tools*, Yourdon Press (Prentice Hall), New York, 1985.

Specialty Areas

Baudin, M., *Manufacturing Systems Analysis*, Yourdon Press Computing Series, New York, 1990.

Hammond, J. S., R. L. Keeney, and H. Raiffa, *Smart Choices: A Practical Guide to Making Better Decisions*, Broadway Books, New York, 2002.

Hayes, Robert H., S. C. Wheelwright, and K. B. Clark, *Dynamic Manufacturing*, The Free Press, New York, 1988.

Keeney, R. L., *Value Focused Thinking*, Harvard University Press, Cambridge, MA, 1992.

Miller, J. G., *Living Systems*, McGraw-Hill, New York, 1978.

Simon, H. A., *Sciences of the Artificial*, MIT Press, Cambridge, MA, 1981.

Thome, B., editor, *Systems Engineering: Principles and Practice of Computer-Based Systems Engineering*, John Wiley, Chichester, Wiley Series on Software Based Systems, 1993.

Software

Boehm, B., *Software Engineering Economics*, Prentice Hall, Englewood Cliffs, NJ, 1981.

Brooks, F. P. Jr., *The Mythical Man-Month, Essays on Software Engineering, 20th Anniversary Edition*, Addison-Wesley, Reading, MA, 1995.

Deutsch, M. S., and R. R. Willis, *Software Quality Engineering*, Prentice-Hall, Upper Saddle River, NJ, 1988.

Gajski, D. D., V. M. Milutinovic, H. J. Siegel, and B. P. Furht, *Computer Architecture*, The Computer Society of the IEEE, Piscataway, NJ, 1987 (Tutorial).

Gamma, E. et al., *Design Patterns: Elements of Reusable Object-Oriented Software Architecture*, Addison-Wesley, Reading, MA, 1994.

Shaw, M., and D. Garlan, *Software Architecture: Perspectives on an Emerging Discipline*, Prentice Hall, Upper Saddle River, NJ, 1996.

Software Productivity Consortium, ADARTS Guidebook, SPC-94040-CMC, Version 2.00.13, Vols. 1–2, September, 1991.

Yourdon, E., and L. L. Constantine, *Structured Design: Fundamentals of a Discipline of Computer Program and Systems Design*, Yourdon Press, New York, 1979.

Systems Sciences

Flood, R. L., and E. R. Carson, *Dealing with Complexity, an Introduction to the Theory and Application of System Sciences,* Plenum Press, New York, 1988.

Genesereth, M. S., and N. J. Nilsson, *Logical Foundations of Artificial Intelligence,* Morgan Kaufmann, San Francisco, CA, 1987.

Gerstein, Dean R. et al., Editors, *The Behavioral and Social Sciences, Achievements and Opportunities,* National Academy Press, Washington, 1988.

Klir, G. J., *Architecture of Systems Problem Solving,* Plenum Press, New York, 1985.

Systems Thinking

Arbib, M. A., *Brains, Machines, and Mathematics,* 2nd edition, Springer-Verlag, Heidelberg, 1987.

Beam, Walter R., *Systems Engineering, Architecture and Design,* McGraw-Hill, New York, 1990.

Boorstin, Daniel J., *The Discoverers,* Vintage Books, New York, 1985.

Boyes, J. L., editor, *Principles of Command and Control,* AFCEA International Press, Fairfax, VA, 1987.

Davis, S. M., *Future Perfect,* Addison-Wesley, Reading, MA, 1987.

Gause, Donald C., and G. M. Weinberg, *Exploring Requirements, Quality Before Design,* Dorset House, New York, 1989.

Hofstadter, D. R., *Gödel, Escher, Bach: An Eternal Golden Braid,* Vintage Books, New York, 1980.

Norman, Donald A., *The Psychology of Everyday Things,* Basic Books, New York, 1988.

Pearl, Judea, *Heuristics,* Addison-Wesley, Reading, MA, 1984.

Rechtin, E., *Systems Architecting: Creating and Building Complex Systems,* Prentice Hall, Englewood Cliffs, NJ, 1991.

Rubinstein, Moshe F., *Patterns of Problem Solving,* Prentice Hall, Englewood Cliffs, NJ, 1975.

Weinberg, Gerald M., *Rethinking Systems Analysis and Design,* Dorset House, New York, 1988.

Appendix C:
On Defining Architecture
and Other Terms

This appendix is for those who need to come to a consensus in a group on a definition for architecture or other major terms used in this book. There are many who might have such a need, and for those who have a need, this appendix might be very useful. Deciding on formal definitions is commonly part of setting up an official corporate training course or documenting a standard (public or corporate). In these situations, an inordinate amount of time can be spent arguing about fine details of definitions. It may be hard to pick and choose among the definitions offered by different standards because they usually do not record the reasoning that brought them to a decision. This appendix is a record of some of the definition-related discussions one of the authors (Maier) has been involved in over several years. It is offered to help others who need to arrive at a group consensus on definitions a ready-made set of choices and reasoning.

Defining "Architecture"

One might think that, with 5,000 years of history, the notion of architecture in buildings would be clearly and crisply defined. Presumably then, the definition could be extended to give a clear and crisp definition to architecture in other fields. However, this is not the case. A formal definition of architecture is elusive even in the case of buildings. And if the definition is elusive in its original domain, is it surprising that a wholly satisfactory definition is elusive in more general domains?

The communities involved in architecture in systems, software, hardware, and other domains have struggled with finding a formal definition. Each group that has set out a formal definition has usually made a unique choice. The choices are often similar, but reflect significantly different ideas. The sections to follow review some of the more distinctive choices. Of course, there are many small variations on each one.

To make sense of the different definitions, it is important to review them with some criteria in mind. In reviewing these definitions, try to answer the following questions with respect to each definition:

1. How does the definition establish what is the concern of the architect and what is not?
2. What is the purpose of the definition? Some purposes might be defining an element of design, education, organizational survival or politics, setting legal boundaries, or even humor.
3. Choose a building you are familiar with. What is its architecture, according to the definition? How well does the definition implied architecture match what you would expect to be the building architect's scope of work?
4. Choose a system you are familiar with. What is its architecture, according to the definition? What things are uniquely determined about the system from the application of that definition?
5. What is the architecture of the Internet, according to the definition?

Webster's Dictionary *Definition*

We begin with the dictionary's definition.[1]

> Architecture: 1. The art or science of building; specifically, the art or practice of designing and building structures and esp. habitable ones. 2a. Formation or construction as or as if the result of conscious act <the ~ of the garden> b. a unifying or coherent form or structure <the novel lacks ~> 3. Architectural product or work 4. A method or style of building 5. The manner in which the components of a computer or computer system are organized and integrated.

The interesting part of this definition, for our purposes, is part 2. The first definition uses architecture in the sense of the profession, not what we are looking for here. This definition says to speak of the architecture of a thing is to speak of its "unifying or coherent form." Unfortunately, it is not obvious what aspect of form is "unifying or coherent." It is something that can be judged, but is hard to define crisply. The civil building example suggests several other ideas about architecture:

1. Architecture is tied to the structure of components, but if a novel can have an architecture, the notion of components is relatively abstract. Components may need to be interpreted broadly in some contexts. No one would confuse the structure of a novel with its organization into chapters — which is the "packaging" of that structure, and is analogous to confusing the architecture of a system with its module structure.
2. The distinction between an architectural level of description and some other level of design description is not crisp. Architectural description is concerned with unifying characteristics or style, and an engineering description is concerned with construction or acquisition.
3. In common use, "architecture" can mean a conceptual thing, the work of architects, and architectural products. Other definitions make sharper distinctions.

This Book

The definition of architecture given in the glossary of this book is as follows:

> **Architecture:** The structure — in terms of components, connections, and constraints — of a product, process, or element.

This definition is specific, it is talking about structure (although that term is open to some interpretation). Components, connections, and constraints are the descriptive terms for architecture. And we can talk about the architecture of a wide variety of things. This book is primarily about architecting, rather than architecture. The reason is that the most important constraints come from the process of doing the architect's role. The most important things come from working with clients to understand purpose and limitations. Architecture should, by the tenets of this book, proceed from the client's needs rather than a presupposed notion of what constitutes an architectural-level definition of a system.

IEEE Architecture Working Group (AWG)

After extended discussion in 1995–1996 in association with developing ANSI/IEEE 1471 Recommended Practice for Architectural Description of Software-Intensive Systems, the Institute of Electrical and Electronics Engineers (IEEE) Working Group chose the following definition:

An Architecture is the highest-level concept of a system in its environment.

"System" in this definition refers back to the official IEEE definition, "a collection of components organized to accomplish a specific function or set of functions." This definition of architecture was intended to capture several ideas:

1. An architecture is a property of a thing or a concept, not a structure. The term "structure" is avoided specifically to avoid any connotation that architecture was solely a matter of physical structure. Concept, which is obviously much more generic, is used instead.
2. The term "highest-level" is used to indicate that architecture is an abstraction, and that it is a fundamental abstraction. A major defect of this definition is that highest-level carries a connotation of levels of hierarchy, and in particular a single hierarchy, which exactly is one of the connotations to be avoided. Also, "highest-level concept" leaves a great deal of room for interpretation.
3. The definition says that architecture is not a property of the system alone, but that the system's environment must be included in a definition of the system's architecture. This has often been referred to as "architecture in context" as opposed to "architecture as structure." It was there to capture the idea that architecture has to encompass purpose and the relationship of the system to its stakeholders. The reader must judge whether or not that interpretation is clear.

This definition was used in several drafts of the 1471 standard but was replaced in the final balloted version. The definition in the final balloted version was as follows:

> **Architecture:** the fundamental organization of a system embodied in its components, their relationships to each other and to the environment and the principles guiding its design and evolution.

This definition is a refinement of definitions from the software engineering community, as discussed below. Those who do not like it might be more inclined to say it was a compromise between conflicting points of view that suffers from the usual problems of a committee decision. The definition starts with the software communities definitions (discussed shortly) and then adds back some of the ideas of the original 1471 definition. The primary refinement is the deemphasis on physical structure and to say that architecture is "embodied" in components, relationships, and principles. Put another way, the definition tries to recognize that, for most systems, most of the time, the architecture is in the arrangement of physical components and their relationships. But, sometimes, the fundamental organization is on a more abstract level.

INCOSE SAWG

The International Council on Systems Engineering (INCOSE) Systems Architecture Working Group (SAWG) adopted a definition for systems architecture. It could as well be read as a definition for "Architecture, of a system." It is as follows:

> Systems Architecture: The fundamental and unifying system structure defined in terms of system elements, interfaces, processes, constraints, and behaviors.

This definition borrows the core of the dictionary definition that architecture represents fundamental, unifying, or essential structure. Exactly what constitutes fundamental, unifying, or essential is not easily defined. It is presumed that recognizing it is partially art and up to the participants. In this definition, the role of multiple aspects making up the architecture is made explicit through the listing of elements, interfaces, processes, constraints, and behaviors. This definition makes, or facilitates making, a sharper separation between an architecture as a conceptual object, an architecture description as concrete object, and the process or act of creating architectures (architecting).

MIL-STD-498

MIL-STD-498, now canceled, had a definition of architecture that specifically pertained to a designated development task.

> Architecture: The organizational structure of a system or CSCI, identifying its components, their interfaces, and a concept of execution among them.

Here architecture is described specifically in three parts: components, interfaces, and a concept of execution. In this sense, it supports the idea of architecture as inherently multiview, although it specifically defines the views where others leave them open. The meaning of "organizational structure" as opposed to some other structure (conceptual, implementation, detailed, and so forth) is not made clear, although the idea is congruent of the common usage of architecture. It also uses "concept" within the definition, but only in referring to execution. Like most definitions, it does not clearly make a distinction between architectural and design concerns.

This definition is also "structuralist" in the sense that it emphasizes the structure of the system rather than its purposes or other relationships. One could interpret the definition to mean that the architect was not concerned with the system's purpose, that architecture came after requirements were fully defined. In fact, that is exactly the interpretation it should be given, at least in the way the associated standards envisioned the systems engineering process executing.

The original IEEE definition (in IEEE 610.12-1990) is a shorter version of this:

> The organizational structure of a system or component.

Perry-Garlan

A widely used definition in the software community is due to Perry and Garlan, although the exact place it first appeared is somewhat obscure.

> The structure of the components of a system, their interrelationships, and principles and guidelines governing their design and evolution over time.

An almost identical definition is used as the definition of architecture in the U.S. DoD C4ISR Architecture Framework, where it is incorrectly credited to the IEEE 610.12 standard for terminology. This definition is another three-part specification: components, interrelationships, and principles-guidelines. As this definition is commonly used, components and interrelationships usually refer to physically identifiable elements. This definition is mostly used in the software architecture community, and there it is common to see components identified as code units, classes, packages, tasks, and other code abstractions. The interrelationships would be calls or lines of inheritance.

The two basic objections to this definition are that it implies (if primarily through use rather than the words) that architecture is the same as physical structure, and that it makes no distinction in level of abstraction. The common usage of architecture is in reference to abstracted properties of things, not to the details. The Perry-Garlan definition can presumably apply to the structure of components at any level of abstraction. Although applicability to multiple levels is, in part, desirable, it is also desirable to distinguish between what constitutes an architectural-level description (whether of a whole system or of a component) from descriptions at lower levels of abstraction.

Maier's Tongue-in-Cheek Rule-of-Thumb

A slightly flip, but illustrative way of defining architecture is to go back to what architects are supposed to do.

> An Architecture is the set of information that defines a systems value, cost, and risk sufficiently for the purposes of the systems sponsor.

Obviously, this definition reflects the issue back to architecting, when the definition of architecture reflects back to architecture. The point of this definition is that architecture is what architects produce, and that what architects do is help clients make decisions about building systems. When the client makes acquisition decisions, architecture has been done (perhaps unconsciously, and perhaps very badly, but done).

Internet Discussion

One of the questions given at the beginning was "What is the architecture of the Internet?" The point of the question is that no reasonable notion of unifying, organizing, or coherent form will produce a physical description of the Internet. The specific pattern of physical links is continuously changing and of little interest. However, there is a very clear unifying structure, but it is a structure in protocols. It is not even a structure in software components, as exactly what software components implement the protocols is not known even to the participating elements. The point about protocols being the organizing structure of the Internet, and in particular the Internet Protocol (IP), was made in Chapter 7 and Figure 7.1.

Summary

Those who must choose definitions have a lot to work with, probably more than they would want. The precise form of the definition is less important than the background of what architecture should be about. What architecting should be was discussed at length in Chapter 1. The specifics of what architects will produce — that is, what an architecture actually looks like — will differ from domain to domain. Ideally, the definition for a given organization should come from that knowledge, the knowledge of what is needed to successfully define a system concept and take it through development. If the organization has that knowledge, it should be able to choose a formal definition that encapsulates it. If the organization does not have that knowledge, then no formal definition will produce it.

Models, Viewpoints, and Views

The terms model, view, and viewpoint are important in setting architecture description standards, or architecture frameworks using the community terminology of Chapter 11. The meaning of these terms changes from standard to standard. The discussion below is intended to capture an argument for a distinction between the two meanings. The distinction can be useful in writing standards, though it is not important in writing architecture descriptions nor is it extensively used in this book.

Why do we need some organizing abstraction beyond just models? Experience teaches that particular collections of models are logically related by the kinds of issues or concerns they address. The idea of a view comes from architectural drawings. In a drawing we talk about the top view or the side view of an object in referring to its physical representation as seen from a point. A view is the representation of a system from a particular point or perspective. A view is a representation of the whole system with respect to a set of related concerns. A viewpoint is the abstraction of many related views; it is the idea of viewing something from "the front," for example.

A view need not correspond to physical appearance. A functional view is a representation of a system abstracting away all nonfunctional or nonbehavioral details. A cutaway view shows some mixture of internal and external physical features in a mixture defined by the illustrator.

A view can be thought of both projectively and constructively. In the projective sense, a view is formed by taking the system and abstracting away all the details unnecessary to the view. It is analogous to taking a multidimensional object and projecting it onto a lower-dimensional space (like a viewing plane). So, for example, a behavioral view is the system pared down to only its behaviors, its set of input to output traces.

In the constructive sense, we build a complete model of the system by building a series of views. Each represents the system from one perspective, and with enough the system should be "completely" defined. It is like sketching a front view, a side view, a top view, and then inferring the structure of the whole object. In more general systems, we might build a functional view, then a physical view, then a data view, then return to the functional view, and so forth, until a complete model is formed from the joint set of views.

In practice, it usually takes several models to represent that whole system relative to typical concerns, at least for high-technology systems. So, a view is usually a collection of models. For example, physical representation seems simple enough, but how many different models are needed to represent the components of an information-intensive system? A complete physical view might need conventional block diagrams of information flow, block diagrams of communication interconnection, facilities layouts, and software component diagrams.

Viewpoints are motivated noticing that we build similar views, using similar methods, for many systems. By analogy, we will want to draw a top view of most systems we build. The civil architect always draws a set of elevations, and elevation drawings share common rules and structures. And an information systems architect will build information models using standard methods for each system. This similarity is because related systems will typically have similar stakeholders, and these stakeholders find their concerns consistently addressed by particular types of models and analysis methods. Hence, a viewpoint can be thought of as a set of modeling or analysis methods together with the concerns those methods address and the stakeholders possessing those concerns.

Working Definitions

These are summarizing definitions, augmented with the notions of consistency and completeness. The concepts here refer to the 1471 information model in Figure 11.1.

Model: An approximation, representation, or idealization of selected aspects of the structure, behavior, operation, or other characteristics of a real-world process, concept, or system (IEEE 610.12-1990).

Viewpoint: A template, pattern, or specification for constructing a view (IEEE 1471-2000).

View: A representation of a system from the perspective of related concerns or issues (IEEE 1471-2000).

Consistency, of views: Two or more views are consistent if at least one can exist that possesses the given views.

Completeness, of view: A set of views is complete if they satisfy (or "cover") all of the concerns of all stakeholders of interest.

Consistency and Completeness

Given multiple views (like top, front, and side) of a physical object, the ideas of consistency and completeness clear. A set of views is consistent if they are abstractions of the same object. A little more generally, they are consistent if at least one real object exists that has the given views. Consistency for physical object and views can be checked through solid geometry. Figure C.1 illustrates the point. The views are consistent if the geometrical object produces them when projected onto the appropriate subspace. Even without the actual object, we can perform geometric checks on the different views.

We cannot (yet) treat consistency in the same rigorous manner if the views are functional and physical and of a complex system. As we

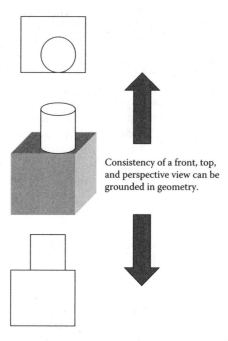

Consistency of a front, top, and perspective view can be grounded in geometry.

Figure C.1 A geometric illustration of the concept of consistency in views.

employ more complex views it is useful to return to the heuristic notion of consistency. Given a few models of a system being architected, we say they are consistent if at least one implementation exists that has the models as abstractions of itself.

Completeness can also be heuristically understood through the geometric analogy. Suppose we have set of visual representations of a material object. What does it mean to claim that the set of representations (views) is "complete"? Logically, it means that the views completely define the object. But, any set of external visual representations can only define the external shape of the object; it cannot define the internal structure, if any. This trivial observation is actually extremely important for understanding architecture. No set of representations is *ever* truly complete. A set of representations can be complete only with respect to something, say with respect to some set of concerns. If the concerns are external shape, then some set of external visual representations can be complete. If the concerns are extended to include internal structures, or strength properties, or weight, or any number of other things, then the set of views must likewise be extended to be "complete."

Reference

1. *Merriam Webster's Collegiate Dictionary*, 10th edition, p. 61.

Glossary

The fields of systems engineering and systems architecting are sufficiently new that many terms have not yet been standardized. Common usage is often different among different groups and in different contexts. However, *for the purposes of this book*, the meanings of the following terms are as follows:

Abstraction: A representation in terms of presumed essentials, with a corresponding suppression of the nonessential.

ADARTS: Ada-Based Design Approach for Real-Time Systems. A software development method (including models, processes, and heuristics) developed and promoted by the Software Productivity Consortium.

Aggregation: The gathering together of closely related elements, purposes, or functions.

Architecting: The processing of creating and building architectures; depending on one's perspective, architecting may or may not be seen as a separable part of engineering. Those aspects of system development most concerned with conceptualization, objective definition, and certification for use.

Architectural style: A form or pattern of design with a shared vocabulary of design idioms and rules for using them (See Shaw and Garlan, 1996, page 19).

Architecture: The structure — in terms of components, connections, and constraints — of a product, process, or element.

Architecture, open: An architecture designed to facilitate addition, extension, or adaptation for use.

Architecture (communication, software, or hardware): The architecture of the particular designated aspect of a large system.

ARPANET/INTERNET: The global computer internetwork, principally based on the TCP/IP packet communications protocol. The ARPANET was the original prototype of the current INTERNET.

Certification: A formal, but not necessarily mathematical, statement that defined system properties or requirements that have been met.

Client: The individual or organization that pays the bills. May or may not be the user.

Complexity: A measure of the numbers and types of interrelationships among system elements. Generally speaking, the more complex a system, the more difficult it is to design, build, and use.

Deductive reasoning: Proceeding from an established principle to its application.

Design: The detailed formulation of the plans or instructions for making a defined system element; a follow-on step to systems architecting and engineering.

Domain: A recognized field of activity and expertise, or of specialized theory and application.

Engineering: Creating cost-effective solutions to practical problems by applying scientific knowledge to building things in the service of mankind (Shaw and Garlan, 1996, page 6). May or may not include the art of architecting.

Engineering, concurrent: Narrowly defined (here) as the process by which product designers and manufacturing process engineers work together to create a manufacturable product.

Heuristic: A guideline for architecting, engineering, or design. Lessons learned expressed as a guideline. A natural language abstraction of experience that passes the tests of Chapter 2.

Heuristic, descriptive: A heuristic that describes a situation.

Heuristic, prescriptive: A heuristic that prescribes a course of action.

IEEE P 1220: An Institute of Electrical and Electronic Engineers standard for systems engineering.

Inductive reasoning: Extrapolating the results of examples to a more general principle.

Manufacturing, flexible: Creating different products on demand using the same manufacturing line. In practice, all products on that line come from the same family.

Manufacturing, lean: An efficient and cost-effective manufacturing or production system based on ultraquality and feedback. (See Womack et al., 1990.)

MBTI: Meyer-Briggs Type Indicator. A psychological test for indicating the temperaments associated with selected classes of problem solving. (See Meyers, Briggs, and McCaulley, 1989.)

Metaphor: A description of an object or system using the terminology and properties of another. For example, the desktop metaphor for computerized document processing.

MIL-STD: Standards for defense system acquisition and development.

Model: An abstracted representation of some aspect of a system.

Model, satisfaction: A model that predicts the performance of a system in language relevant to the client.

Modeling: Creating and using abstracted representations of actual systems, devices, attributes, processes, or software.

Models, integrated: A set of models, representing different views, forming a consistent representation of the whole system.

Normative method: A design or architectural method based on "what should be" — that is, on a predetermined definition of success.

OMT: Object Modeling Technique. An object-oriented software development method. (See Rumbaugh et al., 1991.)

Objectives: Client needs and goals, however stated.

Paradigm: A scheme of things, a defining set of principles, a way of looking at an activity, for example, classical architecting.

Participative method: A design method based on wide participation of interested parties. Designing through a group process.

Partitioning: The dividing up of a system into subsystems.

Progressive design: The concept of a continuing succession of design activities throughout product or process development. The succession progressively reduces the abstraction of the system through models until physical implementation is reached and the system used.

Purpose: A reason for building a system.

Rational method: A design method based on deduction from the principles of mathematics and science.

Requirement: An objective regarded by the client as an absolute — that is, either passed or not.

Scoping: Sizing; defining the boundaries and defining the constraints of a process, product, or project.

Spiral: A model of system development that repeatedly cycles from function to form, build, test, and back to function. Originally proposed as a risk-driven process, particularly applicable to software development with multiple release cycles.

System: A collection of things or elements that, working together, produce a result not achievable by the things alone.

Systems, builder-architected: Systems architected by their builders, generally without a committed client.

Systems, feedback: Systems that are strongly affected by feedback of the output to the input.

Systems, form first: Systems that begin development with a defined form (or architecture) instead of a defined purpose. Typical of builder-architected systems.

Systems, politicotechnical: Technological systems, the development and use of which are strongly influenced by the political processes of government.

Systems, sociotechnical: Technological systems, the development and use of which are strongly affected by diverse social groups. Systems in which social considerations equal or exceed technical ones.

Systems architecting: The art and science of creating and building complex systems. That part of systems development most concerned with scoping, structuring, and certification.

Systems architecting, the art of: That part of systems architecting based on qualitative heuristic principles and techniques — that is, on lessons learned, value judgments, and unmeasurables.

Systems architecting, the science of: That part of systems architecting based on quantitative analytic techniques — that is, on mathematics and science and measurables.

Systems engineering: A multidisciplinary engineering discipline in which decisions and designs are based on their effect on the system as a whole.

Technical decisions: Architectural decisions based on engineering feasibility.

Ultraquality: Quality so high that measuring it in time and at reasonable cost is impractical. (See Rechtin, 1991, Chapter 8.)

Value judgments: Conclusions based on worth (to the client and other stakeholders).

View: A perspective on a system describing some related set of attributes. A view is represented by one or more models.

Waterfall: A development model based on a single sequence of steps; typically applied to the making of major hardware elements.

Zero defects: A production technique based on an objective of making everything perfectly. Related to the "everyone a supplier, everyone a customer" technique for eliminating defects at the source. Contrasts with acceptable quality limits in which defects are accepted providing they do not exceed specified limits in number or performance.

Author Index

427

Subject Index

C

C language, 300
C4I systems, 150, 196, 210
C4ISR framework (CAF), 316, 418
CAF; *see* C4ISR framework
Call distribution systems, 160
Carnegie Mellon University, 171, 377, 387, 389
CASE tools; *see* Computer-aided system engineering (CASE) tools
CD-ROMs, 154
Cellular telephones, 129–130, 150, 214
Certification, 17–18, 27, 94–96
Chaotic behavior, 73, 103, 236–237
Chernobyl, 133
Chunking, 402
Circle-to-spiral model, 97–98
Civil architecture, 20, 217
Civil engineering, 255
Civil works architecture, 5, 125
COCOMO; *see* Constructive cost model
Coding theory, 73
Cold War, 64, 363, 380
Collaborative assembly, 25
Collaborative formation, 350
Collaborative systems, 195–214
    analogies for architecting, 202–203
    collaboration as category, 195–196
    examples of, 197–202
    heuristics, 203–206
    misclassification, 208–211
    standards and, 211–213
    variations on collaborative theme, 207–208
Command relationships model (OV-4), 317
Commercial off-the-shelf (COTS) units, 111
Commercial standards, 382–384
CompressedSize, 302
Compression, 177, 205, 302, 331–332
Computational frameworks, 236
Computational specification, 323
Computer-aided system engineering (CASE) tools, 245, 379
Concept formulation, 179, 245, 400
Concurrent engineering, 3, 98–100
Concurrent progressions, 253–254
Conflict of interest, 18, 76, 402
Congress, 18, 316, 336, 362–369, 373
Construction blueprints, 218
Constructive cost model (COCOMO), 296–297
Consumer electronics, 7, 72, 92, 349

Cost estimates, 20, 253
COTS; *see* Commercial off-the-shelf (COTS) units
Customers
    acceptance/uncertainty of, 64, 75
    feedback, 100–102, 107
    flexible manufacturing and, 108–109
    incremental development for, 57–58
    -sales-delivery loop, 107

D

Data
    attributes, 302
    communication systems, 155
    entry-system, computer-based, 179
    exchange, 295, 320
    flow, 167, 172, 231–232, 234–235, 285–289, 293–297
Data flow diagrams (DFDs), 234–235, 288
Databases, 115, 148–149, 168–169, 261–262
Decision theory, 261, 271–273, 336
Decompositioning, 403
Defect rate, 14, 84, 107, 109–110
Design
    hierarchy, 293
    layered, 164
    politics as factor in, 362–364
    progression, 248–249
    reviews, 234, 399, 404
Design progression, in systems architecting, 247–282
    architecture and design disciplines, 277–280
    design as evolution of models, 250
    design concepts for systems architecture, 254–277
    design progression, 248
    evaluation criteria and heuristic refinement, 250–254
    examples, 250
    overview, 247
Desktop publishing, 71, 263
DeSoto story, 91, 109
DFDs; *see* Data flow diagrams
Diagnostic procedures, 92
Digital electronics, 188, 190, 192, 310
Discrete event systems, 102, 235–236, 239, 286–287
DODAF; *see* U.S. Department of Defense Architecture Framework
Domain-specific languages, 229

I